Y0-BRM-479

Race and Ethnicity in Society

Race and Ethnicity in Society

The Changing Landscape

SECOND EDITION

ELIZABETH HIGGINBOTHAM
University of Delaware

MARGARET L. ANDERSEN
University of Delaware

WADSWORTH
CENGAGE Learning™

Australia • Brazil • Japan • Korea • Mexico • Singapore • Spain • United Kingdom • United States

***Race and Ethnicity in Society: The Changing Landscape*, Second Edition**

Elizabeth Higginbotham and Margaret L. Andersen

Acquisitions Editor: Chris Caldeira

Assistant Editor: Tali Beasley

Editorial Assistant: Erin Parkins

Marketing Manager: Michelle Williams

Marketing Assistant: Ileana Shevlin

Marketing Communications Manager: Linda Yip

Project Manager, Editorial Production: Cheri Palmer

Creative Director: Rob Hugel

Art Director: Caryl Gorska

Print Buyer: Paula Vang

Permissions Editor: Bob Kauser

Production Service: Dusty Friedman, Gary Kliewer, The Book Company

Cover Designer: Riezebos Holzbaur Design Group

Cover Image: Jordan Tinker/ © Mary Lee Bendolph

Compositor: Newgen

© 2009, 2006 Wadsworth, Cengage Learning

ALL RIGHTS RESERVED. No part of this work covered by the copyright herein may be reproduced, transmitted, stored, or used in any form or by any means graphic, electronic, or mechanical, including but not limited to photocopying, recording, scanning, digitizing, taping, Web distribution, information networks, or information storage and retrieval systems, except as permitted under Section 107 or 108 of the 1976 United States Copyright Act, without the prior written permission of the publisher.

For product information and technology assistance, contact us at
Cengage Learning Customer & Sales Support, 1-800-354-9706.

For permission to use material from this text or product, submit all requests online at **cengage.com/permissions.**
Further permissions questions can be e-mailed to
permissionrequest@cengage.com.

Library of Congress Control Number: 2008920836

ISBN-13: 978-0-495-50434-4
ISBN-10: 0-495-50434-3

Wadsworth
10 Davis Drive
Belmont, CA 94002-3098
USA

Cengage Learning is a leading provider of customized learning solutions with office locations around the globe, including Singapore, the United Kingdom, Australia, Mexico, Brazil, and Japan. Locate your local office at **international.cengage.com/region.**

Cengage Learning products are represented in Canada by Nelson Education, Ltd.

For your course and learning solutions, visit **academic.cengage.com.**

Purchase any of our products at your local college store or at our preferred online store **www.ichapters.com.**

Printed in Canada
1 2 3 4 5 6 7 12 11 10 09 08

Contents

B. Families, Communities, and Welfare

C. Residential Segregation and Education

D. Social Justice and Social Control, Courts, Crime and the Law

Preface

The study of race and ethnicity is changing, just as the character of race and ethnic relations is changing. The United States is now a multiracial society. New patterns of immigration since the late 1960s have brought new populations to the United States—populations that are changing the racial and ethnic composition of the nation. Whereas the focus of study about race and ethnicity has long assumed a "Black/White" framework, there is now more attention to the different racial and ethnic groups that make up the United States. This is especially obvious in the rather heated national discussions about immigration and immigration policy.

The current generation of college students has come of age not just in a more diverse nation, but at a time when race *appears* to have lost some of its significance in organizing relationships and social institutions. Popular culture makes it appear that interracial relationships are common and that race no longer matters in shaping social relations. The dominant ideology is one of "color blindness," as if recognizing race is the same as being racist.

But more accurate observation and, indeed, volumes of scholarship indicate that race still matters—and it matters a lot. Racial segregation, not integration, is the norm, despite the illusion of inclusion portrayed in popular culture and in the few numbers of people of color who occupy highly visible places in social, economic, and political institutions. As the articles in this book will show, race is still a fundamental part of social structures. Relationships, resources, and identities continue to be shaped by race and ethnicity.

This anthology is intended to introduce students to the study of race by engaging them in the major topics and themes now framing the study of race in the United States. Because society is changing in its racial and ethnic makeup, current scholarship on race is also changing, focusing more on the social construction of race, racial and ethnic identities, and the containing patterns of racial exclusion. But most of the available textbooks have not kept up with these new realities. The dominant framework in most texts on the sociology of race and ethnicity is still a smorgasbord approach, wherein the book documents the different histories of an array of groups, but where ethnicity is seen as only applying to Whites of European ancestry. Within this model, ethnicity has been perceived as mostly "white," race as "color." Such an approach can no longer capture the complexities and dynamism of race and ethnicity in the United States.

We offer this anthology so that those teaching courses on race and ethnicity will have material to use that reflects the current state of scholarship on race and ethnicity in the United States. This book is intended primarily for courses on the sociology of race and ethnicity, although it can be used for courses in other departments and interdisciplinary programs where courses on race are being taught (such as education, political science, ethnic studies, and some humanities departments).

We have organized the second edition to reflect the different themes that underlie the study of race and ethnicity—consciously doing so instead of organizing it around particular groups. The major themes of this book include:

- showing the diversity of experiences that now constitute "race" in the United States;
- teaching students the significance of race as a socially constructed system of social relations;
- showing the connection between different racial identities and the social structure of race;
- understanding how racism works as a belief system rooted in societal institutions;
- providing a social structural analysis of racial inequality;
- providing a historical perspective on how the racial order has emerged and how it is maintained;
- examining how people have contested the dominant racial order;
- exploring current strategies for building a just multiracial society.

NEW TO THE SECOND EDITION

The second edition of *Race and Ethnicity in the United States* has benefited tremendously from the comments of reviewers and many of those who used the first edition. As a result, we have ***reorganized part of the table of Contents*** to clarify how we treat the topics of immigration, state policy, and racial stratification. As a result, there are three new parts in the book. The first *new Part V, "Race, Nation, and Citizenship"* examines how citizenship rights are denied by the state and how critical issues of real and perceived citizenship are to racial and ethnic groups. This part also includes material addressing this issue in the aftermath of 9/11, which is particularly important given the profiling that many Middle Eastern, Muslim, and African communities are now experiencing. The *new Part VI,*

"Immigration, Race, and Ethnicity" allows us to study immigration and immigration policy more directly and thoroughly, which is especially important now that this topic is engaging the public so intensely. The *new Part VII, "Race, Class, and Inequality"* introduces students to some of the basic concepts of racial stratification, including data on income and wealth, as well as examining the connections among social policies, race, and class. This section also includes material on race and poverty, particularly (though not exclusively) in the context of Hurricane Katrina.

There are ***23 new articles*** in the second edition, reflecting some of the new scholarship on race, ethnicity, immigration, and other topics, as well as articles that update material from the first edition. The new articles were selected not only for their importance, but also for their *accessibility to undergraduate readers.* They are on such topics as stereotyping of Latinas (Judith Ortiz Cofer, "The Myth of the Latin Woman: I Just Met a Girl Named María"); race and hip-hop in contemporary youth culture (Cornell West, "The Necessary Engagement with Youth Culture"); Beverly Tatum's work from *Why Are All the Black Kids Sitting Together in the Cafeteria?* on the formation of racial identity; children of immigrant families (Margie K. Shields and Richard E. Behrman, "Children of Immigrant Families: Analysis and Recommendations"); interracial dating and marriage (Zhenchao Qian, "Breaking the Last Taboo: Interracial Marriage in America"); and federal failures in the aftermath of Hurricane Katrina (Harvey Molotch, "Death on the Roof: Race and Bureaucratic Failure"), to name a few.

The second edition has a much stronger focus on immigration than did the first edition, as well as the inclusion of two articles on Katrina—one looking at the connection between poverty in New Orleans and the dynamics of race, class, and gender and the other examining federal policy and bureaucratic failures following Katrina. We also have included a new article that details some of the consequences of recent educational reforms, such as high stakes testing and the No Child Left Behind Act (in Linda Darling-Hammond's, "The Color Line in American Education: Race, Resources, and Student Achievement").

In addition, we have ***revised our introductions*** to include current data and to reflect events that have occurred since the first edition, such as the important Supreme Court decision abolishing the use of race in determining public school placement.

ORGANIZATION OF THIS BOOK

This book is organized in nine parts, beginning with a short introductory section that shows students the significance of understanding race by opening the book with articles that build from the direct experiences of those from diverse racial-ethnic backgrounds. Thus, Part I ("Race: Why It Matters") establishes the importance of examining race as a contemporary social issue. Articles are selected to grab students' attention by showing the importance of race in the United States even when many people tend to think that focusing on race is itself a form of racism. We include some personal narratives to engage students' empathy and understanding.

Part II ("The Social Construction of Race and Ethnicity") establishes the analytical frameworks that are now being used to think about race in society. The section examines the social construction of race as a concept and experience. Together, the articles show the grounding of racial categories in specific historical and social contexts and their fluidity over time.

Part III ("Representations of Race and Group Beliefs: Prejudice and Racism") examines the most immediately experienced dimensions of race: beliefs and ideology. We think that students learn best about race by examining the manifestations of a racially stratified society in people's beliefs and in the images and representations of popular culture. This section also includes material on the current ideology of colorblind thinking and how it perpetuates racial inequality.

Part IV ("Race and Identity") examines racial identities—a topic that we think is especially interesting to students. In a racially diverse and changing society, many people now form racial identities that cross racial and ethnic boundaries.

Part V ("Race, Nation, and Citizenship") retains the focus on conceptualizing citizenship that was present in the first edition, but strengthens the analysis of citizenship and state policy by including new pieces on state policy and American Indians, as well as new material on the suspicions cast on some new immigrant groups as the result of the politics of fear subsequent to 9/11.

Part VI ("Immigration, Race, and Ethnicity") includes articles that examine different dimensions of immigration experiences. There is a review of immigration policy here with a discussion of the implications of trends in policy for the future of the treatment of immigrants, as well as articles that study generational change within immigrant families. The section also addresses comparisons with current and past immigration patterns, as well as discussing the global context of immigrant movement.

Part VII ("Race, Class, and Inequality") updates material on the wealth divide and the current realities of growing class and race inequality. The section includes new articles on the recent rise in poverty among U.S. citizens and includes new material on how poverty affected the residents of New Orleans both before and after the disaster of Katrina.

Part VIII ("Institutional Segregation and Inequality") details the consequences of race and racism as manifested in different social institutions. Within this part, we include subsections on different institutional sites where racial stratification can be seen, including work, family, community, housing, education, and social justice. Each section includes articles examining the outcomes within social institutions that stem from the reality of racial inequality in society. We include articles on work and labor markets, the criminal justice system, housing, education, and social welfare.

Part IX ("Mobilizing for Change: Looking Forward and Learning from the Past") focuses on social movements and social change. One purpose of this section is to teach students how the past matters in understanding how people have mobilized for change (for example, in the Civil Rights Movement). But this section studies activism of the past with an eye toward guiding future social policy, social activism, and social movements. We include articles in this section (and throughout the book) that also show how people have resisted the oppression of race.

PEDAGOGICAL FEATURES

We have included ***Discussion Questions*** for each article in this book, with the goal of helping students grasp the major points of each argument. These questions can also be used for student paper assignments, research exercises, and class discussion.

At the end of each major part, we also include ***Student Exercises*** that can also be used for projects and discussion.

ACKNOWLEDGMENTS

We have benefited from the support and encouragement of many people who have either discussed the contents of the book with us or provided clerical and computer assistance or other forms of help that enabled us to complete/work on this book even in the midst of many other commitments. We thank Maxine Baca Zinn, Victoria Baynes Becker, Bonnie Thornton Dill, Ben Fleury-Steiner, Charles Gallagher, Valerie Hans, Linda Keen, Richard Rosenfeld, and Judy Watson for all they have done to help us. Special thanks go to Michelle Wilcox for her research assistance. We also appreciate the support provided by the University of Delaware and the Center for the History of Business, Technology, and Society at the Hagley Museum and Library. The editors at Wadsworth have also been enthusiastic about this project, so we thank Chris Caldeira for her enthusiasm and guidance, as well as Eve Howard for supporting this project. The suggestions of those who carefully reviewed the first drafts of this book were extremely valuable and have helped make this a stronger anthology, thus we thank Julius H. Bailey, University of Redlands; Marianne Cutler, East Stroudsburg University; Karen Hayden, Merrimack College; Karen Tejada, State University of New York–University at Albany; Lisa Wade, Occidental College; Pamela Williams-Paez, College of the Canyons; George Wilson, University of Miami.

About the Editors

Elizabeth Higginbotham (B.A., City College of the City University of New York; M.A., Ph.D., Brandeis University) is Professor of Sociology, Black American Studies, and Women's Studies at the University of Delaware. She is the author of *Too Much to Ask: Black Women in the Era of Integration* (University of North Carolina Press, 2001) and co-editor of *Women and Work: Exploring Race, Ethnicity, and Class* (Sage Publications, 1997; with Mary Romero). She has also authored many articles in journals and anthologies on the work experiences of African American women, women in higher education, and curriculum transformation. While teaching at the University of Memphis, she received the Superior Performance in University Research Award for 1991–92 and 1992–93. Along with colleagues Bonnie Thornton Dill and Lynn Weber, she is a recipient of the American Sociological Association Jessie Bernard Award (1993) and Distinguished Contributions to Teaching Award (1993) for the work of the Center for Research on Women at the University of Memphis. She also received the 2003–2004 Robin M. Williams Jr. Award from the Eastern Sociological Society, given annually to one distinguished sociologist. Elected in 2006, she serves a term as Vice President of the Eastern Sociological Society from 2007 to 2008.

Margaret L. Andersen (B.A., Georgia State University; M.A., Ph.D. University of Massachusetts, Amherst) is the Edward F. and Elizabeth Goodman Rosenberg Professor of Sociology at the University of Delaware. She is the author of *On Land and On Sea: A Century of Women in the Rosenfeld Collection; Thinking about Women; Race, Class, and Gender: An Anthology* (co-edited with Patricia Hill Collins); *Sociology: Understanding a Diverse Society* (with Howard F. Taylor); *Sociology: The Essentials* (with Howard F. Taylor); and *Understanding Society: An Introductory Reader* (co-edited with Kim Logio and Howard F. Taylor). Professor Andersen was the Vice President of the American Sociological Association (ASA) for 2008-09 and received the ASA's prestigious Jessie Bernard Award (2006). She also received the 2004 Sociologists for Women in Society Feminist Lecturer Award and the 2007-2008 Eastern Sociological Society Robin F. Williams

Lecturer Award. She has received two teaching awards at the University of Delaware: the University Excellence in Teaching Award and the College of Arts and Sciences Outstanding Teacher Award. She currently chairs the National Advisory Board for the Stanford University Center for Comparative Studies in Race and Ethnicity and is the former editor of *Gender & Society*.

Introduction

ELIZABETH HIGGINBOTHAM AND
MARGARET L. ANDERSEN

In the summer of 2007, the Gallup Organization released the results of two separate public opinion polls. One showed that two-thirds of Black Americans are dissatisfied with how Black Americans are treated in society—the highest dissatisfaction recorded since 2001. Two-thirds of Hispanics also said they were dissatisfied with how Hispanics are treated (Saad 2007a).

During the same week, another survey asked a national sample of Black, Hispanic, and White Americans whether they believed that Black children have as good a chance for an education as do White children. Eighty percent of White Americans believe that Black children have as good a chance as White children to get a good education in their community, compared to 49 percent of Black Americans who believe so and 73 percent of Hispanics (Saad 2007b). How in the world can there by such divergent views about race in America?

One society, two very different views about the state of race relations. If, as Whites believe, Black children have an equal chance for an education, why would Black Americans and Hispanics both be reporting a downturn in how they see themselves treated? As Robert Blauner discusses in the opening article of this book, people seem to be talking past one another when it comes to race. There is a widespread belief that race no longer matters when it comes to people's life chances and opportunities. Yet, race obviously does matter in shaping people's perceptions of reality and, as we shall see, the real experiences that people have living in this society—a society that, despite widespread belief, is structured along lines of racial inequality.

The reality of race in the United States is changing. Whereas once there were laws that formally segregated people from one another, now the United States is supposed to be an open society where people have equal opportunities, are free to mix with one another, and enjoy freedoms unfettered by their racial or ethnic background. Formal, state-sanctioned segregation is illegal. There are laws that protect civil rights based on race, sex, national origin, and religion. Most people believe that there should be equal opportunity for all and that no one should be barred from full participation in society because of race or ethnic background.

Yet, racial and ethnic inequality continue. Communities and schools are racially segregated. Police, shopping clerks, airport security officials, and others engage in *racial profiling*, defined as the singling out of individuals because of their presumed identification with a particular group, such as Arab Americans or Black Americans. Disparities in levels of income and wealth persist despite several decades of equal employment legislation. Poverty rates are highest among African American and Latina single-parent families. Many people, including people of color, hold racial and ethnic stereotypes about other groups. Indeed, a "we/they" dichotomy characterizes much of the world conflict in which the United States is now engaged.

At the same time, the U.S. population is becoming more diverse. Population projections indicate that White people will soon be a numerical minority. In some states (notably, California, Texas, and Hawaii), that is nearly the case now. Immigration to the United States is also bringing more diversity to the nation. Whereas in the past most immigrants came from western Europe, today's immigrants are most likely to be from Asia, Latin America, and the Caribbean. This increased diversity is bringing new forms of culture and new relationships to this nation. Young people, regardless of their race or ethnic identity, listen to music and buy clothing inspired by urban African American, Latina/Latino, and Asian cultures. People of different racial and ethnic identities meet and fall in love, resulting in an increase of so-called biracial or multiracial people. And, whereas in the past, the nation's largest minority group was African American, now Hispanics are the largest "minority" group.

All of these changes are transforming the social landscape of what it means to be American. While some people in America might not think much about their racial identity, others proudly assert they are from Korea or El Salvador, that they are American-born Chinese, fifth generation African American, or a member of the Choctaw nation. How does this population diversity change how people think about race and ethnicity in the United States?

Thinking about race was once framed by a presumed "Black-White" dichotomy. That way of thinking about race is no longer adequate. Certainly African American experience is a central part of how U.S. social institutions emerged, but the landscape of race in the United States is more complex than thinking in Black-White terms can capture. Diversity both within and across racial groups cannot be understood in simple or one-dimensional frameworks. The multiracial character of the United States demands that we think more broadly and more inclusively about the meaning of race and its sister concept, ethnicity.

Race and ethnicity are social categories that define a complex set of social relationships. They are ingredients in the social institutions of our nation. Social scientists have pondered the meaning and significance of race and ethnicity for years. Their analyses have had to change as the society itself changes and as groups shift in their positions relative to each other. Fundamentally, U.S. society is marked by racial hierarchy, where groups are ranked and assigned different rights and values based on their race and ethnicity, just as other social hierarchies exist based on gender, social class, sexuality, and age. These hierarchies are based on power differences among groups. These power differences and the structures in which they play out are supported by ideas that define race as a "natural" trait of groups—an idea we examine in Part II. Who decides what racial categories are important and what they mean? What are the consequences for various individuals in the society? These are the kinds of questions you'll begin to ask once you no longer take race for granted and begin to see its underlying social basis.

At different points in history *race* has taken on different meanings. Many of the people currently considered White and thought of as the majority group are descendants of immigrants who at one time were believed to be racially distinct from native-born White Americans, the majority of whom were Protestants. Protestants were hesitant to welcome German Catholics and Jews from western Europe in the early nineteenth century. They were even more distrusting of Chinese immigrants in the 1850s and 1860s and, later in the 1880s, of the Europeans from eastern and southern Europe, including Poland, Russia, and Italy. Powerful White leaders at the time used the law and other means to push for greater control of the national borders, attempting to shape the racial composition of the nation. Asians, Jewish people, and Italians were viewed as strangers who were not as "worthy" as the descendants of the presumed founders of the nation.

Race has now shifted in meaning, but racial borders between groups are still contested and changing. How we think about race and ethnicity now can help us prepare for these continuing social changes. Thus, this book is about *race and ethnicity in society*. It is not just about people of color, although understanding the experience of people of color—as well as dominant groups—is critical to understanding the social reality of race. Even though many are reluctant to acknowledge the social reality of race, race and ethnicity are social facts. They are fundamental to how we think about ourselves. We all walk around in bodies that are assigned racial meanings—meanings we give to ourselves and that others give to us. Race shapes basic social institutions, and racial politics are a major dimension of social change—even when not explicitly acknowledged as such. Even though many deny that race is "real," it is a lens through which we often view our lives and those of others.

Our task in this book is to examine the changing landscape of race and ethnicity in the United States. We do so by examining the *social meaning of race and ethnicity*—how they form our identities and beliefs about each other; how they become the basis for the distribution of social and economic resources, and how they structure social institutions, including work, families, communities, education, and the law. But we also emphasize how people have organized to resist the inequities that exist in a racially divided society. Underlying this book is the idea that racial hierarchies, like other social hierarchies, are unstable, fluid, and constantly changing. Systems of inequality are usually challenged by those who are disadvantaged by them, adding to the changes in society that define race and ethnic relations.

We also note that the study of race has changed dramatically in recent years. It has moved from a focus on the Black-White dichotomy to an understanding that many different racial-ethnic groups make up the United States. The current generation of college students has come of age in a more diverse nation, but also at a time when issues of race continue to be contested. Current scholarship on race has also changed. It now focuses on how race is socially constructed and is a fundamental part of social structure. This anthology is intended to introduce students to the study of race by engaging them in the major topics and themes now framing the study of race in the United States.

Several themes guide the articles selected for this book:

- the diversity of experiences that now constitute race and ethnicity in the United States;
- the significance of race and ethnicity as socially constructed systems of social relations;
- the connection between diverse racial identities and the social structure of race and ethnicity;
- how racism works as a belief system rooted in societal institutions;
- how the racial order has emerged historically and how it is maintained;
- how people have contested the dominant racial order; and,
- current strategies for building a more just multiracial society.

ORGANIZATION OF THE BOOK

The book is divided into nine parts, beginning with this introductory essay by the editors discussing just how much race matters in U.S. society. Part I establishes the importance of examining race as a contemporary social issue and introduces you to some of the diverse experiences that people have as the result of their different racial-ethnic backgrounds. The articles in this part show how the U.S. population is becoming more racially and ethnically diverse; some of the articles use personal narratives to help you gain empathy and understanding.

The articles in Part II, "The Social Construction of Race and Ethnicity," show how thinking about race and racism has emerged over time in Western thought. In this part, you will learn to be critical of purely biological constructions of race, learning instead how race is defined through society, history, and culture. Together, the articles show both the fluidity of racial categories and their grounding in specific historical and social contexts.

Part III, "Representations of Race and Group Beliefs: Prejudice and Racism," examines the most immediately experienced dimensions of race: racial beliefs and the representation of race in popular culture. We think you will learn best about race by examining how people's beliefs about it emerge in a racially stratified society. Images in popular culture are central to this process. Because society is so strongly influenced by the media, we include articles on film and music to help you see from the beginning how much thinking about race is influenced by popular culture. This part also includes material on the current ideology of color-blind thinking and how it cloaks the persistence of institutional racism.

In Part IV, "Race and Identity," we examine the complexity of racial identity, a topic that we think is especially interesting to you as students. Everyone develops a racial identity, including White people, but we vary in our awareness of our identities. Furthermore, those identities represent varying degrees of social acceptance in the society. In a racially diverse and changing society, many people now form racial identities that cross racial boundaries. This section explores current issues of racial identity.

Part V, "Race, Nation, and Citizenship," shows how the state has denied full rights of citizenship to different groups based on their racial and/or ethnic group membership. The rights of citizenship can include formal rights, such as the right to vote, but they can include more subtle dimensions of citizenship, such as whether one is perceived as belonging to American society. Over the course of history, numerous groups—including immigrant groups—have been denied such rights, both formally and informally—as, for example, in the contemporary example of racial and ethnic profiling.

Part VI, "Immigration, Race, and Ethnicity," explores the currently important topic of immigration. Perhaps not since the early twentieth century when immigration was occurring at such a high rate has the subject of immigration garnered so much national attention. This section reviews contemporary trends in immigration and immigration policies, as well as compares these current trends to those of the past. The section also analyzes immigration in the context of global changes—changes that have introduced new forms of social relationships, such as transnational families and communities.

Part VII, "Race, Class, and Inequality," focuses on racial stratification, including the connection between race and class. The section includes current data on income and wealth differences among and within different racial and ethnic groups. This section also includes articles on the connection between race and poverty, as was made clearly evident in the aftermath of Hurricane Katrina. The articles in this section also include a discussion of social policy in the context of rising inequality.

In Part VIII, "Institutional Segregation and Inequality," we detail the consequences of race and racism as manifested in different social institutions, including work, family, health, housing, education, and social justice. This part is longer than the others because we wanted to examine in detail the different social institutions that reflect racial stratification. Institutional racism means that different groups have unequal access to institutional resources. In this part we also emphasize what groups have done to withstand institutional exploitation. The subsections include work and labor markets; families, communities, and welfare; residential segregation and education; and social justice and social control.

Part IX, "Mobilizing for Change: Looking Forward and Learning from the Past," focuses on social movements and social change. One purpose of this part is to illustrate how the past matters in understanding how people have mobilized for change (for example, in the Civil Rights Movement). We also include articles in this part that show the different ways people have resisted racial oppression.

Throughout the book we include the experiences of diverse groups, showing the complexity, pervasiveness, and importance of race and ethnicity in people's lives. As the United States moves through the twenty-first century, we know that the racial landscape will continue to change, but we hope that it will do so for the better as people become better informed about how race and ethnicity continue to shape the experience of diverse groups in society.

REFERENCES

Saad, Lydia. 2007a (July 6). "A Downturn in Black Perceptions of Racial Harmony." *The Gallup Poll.* Princeton, NJ: The Gallup Organization.

Saad, Lydia. 2007b (July 2). "Black-White Educational Opportunities Widely Seen as Equal." *The Gallup Poll.* Princeton, NJ: The Gallup Organization.

PART I

INTRODUCTION

Race: Why It Matters

Supreme Court Justice Ruth Bader Ginsburg wrote:

> In the wake of a system of racial caste only recently ended, large disparities endure. Unemployment, poverty, and access to health care vary disproportionately by race. Neighborhoods and schools remain racially divided.... Irrational prejudice is still encountered in real estate markets and consumer transactions. Bias both conscious and unconscious, reflecting traditional and unexamined habits of thought, keeps up barriers that must come down if equal opportunity and nondiscrimination are even genuinely to become this country's law and practice.[1]

In other words, race matters, as philosopher Cornel West reminded the public in his best-selling book entitled *Race Matters* (1993).

Race matters because it is one of the most significant ways that U.S. society distributes economic, social, political, and cultural resources. Race matters because it is one of the ways we define ourselves and other people. Race matters because it segregates our neighborhoods, our schools, our churches, and our relationships. Race matters because it is often a matter of heated political debate, and the dynamics of race lie at the heart of the systems of justice and social welfare.

Yet, many people claim that the United States is now a colorblind society where race no longer matters. Personal attitudes, not societal factors, are seen as causes for success or failure. People are often blamed when they do not succeed, and that blame is often associated with race.

Some think that as a society we can overcome racial problems by abolishing race as a category of identification. This thinking was reflected in the 2007 Supreme Court case *Parents Involved in Community Schools v. Seattle School District*, where in five of the nine Supreme Court justices ruled that school districts could

1. *Gratz v. Bollinger*, 123 S. Ct. 2411.

not use race as a factor in determining which schools children attend. Their decision was based on the argument that, in the words Chief Justice John Roberts, "the way to stop discrimination on the basis of race is to stop discriminating on the basis of race."

This is a very different conclusion than the earlier view of Supreme Court Justice Harry Blackmun, who argued (in the important Bakke decision on affirmative action) that "in order to get beyond racism, we must first take account of race. There is no other way, and in order to treat some persons equally, we must treat them differently" (Blackmun 1978). But now, the U.S. Supreme Court—in a narrowly divided opinion—is saying that one should not take race into account in order to halt racial inequality—as if race should no longer matter.

The problem is that race does matter—and it matters not only in education but also in workplaces, in communities, in health care systems, in courts and policing, in everyday interactions—in fact, in every single incidence of day-to-day life, race matters and it matters a lot. Race is embedded in social institutions and in social relationships. And it has real and recurring social and historical consequences.

Consider these facts:

- There is a 61 percent gap between Black and White household income (DeNavas-Walt, Proctor, and Lee 2006).
- U.S. schools are now more segregated than thirty years ago, and the most successful plans for desegregation are being dismantled (Frankenberg, Lee, and Orfield 2003).
- Despite high levels of poverty, there is also a substantial growth of the Black, Asian American, Native American and Latino middle class (Pattillo-McCoy 1999).
- Interracial marriage, although increasing, is still rare—constituting only 4 percent of all marriages (U.S. Census Bureau 2007).

The official statement of the American Sociological Association (ASA) reprinted in Part I shows the many ways that race matters. The statement was collectively developed by a group of sociological experts on race who reviewed the many social scientific studies of race. The ASA concluded that, despite beliefs to the contrary, race is still a mechanism for sorting people in society, in their words—a "stratifying practice." Disparities in jobs, housing, health, education, and other facets of life remain, and these can be attributed to race, along with its relationship to other social factors such as social class, gender, age, nationality, and so on. Because of the strong sociological significance of race, we cannot just make it go away by denying that it is there.

The social reality of race is becoming even more evident as the nation becomes more racially and ethnically diverse. Sociologists often frame discussions of race and ethnicity in terms of "minority/majority group." These terms are used not in a numerical sense, but as a way to represent who holds power in society,  namely, the "majority." Although the nomenclature is intended to refer to the power relations associated with race and ethnicity, it can be misleading in the context of greater diversity. Indeed, it may seem to neutralize the relations of power that exist between dominant and subordinate groups. "Dominant and subordinate" are stronger in their connotation and better represent how power shapes relations between race and ethnic groups in society. As society changes, so does the language that social scientists use to describe and analyze it.

Anyone who thinks that race does not matter much might want to step into the shoes of those who know it does. Robert Blauner ("Talking Past One Another") shows how a racial fault line has developed, with White and Black people having different worldviews about the reality of race. This racial fault line inhibits cross-race communication and understanding, especially if people deny that racism still matters. And clearly it does. Something as simple as the act of shopping elicits different experiences for White people and people of color. Blauner also points out that racism can sometimes be subtle, sometimes not; it is not always overt, nor is it always intentional. Racism is institutionalized—that is, built into the very fabric of society. As the nation is becoming more racially and ethnically diverse, the racial fault lines may be unfolding into multiple worldviews—each of them founded in a different racial-ethnic social location.

We include as the second piece in this section a personal narrative by Berta Esperanza Hernández-Truyol who questions the way that racial-ethnic categories can lump together people of very different backgrounds and experiences. Still, these categories (like Hispanic, Latino, Black, Asian American)—even when diverse within—shape one's identity and experiences in the world. Hernández-Truyol recalls a job interview where the stereotype of Latinas leads others to think that she is not qualified for a job. Such experiences are common to many racial-ethnic groups and shows the continuing and persistent ways that race matters in the United States.

Judith Ortiz Cofer's essay ("The Myth of the Latin Woman: I Just Met a Girl Named Maria") also shows the hurt that racial-ethnic stereotypes produce when targeted at people of color. Her narrative shows not just the continuing power of stereotyping to do harm, but the linkage between racial stereotyping and gender stereotypes as well. For Latinas, this has meant challenging social attitudes that associate sexuality with women of color, Latinas in particular. Cofer's essay also discusses some of the generational changes that emerge as groups encounter new social expectations and new social and cultural realities.

In a racially stratified society, people of color become objects of suspicion. This social reality has become familiar to Arab Americans in the aftermath of the devastating events of September 11, 2001. Many Arab Americans are now experiencing racial profiling in their daily lives. They join other people of color who appear "foreign" and who are subjected to racial and ethnic stereotyping. Once you realize that profiling happens, you can see how resentment, hostility, and suspicion characterize human interaction under conditions of racial inequality. But, we can learn to challenge these ways of interacting by becoming more aware of the reality of race in everyday life and working to change it—in our personal lives, as well as within larger societal structures.

Moustafa Bayoumi ("What Does It Feel Like to Be a Problem?") asks us to think about race as something other than a social problem. Yes, race does generate social problems, but seeing people only in the framework of a problem diminishes their humanity and makes us lose sight of the human creativity and adaptation that takes place even when people face harsh conditions. Reflecting on 9/11, Bayoumi wonders if this event will further divide people or provide an opportunity through which people can challenge the commission of hate crimes, the growth of exclusionary nationalism, and "we/they" thinking. The tragedy of 9/11 may have been that this was the first time that many White Americans had felt someone hated them simply because of the group they belonged to—a feeling that many people of color have repeatedly experienced.

In this first part, we examine some of the ways that race matters, thus setting the stage for understanding the many dimensions of race in society explored in subsequent parts.

REFERENCES

Blackmun, Justice Harry. 1978. *Regents of the University of California v. Bakke*, U.S. Supreme Court, No. 76–811.

DeNavas-Walt, Carmen, Bernadette D. Proctor, and Cheryl Hill Lee. 2006. *Income, Poverty, and Health Insurance Coverage in the United States 2005*. Washington, DC: U.S. Department of Commerce.

Frankenberg, Erica, Chungmei Lei, and Gary Orfield. 2003. "A Multiracial Society with Segregated Schools: Are We Losing the Dream?" The Civil Rights Project, Harvard University, Web site: www.civilrightsproject.harvard.edu

Pattillo-McCoy, Mary. 1999. *Black Picket Fences: Privilege and Peril among the Black Middle Class*. Chicago: University of Chicago Press.

U.S. Census Bureau, 2007. *Statistical Abstract of the United States 2006*. Washington, DC: U.S. Census Bureau.

West, Cornel. 1993. *Race Matters*. Boston: Beacon Press.

1

Talking Past One Another

Black and White Languages of Race

ROBERT BLAUNER

This article shows the significance of the racial divide, as reflected in the different understandings of race generally held by Black and White Americans. Many White Americans think of racism as a thing of the past, not understanding how embedded racism is in social institutions. Blauner emphasizes the need for interracial dialogue to overcome this racial divide.

... I want to advance the proposition that there are two languages of race in America. I am not talking about black English and standard English, which refer to different structures of grammar and dialect. "Language" here signifies a system of implicit understandings about social reality, and a racial language encompasses a worldview.

Blacks and whites differ on their interpretations of social change ... because their racial languages define the central terms, especially "racism," differently. Their racial languages incorporate different views of American society itself, especially the question of how central race and racism are to America's very existence, past and present. Blacks believe in this centrality, while most whites, except for the more race-conscious extremists, see race as a peripheral reality. Even successful, middle-class black professionals experience slights and humiliations—incidents when they are stopped by police, regarded suspiciously by clerks while shopping, or mistaken for messengers, drivers, or aides at work—that remind them they have not escaped racism's reach. For whites, race becomes central on exceptional occasions: collective, public moments ... when the veil is lifted, and private ones, such as a family's decision to escape urban problems with a move to the suburbs. But most of the time European-Americans are able to view racial issues as aberrations in American life, much as Los Angeles Police Chief Daryl Gates used the term "aberration" to explain his officers' beating of Rodney King in March 1991.

Because of these differences in language and worldview, blacks and whites often talk past one another, just as men and women sometimes do. I first noticed this in my classes, particularly during discussions of racism. Whites locate racism in color consciousness and its absence in color blindness. They regard it as a kind

SOURCE: Reprinted with permission from Bob Blauner, "Talking Past Each Other," *The American Prospect* Vol. 3 No. 10, June 10, 1992. *The American Prospect*, 11 Beacon Street, Suite 1120, Boston, MA 02108. All rights reserved.

of racism when students of color insistently underscore their sense of differences, their affirmation of ethnic and racial membership, which minority students have increasingly asserted. Many black, and increasingly also Latino and Asian, students cannot understand this reaction. It seems to them misinformed, even ignorant. They in turn sense a kind of racism in the whites' assumption that minorities must assimilate to mainstream values and styles. Then African Americans will posit an idea that many whites find preposterous: Black people, they argue, cannot be racist, because racism is a system of power, and black people as a group do not have power.

In this and many other arenas, a contest rages over the meaning of racism. Racism has become the central term in the language of race. From the 1940s through the 1980s new and multiple meanings of racism have been added to the social science lexicon and public discourse. The 1960s were especially critical for what the English sociologist Robert Miles has called the "inflation" of the term "racism." Blacks tended to embrace the enlarged definitions, whites to resist them. This conflict, in my view, has been at the very center of the racial struggle during the past decade.

THE WIDENING CONCEPTION OF RACISM

The term "racism" was not commonly used in social science or American public life until the 1960s. "Racism" does not appear, for example, in the Swedish economist Gunnar Myrdal's classic 1944 study of American race relations, *An American Dilemma.* But even when the term was not directly used, it is still possible to determine the prevailing understandings of racial oppression.

In the 1940s racism referred to an ideology, an explicit system of beliefs postulating the superiority of whites based on the inherent, biological inferiority of the colored races. Ideological racism was particularly associated with the belief systems of the Deep South and was originally devised as a rationale for slavery. Theories of white supremacy, particularly in their biological versions, lost much of their legitimacy after the Second World War due to their association with Nazism. In recent years cultural explanations of "inferiority" are heard more commonly than biological ones, which today are associated with such extremist "hate groups" as the Ku Klux Klan and the White Aryan Brotherhood.

By the 1950s and early 1960s, with ideological racism discredited, the focus shifted to a more discrete approach to racially invidious attitudes and behavior, expressed in the model of prejudice and discrimination. "Prejudice" referred (and still does) to hostile feelings and beliefs about racial minorities and the web of stereotypes justifying such negative attitudes. "Discrimination" referred to actions meant to harm the members of a racial minority group. The logic of this model was that racism implied a double standard, that is, treating a person of color differently—in mind or action—than one would a member of the majority group.

By the mid-1960s the terms "prejudice" and "discrimination" and the implicit model of racial causation implied by them were seen as too weak to explain the sweep of racial conflict and change, too limited in their analytical power, and for some critics too individualistic in their assumptions. Their original meanings tended to be absorbed by a new, more encompassing idea of racism. During the 1960s the referents of racial oppression moved from individual actions and beliefs to group and institutional processes, from subjective ideas to "objective" structures or results. Instead of intent, there was now an emphasis on process: those more objective social processes of exclusion, exploitation, and discrimination that led to a racially stratified society.

The most notable to these new definitions was "institutional racism." In their 1967 book *Black Power*, Stokely Carmichael and Charles Hamilton stressed how institutional racism was different and more fundamental than individual racism. Racism, in this view, was built into society and scarcely required prejudicial attitudes to maintain racial oppression.

This understanding of racism as pervasive and institutionalized spread from relatively narrow "movement" and academic circles to the larger public with the appearance in 1968 of the report of the commission on the urban riots appointed by President Lyndon Johnson and chaired by Illinois governor Otto Kerner. The Kerner Commission identified "white racism" as a prime reality of American society and the major underlying cause of ghetto unrest. America, in this view, was moving toward two societies, one white and one black (it is not clear where other racial minorities fit in). Although its recommendations were never acted upon politically, the report legitimated the term "white racism" among politicians and opinion leaders as a key to analyzing racial inequality in America.

Another definition of racism, which I would call "racism as atmosphere," also emerged in the 1960s and 1970s. This is the idea that an organization or an environment might be racist because its implicit, unconscious structures were devised for the use and comfort of white people, with the result that people of other races will not feel at home in such settings. Acting on this understanding of racism, many schools and universities, corporations, and other institutions have changed their teaching practices or work environments to encourage a greater diversity in their clientele, students, or workforce.

Perhaps the most radical definition of all was the concept of "racism as results." In this sense, an institution or an occupation is racist simply because racial minorities are underrepresented in numbers or in positions of prestige and authority.

Seizing on different conceptions of racism, the blacks and whites I talked to in the late 1970s had come to different conclusions about how far America had moved toward racial justice. Whites tended to adhere to earlier, more limited notions of racism. Blacks for the most part saw the newer meanings as more basic. Thus African Americans did not think racism had been put to rest by civil rights laws, even by the dramatic changes in the South. They felt that it still pervaded American life, indeed, had become more insidious because the subtle forms were harder to combat than old-fashioned exclusion and persecution.

Whites saw racism largely as a thing of the past. They defined it in terms of segregation and lynching, explicit white supremacist beliefs, or double standards in hiring, promotion, and admissions to colleges or other institutions. Except for affirmative action, which seemed the most blatant expression of such double standards, they were positively impressed by racial change. Many saw the relaxed and comfortable relations between whites and blacks as the heart of the matter. More crucial to blacks, on the other hand, were the underlying structures of power and position that continued to provide them with unequal portions of economic opportunity and other possibilities for the good life.

The newer, expanded definitions of racism just do not make much sense to most whites. I have experienced their frustrations directly when I try to explain the concept of institutional racism to white students and popular audiences. The idea of racism as an "impersonal force" loses all but the most theoretically inclined. Whites are more likely than blacks to view racism as a personal issue. Both sensitive to their own possible culpability (if only unconsciously) and angry at the use of the concept of racism by angry minorities, they do not differentiate well between the racism of social structures and the accusation that they as participants in that structure are personally racist.

The new meanings make sense to blacks, who live such experiences in their bones. But by 1979 many of the African Americans in my study, particularly the older activists, were critical of the use of racism as a blanket explanation for all manifestations of racial inequality. Long before similar ideas were voiced by the black conservatives, many blacks sensed that too heavy an emphasis on racism led to the false conclusion that blacks could only progress through a conventional civil rights strategy of fighting prejudice and discrimination. (This strategy, while necessary, had proved very limited.) Overemphasizing racism, they feared, was interfering with the black community's ability to achieve greater self-determination through the politics of self-help. In addition, they told me that the prevailing rhetoric of the 1960s had affected many young blacks. Rather than taking responsibility for their own difficulties, they were now using racism as a "cop-out."

In public life today this analysis is seen as part of the conservative discourse on race. Yet I believe that this position originally was a progressive one, developed out of self-critical reflections on the relative failure of 1960s movements. But perhaps because it did not seem to be "politically correct," the left-liberal community, black as well as white, academic as well as political, has been afraid of embracing such a critique. As a result, the neoconservatives had a clear field to pick up this grass-roots sentiment and to use it to further their view that racism is no longer significant in American life. This is the last thing that my informants and other savvy African Americans close to the pulse of their communities believe.

By the late 1970s the main usage of racism in the mind of the white public had undoubtedly become that of "reverse racism." The primacy of "reverse racism" as "the really important racism" suggests that the conservatives and the liberal-center have, in effect, won the battle over the meaning of racism.

Perhaps this was inevitable because of the long period of backlash against all the progressive movements of the 1960s. But part of the problem may have been the inflation of the idea of racism. While institutional racism exists, such a concept loses practical utility if every thing and every place is racist. In that case, there is effectively nothing to be done about it. And without conceptual tools to distinguish what is important from what is not, we are lost in the confusion of multiple meanings....

The question then becomes what to do about these multiple and confusing meanings of racism and their extraordinary personal and political charge. I would begin by honoring both the black and white readings of the term. Such an attitude might help facilitate the interracial dialogue so badly needed and yet so rare today.

Communication can only start from the understandings that people have. While the black understanding of racism is, in some sense, the deeper one, the white views of racism (ideology, double standard) refer to more specific and recognizable beliefs and practices. Since there is also a cross-racial consensus on the immorality of racist ideology and racial discrimination, it makes sense whenever possible to use such a concrete referent as discrimination rather than the more global concept of racism. And reemphasizing discrimination may help remind the public that racial discrimination is not just a legacy of the past.

The intellectual power of the African American understanding lies in its more critical and encompassing perspective.

DISCUSSION QUESTIONS

1. What different definitions of racism does Blauner identify as having emerged at different points in time in this country? How does he now define racism?
2. What different views of racism does Blauner say White people and Black people hold? How does this result in Black people and White people "talking past one another?"

2

Building Bridges

Latinas and Latinos at the Crossroads

BERTA ESPERANZA HERNÁNDEZ-TRUYOL

In this article Hernández-Truyol shows how lumping diverse people together into a single racial or ethnic category can hide the various experiences that people of different racial-ethnic origins have. Nonetheless, her experience in a job interview also shows how stereotypes—fixed within racial and ethnic categories—continue to guide people's social interactions and understandings of one another.

Can one lump persons in a generic "hispanic" category? Only by ignoring the diversity of the latina/o population. For example, federal forms usually provide the following options: black (not of hispanic origin); white (not of hispanic origin); hispanic. As the forms seek information in the conjunctive, implicitly recognizing that ethnic identity and racial identity are two separate, co-existing traits, it is particularly ironic that latinas/os are deprived of the opportunity to identify as ethnic, i.e., latina/o, including subcategory identification such as Cuban, Mexican, Puerto Rican, as well as to identify by race. As multiple-layered selves we are denied part of our personhood when we have to deny part of who we are. Our experience simply cannot be sanitized to fit a mold in the creation of which we were not considered.

Two specific points are noteworthy. First, the disjunctive nature of the categories with which latinas/os are expected to identify collapses and simultaneously excises latina/o ethnicity from the black or white races and places latinas/os as separate from both. Second, it prevents latinas/os from claiming their racial identification, black or white, and renders invisible latinas/os of other racial and ethnic backgrounds such as Asian, Indios, Mestizos, and so on. Certainly, the insensitivity and the under- and over-inclusiveness of any generic latina/o categorization and the homogenization it engenders further the myth of a monolithic latina/o identity. What is tragically wrong with this picture is that latinas/os, in reality, are a racially and culturally diverse group. These experiences inform our perceptions differently. Consider who we are.

I moved to Albuquerque back in the summer of '82. That is the year when I started teaching at the University of New Mexico. I had been out there to find a

SOURCE: From "Building Bridges—Latinas and Latinos at the Crossroads: Realities, Rhetoric and Replacement." 25 *Colum. Hum. Rts. L. Rev.* 369 (1994). Originally published in the *Columbia Human Rights Law Review*, Reprinted by permission.

place to live in the spring. I fell in love with New Mexico when I first visited the university to interview for the teaching slot. It felt like home, the familiar Spanish influence, the rice and beans, the sunlight and the bright clothing. That summer I arrived the day before my house closing—late, with my dog in tow, and hungry. Starving, really. And when I am hungry I have to eat. But with being in a new place and all, and the excitement of the closing, the furniture arriving, etc., I figured a light meal would do. So I went into the only place I found open and ordered a tortilla, a plain tortilla. There, I was so happy, I could even order food in Spanish. The waitress looked at me kind of funny and asked, simply, "Are you sure all you want is a tortilla?" "Yes," I said. "Plain?" she asked. "Yes," I said, "it's late." So with a shrug of the shoulders she disappeared and promptly returned and put this plate in front of me. Sitting on the plate was this flat thing, white, warm, soft. My turn to ask, "And what is this?" "Your order ma'am." And we stared at each other. I ate this thing, although I did not quite know how I was supposed to do that. I got funny looks when I went at it with fork and knife. I ate, I paid, I left—still hungry and now confused. Clarity occurred several days later when I was formally introduced to the tortilla—that Mexican tortilla anyway, one that I learned to love even plain (well, with some melted butter). But that was not the tortilla I had in mind. That first night I wanted my tortilla, a simple omelette to those of us from the Caribbean.

This is a simple story about complex and diverse peoples. Here we were dealing with the same word: tortilla, and the same language: Spanish. Yet our different cultures give the word different meanings. The existing approach, one that would consider us both the same notwithstanding our common "language" and other cultural similarities, excludes. It denies the different cultural experiences of latinas/os and falsely homogenizes by making us the "same" when we are not. It creates a generic "hispanic" that in reality does not exist. Such a single-trait perspective (and whose point-of-view is it anyway?) does everyone a disservice by misinforming. A latina/o can be black, and no less latina/o, just as a woman can be latina/o and no less woman. Prevalent single-trait, uni-perspective methodologies prevent the constructive bridge-building that could occur if the focus were on universality. Yet, our differences need not make us adversaries. A multiple-perspective approach promotes understanding among different peoples by affording them the comfort to talk with each other armed with the knowledge that they will have different perspectives based on their life experiences. A multiple-perspective approach therefore promotes understanding instead of generating conflict.

TYPECASTING

Stereotyping is one of the greatest problems latinas/os must battle to debunk the myth of a monolithic latina/o. Stereotyping puts people in boxes and creates images that result in false presumptions accepted as incontrovertible truths. To be sure, we have seen how this problem plagues all women and men of color. Latinas/os can not escape its trap.

The Meat Market

I had an interview I will never forget. It was back in 1981, my first participation in the recruitment conference, back in the days when it was held in Chicago. I had been running around, up and down, a day full of thirty-minute interviews scheduled back to back. My 2:00 P.M. interview was with one of the good, progressive schools. My interviewers were four, what I would then have described as older (but today would describe as middle-aged), white men in lawyer uniform: wing-tips, pin-striped suits, white shirts, red and blue striped ties, dark socks, grey hair parted to the side, and wire-rimmed glasses. I arrived at the door, wearing my costume: a camel hair suit, blouse, pumps, leather briefcase. I knocked. El Jefe (read: dean) answered my knock. There was a pause as el Jefe and his three colleagues first looked at each other and briefly stared at me, in silence. Then they looked me up and down once, twice, three times. Silence. "Uh," said el Jefe as he proceeded to the door, "you must be at the wrong room. We are scheduled to see a Miss Ber … uh Ber … uh … Ber (mumble, groan) HERnandez." (Unpronounceable name indeed!) My turn. Pause. Extend hand, grip firmly. "I am Berta Esperanza Hernández," I introduced myself and proceeded to enter the interview suite and shake hands with the other three interviewers. Once everyone was seated we engaged in the usual obligatory preliminary chit-chat. We talked law, how exactly would I teach labor law, as I recall. Some ten of fifteen minutes into the interview (which considering its beginnings was proceeding unexpectedly smoothly) el Jefe asked, "Excuse me, do you mind telling us if you are from an academic background?" Young, yes; naive, yes; I still had radar. So I pointed to their hands (each one was holding a copy of my resume), and noted that the resume fully covered my educational experience. "No, no," el Jefe said, speaking for all of them, "we mean are you from an academic (emphasis here with arched eyebrows) background?" I too can play, I thought. "I am afraid I don't understand. I went to law school at …" Again the question (with very arched eyebrows). Again my answer, adding, "High School I attended in San Juan, Puerto Rice, St. John's Prep." Until finally el Jefe asked, "Well, er, I, what we mean, is, er, well is your father a professional"? There it was, he said it. "My father is a banker, my mother is a lawyer," I replied. Pause. Long Pause. "Um, Er, Well," said el Jefe, "I am afraid that with your background our Chicano students would not be able to relate to you. We are afraid they would consider you elite." Pause, again; long pause again, but this time it was mine. At which point I calmly (I think) stood up and, while shaking their startled hands and doing goodbye I said, "Well, I guess we do not have anything else to talk about then."

I can only speculate why the four interviewers concluded, erroneously, first that I was not me; and second, that their Chicano students would be unable to relate to me or I to them. However, it is plain that their conclusions, incorrect as they may have been, were driven initially by their pre-conceived image of what a latina law professor should look like and by their presuppositions as to what her family should be like. One thing was clear to me: their view depended upon coded ethnicity-gender and class assumptions. Fist, their image of latinas was one that did not look like me, and I hate to disappoint any who might infer I look "different," I am really rather typical looking, if a bit on the tall side: 5'8", dark eyes, olive skin. Maybe it was my lawyer uniform that threw them off; although, I must confess, almost everyone who goes to the recruitment conference wears some variant of this uniform. Moreover, latinas/os from educated, professional families did not quite fit their image of who latinas/os are supposed to be. And perhaps because I did not fit their image on either count they decided my "difference" was that I was not Chicana. This is the tragic flaw of homogenizing and stereotyping: somebody else's image of who we are, what our families are like, what we do and what we look like makes us the image. It is this imagery—gender, race, ethnic, color and class stereotyping—that falsely imprisons all of us. And I mean all. Latinas/os too.

DISCUSSION QUESTIONS

1. Why does Hernández-Truyol think that generic categories referring to racial and ethnic groups are problematic?
2. How does the use of the category "Hispanic" produce misleading assumptions about different individuals and their abilities? How was this reflected in Hernández-Truyol's job intereview?

3

The Myth of the Latin Woman

I Just Met a Girl Named María

JUDITH ORTIZ COFER

Cofer's narrative about her experience in everyday encounters shows the continuing significance of racial and ethnic stereotyping and how it feels when directed at you. She also shows the role of the media in constructing some of these stereotypes, as well as highlighting some of the generational changes that have occurred for Puerto Ricans.

On a bus trip to London from Oxford University, where I was earning some graduate credits one summer, a young man, obviously fresh from a pub, spotted me and as if struck by inspiration went down on his knees in the aisle. With both hands over his heart he broke into an Irish tenor's rendition of "María" from *West Side Story*. My politely amused fellow passengers gave his lovely voice the round of gentle applause it deserved. Though I was not quite as amused, I managed my version of an English smile: no show of teeth, no extreme contortions of the facial muscles—I was at this time of my life practicing reserve and cool. Oh, that British control, how I coveted it. But María had followed me to London, reminding me of a prime fact of my life: you can leave the Island, master the English language, and travel as far as you can, but if you are a Latina, especially one like me who so obviously belongs to Rita Moreno's gene pool, the Island travels with you.

This is sometimes a very good thing—it may win you that extra minute of someone's attention. But with some people, the same things can make *you* an island—not so much a tropical paradise as an Alcatraz, a place nobody wants to visit. As a Puerto Rican girl growing up in the United States and wanting like most children to "belong," I resented the stereotype that my Hispanic appearance called forth from many people I met.

Our family lived in a large urban center in New Jersey during the sixties, where life was designed as a microcosm of my parents' casas on the island. We spoke in Spanish, we ate Puerto Rican food bought at the bodega, and we practiced strict Catholicism complete with Saturday confession and Sunday mass at a

SOURCE: From *The Latin Deli: Prose and Poetry by Judith Ortiz Cofer*, pp. 148-154.
Copyright © 1993. Reprinted by permission of The University of Georgia Press.

church where our parents were accommodated into a one-hour Spanish mass slot, performed by a Chinese priest trained as a missionary for Latin America.

As a girl I was kept under strict surveillance, since virtue and modesty were, by cultural equation, the same as family honor. As a teenager I was instructed on how to behave as a proper señorita. But it was a conflicting message girls got, since the Puerto Rican mothers also encouraged their daughters to look and act like women and to dress in clothes our Anglo friends and their mothers found too "mature" for our age. It was, and is, cultural, yet I often felt humiliated when I appeared at an American friend's party wearing a dress more suitable to a semiformal than to a playroom birthday celebration. At Puerto Rican festivities, neither the music nor the colors we wore could be too loud. I still experience a vague sense of letdown when I'm invited to a "party" and it turns out to be a marathon conversation in hushed tones rather than a fiesta with salsa, laughter, and dancing—the kind of celebration I remember from my childhood.

I remember Career Day in our high school, when teachers told us to come dressed as if for a job interview. It quickly became obvious that to the barrio girls, "dressing up" sometimes meant wearing ornate jewelry and clothing that would be more appropriate (by mainstream standards) for the company Christmas party than as daily office attire. That morning I had agonized in front of my closet, trying to figure out what a "career girl" would wear because, essentially, except for Marlo Thomas on TV, I had no models on which to base my decision. I knew how to dress for school: at the Catholic school I attended we all wore uniforms; I knew how to dress for Sunday mass, and I knew what dresses to wear for parties at my relatives' homes. Though I do not recall the precise details of my Career Day outfit, it must have been a composite of the above choices. But I remember a comment my friend (an Italian-American) made in later years that coalesced my impressions of that day. She said that at the business school she was attending the Puerto Rican girls always stood out for wearing "everything at once." She meant, of course, too much jewelry, too many accessories. On that day at school, we were simply made the negative models by the nuns who were themselves not credible fashion experts to any of us. But it was painfully obvious to me that to the others, in their tailored skirts and silk blouses, we must have seemed "hopeless" and "vulgar." Though I now know that most adolescents feel out of step much of the time, I also know that for the Puerto Rican girls of my generation that sense was intensified. The way our teachers and classmates looked at us that day in school was just a taste of the culture clash that awaited us in the real world, where prospective employers and men on the street would often misinterpret our tight skirts and jingling bracelets as a come-on.

Mixed cultural signals have perpetuated certain stereotypes—for example, that of the Hispanic woman as the "Hot Tamale" or sexual firebrand. It is a one-dimensional view that the media have found easy to promote. In their special vocabulary, advertisers have designated "sizzling" and "smoldering" as the adjectives of choice for describing not only the foods but also the women of Latin America. From conversations in my house I recall hearing about the harassment that Puerto Rican women endured in factories where the "boss men" talked to them as if sexual innuendo was all they understood and, worse, often gave them the choice of submitting to advances or being fired.

It is custom, however, not chromosomes, that leads us to choose scarlet over pale pink. As young girls, we were influenced in our decisions about clothes and colors by the women—older sisters and mothers who had grown up on a tropical island where the natural environment was a riot of primary colors, where showing your skin was one way to keep cool as well as to look sexy. Most important of all, on the island, women perhaps felt freer to dress and move more provocatively, since, in most cases, they were protected by the traditions, mores, and laws of a Spanish/Catholic system of morality and machismo whose main rule was: *You may look at my sister, but if you touch her I will kill you.* The extended family and church structure could provide a young woman with a circle of safety in her small pueblo on the island; if a man "wronged" a girl, everyone would close in to save her family honor.

This is what I have gleaned from my discussions as an adult with older Puerto Rican women. They have told me about dressing in their best party clothes on Saturday nights and going to the town's plaza to promenade with their girlfriends in front of the boys they liked. The males were thus given an opportunity to admire the women and to express their admiration in the form of *piropos:* erotically charged street poems they composed on the spot. I have been subjected to a few piropos while visiting the Island, and they can be outrageous, although custom dictates that they must never cross into obscenity. This ritual, as I understand it, also entails a show of studied indifference on the woman's part; if she is "decent," she must not acknowledge the man's impassioned words. So I do understand how things can be lost in translation. When a Puerto Rican girl dressed in her idea of what is attractive meets a man from the mainstream culture who has been trained to react to certain types of clothing as a sexual signal, a clash is likely to take place. The line I first heard based on this aspect of the myth happened when the boy who took me to my first formal dance leaned over to plant a sloppy overeager kiss painfully on my mouth, and when I didn't respond with sufficient passion said in a resentful tone: "I thought you Latin girls were supposed to mature early"—my first instance of being thought of as a fruit or vegetable—I was supposed to *ripen,* not just grow into womanhood like other girls.

It is surprising to some of my professional friends that some people, including those who should know better, still put others "in their place." Though rarer, these incidents are still commonplace in my life. It happened to me most recently during a stay at a very classy metropolitan hotel favored by young professional couples for their weddings. Late one evening after the theater, as I walked toward my room with my new colleague (a woman with whom I was coordinating an arts program), a middle-aged man in a tuxedo, a young girl in satin and lace on his arm, stepped directly into our path. With his champagne glass extended toward me, he exclaimed, "Evita!"

Our way blocked, my companion and I listened as the man half-recited, half-bellowed "Don't Cry for Me, Argentina." When he finished, the young girl said: "How about a round of applause for my daddy?" We complied, hoping this would bring the silly spectacle to a close. I was becoming aware that our little group I was attracting the attention of the other guests. "Daddy" must have perceived this too, and he once more barred the way as we tried to walk past him. He began to shout-sing a ditty to the tune of "La Bamba"—except the lyrics were about a girl named María whose exploits all rhymed with her name and gonorrhea. The girl kept saying "Oh, Daddy" and looking at me with pleading eyes. She wanted me to laugh along with the others. My companion and I stood silently waiting for the man to end his offensive song. When he finished, I looked not at him but at his daughter. I advised her calmly never to ask her father what he had done in the army. Then I walked between them and to my room. My friend complimented me on my cool handling of the situation. I confessed to her that I really had wanted to push the jerk into the swimming pool. I knew that this same man—probably a corporate executive, well educated, even worldly by most standards—would not have been likely to regale a white woman with a dirty song in public. He would perhaps have checked his impulse by assuming that she could be somebody's wife or mother, or at least *somebody* who might take offense. But to him, I was just an Evita or a María: merely a character in his cartoon-populated universe.

Because of my education and my proficiency with the English language, I have acquired many mechanisms for dealing with the anger I experience. This was not true for my parents, nor is it true for the many Latin women working at menial jobs who must put up with stereotypes about our ethnic group such as: "They make good domestics." This is another facet of the myth of the Latin woman in the United States. Its origin is simple to deduce. Work as domestics, waitressing, and factory jobs are all that's available to women with little English and few skills. The myth of the Hispanic menial has been sustained by the same media phenomenon that made "Mammy" from *Gone with the Wind* America's idea of the black woman for generations; María, the housemaid or counter girl, is now indelibly etched into the national psyche. The big and the little screens have presented us with the picture of the funny Hispanic maid, mispronouncing words and cooking up a spicy storm in a shiny California kitchen.

This media-engendered image of the Latina in the United States has been documented by feminist Hispanic scholars, who claim that such portrayals are partially responsible for the denial of opportunities for upward mobility among Latinas in the professions. I have a Chicana friend working on a Ph.D. in philosophy at a major university. She says her doctor still shakes his head in puzzled amazement at all the "big words" she uses. Since I do not wear my diplomas around my neck for all to see, I too have on occasion been sent to that "kitchen," where some think I obviously belong.

One such incident that has stayed with me, though I recognize it as a minor offense, happened on the day of my first public poetry reading. It took place in Miami in a boat-restaurant where we were having lunch before the event. I was nervous and excited as I walked in with my notebook in my hand. An older woman motioned me to her table. Thinking (foolish me) that she wanted me to autograph a copy of my brand new slender volume of verse, I went over. She ordered a cup of coffee from me, assuming that I was the waitress. Easy enough to mistake my poems for menus, I suppose. I know that it wasn't an intentional act of cruelty, yet with all the good things that happened that day, I remember that scene most clearly, because it reminded me of what I had to overcome before anyone would take me seriously. In retrospect I understand that my anger gave my reading fire, that I have almost always taken doubts in my abilities as a challenge—and that the result is, most times, a feeling of satisfaction at having won a convert when I see the cold, appraising eyes warm to my words, the body language change, the smile that indicates that I have opened some avenue for communication. That day I read to that woman and her lowered eyes told me that she was embarrassed at her little faux pas, and when I willed her to look up at me, it was my victory, and she graciously allowed me to punish her with my full attention. We shook hands at the end of the reading, and I never saw her again. She has probably forgotten the whole thing but maybe not.

Yet I am one of the lucky ones. My parents made it possible for me to acquire a stronger footing in the mainstream culture by giving me the chance at an education. And books and art have saved me from the harsher forms of ethnic and racial prejudice that many of my Hispanic *compañeras* have had to endure. I travel a lot around the United States, reading from my books of poetry and my novel, and the reception I most often receive is one of positive interest by people who want to know more about my culture. There are, however, thousands of Latinas without the privilege of an education or the entrée into society that I have. For them life is a struggle against the misconceptions perpetuated by the myth of the Latina as whore, domestic, or criminal. We cannot change this by legislating the way people look at us. The transformation, as I see it, has to occur at a much more individual level. My personal goal in my public life is to try to replace the old pervasive stereotypes and myths about Latinas with a much more interesting set of realities. Every time I give a reading, I hope the stories I tell, the dreams and fears I examine in my work, can achieve some universal truth which will get my audience past the particulars of my skin color, my accent, or my clothes.

I once wrote a poem in which I called us Latinas "God's brown daughters." This poem is really a prayer of sorts, offered upward, but also, through the human-to-human channel of art, outward. It is a prayer for communication, and for respect. In it, Latin women pray "in Spanish to an Anglo God / with a Jewish heritage," and they are "fervently hoping / that if not omnipotent, / at least He be bilingual."

DISCUSSION QUESTIONS

1. What stereotypes does Cofer find in her everyday life? How do they show the mingling of race and gender in the way stereotypes are expressed?
2. What generational differences have emerged for Puerto Ricans living in the U.S. mainland?

4

How Does It Feel to Be a Problem?

MOUSTAFA BAYOUMI

Prior to 9/11, few White Americans probably understood what it was like for someone to hate you simply because of the group to which you belonged. Bayoumi's personal account of his post-9/11 feelings not only reveals the extent to which racial-ethnic profiling now shapes the experiences of Arab Americans but also examines what it is like to be a target of prejudice and racism.

(NEW YORK CITY, SEPTEMBER 25, 2001)

Thankfully, I was spared any personal loss. Like so many others in the city which I love I have spent much of the past two weeks reeling from the devastation. Mostly this has meant getting back in touch with friends, frantically calling them on the phone, rushing around the city to meet with them to give them a consoling hug, but knowing that really it was me looking for the hug. I dash off simple one-line emails, "let me know you're okay, okay?"

Old friends from around the world responded immediately. An email from Canada asks simply if I am all right. Another arrives from friends in Germany telling me how they remember, during their last visit to see me, the view from the top of the towers. A cousin in Egypt states in awkward English, "I hope this attack will not affect you. We hear that some of the Americans attack Arabs and Muslims. I will feel happy if you be in contact with me."

I am all right, of course, but I am devastated. In the first days, I scoured the lists of the dead and missing hoping not to find any recognizable names, but I come across the name of a three-year-old child, and my heart collapses. I hear my neighbor, who works downtown, arrive home, and I knock on her door. She tells me how she was chased by a cloud of debris into a building, locked in there for over an hour, and then, like thousands of others, walked home. I can picture her with the masses in the streets, trudging bewildered like refugees, covered in concrete and human dust. Later, I ride the subway and see a full-page picture of the towers on fire with tiny figures in the frame silently diving to their deaths, and I start to cry.

SOURCE: From *Amerasia Journal* 27:3 (2001)/28:1 (2002): 69-77. Reprinted by permission.

In the following days, I cried a lot. Then, with friends, I attended a somber peace march in Brooklyn, sponsored by the Arab community. Thousands, overwhelmingly non-Arab and non-Muslim, show up, and I feel buoyed by the support. A reporter from Chile notices my Arab appearance and asks if she can interview me. I talk to her but am inwardly frightened by her locating me so easily among the thousands. Many people are wearing stickers reading: "We Support Our Arab neighbors," which leaves me both happy and, strangely, crushed. Has it really come to this? Now it has become not just a question of whether we—New Yorkers—are so vulnerable as a city but whether we—in the Arab and Muslim communities—are so vulnerable by our appearances. Is our existence so precarious here? I want to show solidarity with the people wearing the stickers, so how can I possibly explain to them how those stickers scare me?

Before September 11, I used to be fond of saying that the relations between the Muslim world and the West have never been at a lower point since the crusades. They have now sunk lower. The English language lexicon is, once again, degraded by war. President George W. Bush's ignorant use of the word "crusade" is but a manifestation. Why don't we ask the Apache what they think of the Apache helicopter? Is there any phrase more disingenuous in the English language than "collateral damage"? ...

For the first four weeks after the attacks, I felt a bubble of hope in the dank air of New York. The blunt smell of smoke and death that hung in the atmosphere slowed the city down like I had never experienced it before. New York was solemn, lugubrious, and, for once, without a quick comeback. For a moment, it felt that the trauma of suffering—not the exercise of reason, not the belief in any God, not the universal consumption of a fizzy drink, but the simple and tragic reality that it hurts when we feel pain—was understood as the thread that connects all of humanity. From this point, I had hope that a lesson was being learned, that inflicting more misery cannot alleviate the ache of collective pain.

When the bombing began, the bubble burst. Where there was apprehension, now there was relief in the air. It felt like the city was taking a collective sigh, saying to itself that finally, with the bombing, we can get back to our own lives again. With a perverted logic, dropping munitions meant all's right with the world again.

Television, the great mediator, allows the public to feel violence or to abstract it. New Yorkers qualify as human interest. Afghans if they are lucky, get the long shot. In late October, CNN issued a directive to its reporters, for it seems that even a little bit of detached compassion is too much in the media world. "It seems perverse to focus too much on the casualties or hardship of Afghanistan," their leadership explains. "We must talk about how the Taliban is using civilian shields and how the Taliban has harbored the terrorists responsible for killing close to 5,000 innocent people." God forbid, we viewers see the pain ordinary Afghans are forced to endure. CNN must instead issue policy like a nervous state, rather than investigate how cluster bombs, freely dropped in the tens of thousands from the skies, metamorphose into land mines since about 7 percent of these soda-can-sized bombs don't explode on contact. In Canada, in the U.K., across the Arab world, this is becoming an issue. But in the United States, a cluster bomb sounds like a new kind of candy bar.

This is not to say that people in the United States are foolish, but they are by and large woefully underinformed. A study taken during the course of the Gulf War revealed that the more TV one watched, the less one actually knew about the region. In the crash course on Islam that the American public is now receiving, I actually heard a group of well-suited pundits on MSNBC (or its equivalent, I can no longer separate the lame from the loony) ask questions of a Muslim about the basics of the faith, questions like "Now, is there a difference between Moslem and Muslim?" No lie. The USA-Patriot Act, the end of the world as we know it (or at least of judicial review), actually includes the expression "Muslim descent," as if Islam is a chromosome to be marked by the human genome project. About Islam, most people in the United States still know nothing, unlike professional sports, where many are encyclopedias....

For years, the organized Arab American community has been lobbying to be recognized with minority status. The check boxes on application forms have always stared defiantly out at me. Go ahead, try to find yourself, they seem to be taunting me. I search and find that, in the eyes of the government, I am a white man.

It is a strange thing, to be brown in reality and white in bureaucracy. Now, however, it is stranger than ever. Since 1909, when the government began questioning whether Syrian immigrants were of "white" stock (desirable) or Asian stock (excludable), Arab immigrants in this country have had to contend with fitting their mixed hues into the primary colors of the state. As subjects of the Ottoman empire, and thus somehow comingled with Turkish stock (who themselves claim descent from the Caucauses, birthplace of the original white people in nineteenth century thinking, even though the location is Asia Minor), Arabs, Armenians, and other Western Asians caused a good deal of consternation among the legislators of race in this country. Syrians and Palestinians were in 1899 classified as white, but by 1910 they were reclassified as "Asiatics." A South Carolina judge in 1914 wrote that even though Syrians may be white people, they were not "that particular free white person to whom the Act of Congress had donated the privilege of citizenship" (that being reserved exclusively for people of European descent). What is it Du Bois wrote: "How does it feel to be a problem?"

In the twenty-first century, we are back to being white on paper and brown in reality. After the attacks of September 11, the flood to classify Arabs in this country was drowning our community like a break in a dam. This impact? Hundreds of hate crimes, many directed at South Asians and Iranians, whom the perpetrators misidentified as Arab (or, more confusingly, as "Muslim": again as if that were a racial category). In the days following the attack (September 14/15), a Gallup Poll revealed that 49 percent of Americans supported "requiring Arabs, including those who are U.S. citizens, to carry a special ID." Fifty-eight percent also supported "requiring Arabs, including those who are U.S. citizens, to undergo special, more intensive security checks before boarding airplanes in the U.S." Debate rages across the nation as to the legitimacy of using "racial profiling" in these times (overwhelmingly pro). The irony, delicious if it were not so tragic, is that they are racially profiling a people whom they don't even recognize as a race....

From the fall to the fallout, I have been living these days in some kind of limbo. The horrific attacks of September 11 have damaged everyone's sense of security, a principle enshrined in the Universal Declaration of Human Rights, and I wonder if for the first [time] that I can remember in the United States, we can start to reflect on that notion more carefully. All the innocents who have perished in this horrendous crime deserve to be mourned, whether they be the rescue workers, the financiers, the tourists, or the service employees in the buildings. An imam in the city has told me how a local union requested his services for a September 11 memorial of their loss since a quarter of their membership was Muslim. Foreign nationals from over eighty countries lost their lives, and the spectacular nature of the attacks meant that the world could witness the United States' own sense of security crumble with the towers. The tragedy of September 11 is truly of heartbreaking proportions. The question remains whether the United States will understand its feelings of stolen security as an unique circumstance, woven into the familiar narrative of American exceptionalism, or whether the people of this country will begin to see how security of person must be guaranteed for all. Aren't we all in this together? …

Overheard on a city bus, days after the attacks: "They will take them, like they did the Japanese, into camps. I think that's what they're going to do." In *Korematsu v. United States*, the infamous Supreme Court decision on Japanese internment, Justice Black wrote:

> It is said that we are dealing here with the case of imprisonment of a citizen in a concentration camp solely because of his ancestry, without evidence or inquiry concerning his loyalty and good disposition towards the United States. Our task would be simple, our duty clear, were this a case involving the imprisonment of a loyal citizen in a concentration camp because of racial prejudice. Regardless of the true nature of the assembly and relocation centers—and we deem it unjustifiable to call them concentration camps with all the ugly connotations that term implies—we are dealing specifically with nothing but an exclusion order.

In this exercise in rationalizing racism, Justice Black's backward logic is underscored by *his* taking offense at the term "concentration camp."

Of course, the Japanese American experience is not far from everyone's mind these days. Two months after the attack on Pearl Harbor, President Roosevelt signed Executive Order 9066, which led to the internment of over 110,000 Americans of Japanese descent. Without any need for evidence, anyone of Japanese ancestry, whether American-born or not, could be rounded up and placed in detention, all of course in the name of democracy and national security. What is less well-known is that in the 1980s, a multi-agency task force of the government, headed by the INS, had plans to round up citizens of seven Arab countries and Iran and place them in a camp in Oakdale, Louisiana, in the event of a war or action in the Middle East.

Will we see the return of the camps? I doubt it. How do you round up some seven or eight million people, geographically and economically dispersed throughout the society in ways people of Japanese descent were not in the 1940s? I suspect that this time we are not in for such measures, but we are already in the middle of something else. Over 1,000 people, most non-citizens, most Arab and Muslim, have been taken into custody under shadowy circumstances reminiscent of the *disappeared* of Argentina. Targeted for their looks, their opinions, or their associations, not one has yet to be indicted on any charge directly related to the attacks of September 11. Now, being Muslim means you are worthy of incarceration. INS administrative courts are the places where much of this happens, since non-citizens are the weakest segment of the population from a judicial point of view. Islam in this scenario becomes both racial and ideological.

In 1920, Attorney General A. Mitchell Palmer launched a nationwide assault on suspected communists and rounded up thousands without any judicial review (this event, an egregious abuse of authority, launched the ACLU). It too was directed mainly at immigrants to this country, and was covert and indiscriminate. This is what Palmer had to say about it: "How the Department of Justice discovered upwards of 60,000 of these organized agitators of the Trotsky doctrine in the United States, is the confidential information upon which the government is now sweeping the nation clean of such alien filth." John Ashcroft may be more circumspect in his language, but what we are facing now is a combination of both Yellow Peril and the Red Scare. Call it the Green Scare if you will, and recognize it is as a perilous path.

DISCUSSION QUESTIONS

1. How did the events of September 11, 2001, now called simply 9/11, redefine Arab Americans as a minority group?
2. How does their experience compare to that of Japanese Americans during World War II? How is it different?

5

The Importance of Collecting Data and Doing Social Scientific Research on Race

THE AMERICAN SOCIOLOGICAL ASSOCIATION

Some have argued that the concept of race should be abandoned, as if eliminating the term would somehow lessen racism. In the context of this debate, the American Sociological Association assembled a panel of experts who presented this statement to indicate the importance of maintaining the concept of race because of what it reveals about the realities of experience for different groups in U.S. social institutions.

RACIAL CLASSIFICATIONS AS THE BASIS FOR SCIENTIFIC INQUIRY[1]

Race is a complex, sensitive, and controversial topic in scientific discourse and in public policy. Views on race and the racial classification system used to measure it have become polarized.[2] In popular discourse, racial groups are viewed as physically distinguishable populations that share a common geographically based ancestry. "Race" shapes the way that some people relate to each other, based on their belief that it reflects physical, intellectual, moral, or spiritual superiority or inferiority. However, biological research now suggests that the substantial overlap among any and all biological categories of race undermines the utility of the concept for scientific work in this field.

SOURCE: From the American Sociological Association, "The Importance of Collecting Data and Doing Social Scientific Research on Race, 2003." Reprinted by permission of the American Sociological Association.

1. Editor's note: The complete reference list is available in the online version. See http://www.asanet.org/governance/racestmt.html

2. The federal government defines race categories for statistical policy purposes, program administrative reporting, and civil rights compliance, and sets forth minimum categories for the collection and reporting of data on race. The current standards, adopted in October 1997, include five race categories: American Indian or Alaska Native; Asian; Black or African American; Native Hawaiian or Other Pacific Islander; and White. Respondents to federal data collection activities must be offered the option of selecting one or more racial designations. Hispanics or Latinos, whom current standards define as an ethnic group, can be of any race. However, before the government promulgated standard race categories in 1977, some U.S. censuses designated Hispanic groups as race categories (e.g., the 1930 census listed Mexicans as a separate race).

How, then, can it be the subject of valid scientific investigation at the social level? The answer is that social and economic life is organized, in part, around race as a social construct. When a concept is central to societal organization, examining how, when, and why people in that society use the concept is vital to understanding the organization and consequences of social relationships.

Sociological analysis of the family provides an analogue. We know that families take many forms; for example, they can be nuclear or extended, patrilineal or matrilineal. Some family categories correspond to biological categories; other do not. Moreover, boundaries of family membership vary, depending on a range of individual and institutional factors. Yet regardless of whether families correspond to biological definitions, social scientists study families and use membership in family categories in their study of other phenomena, such as well-being. Similarly, racial statuses, although not representing biological differences, are of sociological interest in their form, their changes, and their consequences.

THE SOCIAL CONCEPT OF RACE

Individuals and social institutions evaluate, rank, and ascribe behaviors to individuals on the basis of their presumed race. The concept of race in the United States—and the inevitable corresponding taxonomic system to categorize people by race—has changed, as economic, political, and historical contexts have changed. Sociologists are interested in explaining how and why social definitions of race persist and change. They also seek to explain the nature of power relationships between and among racial groups, and to understand more fully the nature of belief systems about race—the dimensions of how people use the concept and apply it in different circumstances.

Social Reality and Racial Classification

The way we define racial groups that comprise "the American mosaic" has also changed, most recently as immigrants from Asia, Latin America, and the Caribbean have entered the country in large numbers. One response to these demographic shifts has been the effort (sometimes contentious) to modify or add categories to the government's official statistical policy on race and ethnicity, which governs data collection in the census, other federal surveys, and administrative functions. Historically, changes in racial categories used for administrative purposes and self-identification have occurred within the context of a polarized biracialism of Black and White; other immigrants to the United States, including those from Asia, Latin America, and the Caribbean, have been "racialized" or ranked in between these two categories.

Although racial categories are legitimate subjects of empirical sociological investigation, it is important to recognize the danger of contributing to the popular conception of race as biological. Yet refusing to employ racial categories for administrative purposes and for social research does not eliminate their use in daily life, both by individuals and within social and economic institutions. In France, information on race is seldom collected officially, but evidence of systematic racial discrimination remains. The 1988 Eurobarometer revealed that, of the 12 European countries included in the study, France was second (after Belgium) in both anti-immigrant prejudice and racial prejudice. Brazil's experience also is illustrative: The nation's then-ruling military junta barred the collection of racial data in the 1970 census, asserting that race was not a meaningful concept for social measurement. The resulting information void, coupled with government censorship, diminished public discussion of racial issues, but it did not substantially reduce racial inequalities. When racial data were collected again in the 1980 census, they revealed lower socio-economic status for those with darker skin.

THE CONSEQUENCES OF RACE AND RACE RELATIONS IN SOCIAL INSTITUTIONS

Although race is a social construct (in other words, a social invention that changes as political, economic, and historical contexts change), it has real consequences across a wide range of social and economic institutions. Those who favor ignoring race as an explicit administrative matter, in the hope that it will cease to exist as a social concept, ignore the weight of a vast body of sociological research that shows that racial hierarchies are embedded in the routine practices of social groups and institutions.

Primary areas of sociological investigation include the consequences of racial classification as:

- A sorting mechanism for mating, marriage and adoption.
- A stratifying practice for providing or denying access to resources.
- An organizing device for mobilization to maintain or challenge systems of racial stratification.
- A basis for scientifically investigating proximate causes.

Race as a Sorting Mechanism for Mating, Marriage, and Adoption

Historically, race has been a primary sorting mechanism for marriage (as well as friendship and dating). Until anti-miscegenation laws were outlawed in the United States in 1967, many states prohibited interracial marriage. Since then, intermarriage rates have more than doubled to 2.2 percent of all marriages, according to the latest census information. When Whites (the largest racial group in the United States) intermarry, they are most likely to marry Native Americans/American Indians and least likely to marry African Americans. Projections to the year 2010 suggest that intermarriage and, consequently, the universe of people identifying with two or more races is likely to increase, although most marriages still occur within socially designated racial groupings.

Race as a Stratifying Practice

Race serves as a basis for the distribution of social privileges and resources. Among the many arenas in which this occurs is education. On the one hand, education can be a mechanism for reducing differences across members of racial categories. On the other hand, through "tracking" and segregation, the primary and secondary educational system has played a major role in reproducing race and class inequalities. Tracking socializes and prepares students for different education and career paths. School districts continue to stratify by race and class through two-track systems (general and college prep/advanced) or systems in which all students take the same courses, but at different levels of ability. African Americans, Hispanics, American Indians, and students from low socioeconomic backgrounds, regardless of ability levels, are over-represented in lower level classes and in schools with fewer Advanced Placement classes, materials, and instructional resources.

Race as an Organizing Device for Mobilization to Maintain or Challenge Systems of Racial Stratification

Understanding how social movements develop in racially stratified societies requires scholarship on the use of race in strategies of mobilization. Racial stratification has clear beneficiaries and clear victims, and both have organized on racial terms to challenge or preserve systems of racial stratification. For example, the apartheid regime in South Africa used race to maintain supremacy and privilege for Whites in nearly all aspects of economic and political life for much of the 20th century. Blacks and others seeking to overthrow the system often were able to mobilize opposition by appealing to its victims, the Black population. The American civil rights movement was similarly successful in mobilizing resistance to segregation, but it also provoked some White citizens into organizing their own power base (for example, by forming White Citizen's Councils) to maintain power and privilege.

Race and Ethnicity as a Basis for the Scientific Investigation of Proximate Causes and Critical Interactions

Data on race often serve as an investigative key to discovering the fundamental causes of racially different outcomes and the "vicious cycle" of factors affecting these outcomes. Moreover, because race routinely interacts with other primary categories of social life, such as gender and social class, continued examination of these bases of fundamental social interaction and social cleavage is required. In the health arena, hypertension levels are much higher for African Americans than other groups. Sociological investigation suggests that discrimination and unequal allocation of society's resources might expose members of this racial group to higher levels of stress, a proximate cause of hypertension. Similarly, rates of prostate cancer are much higher for some groups of men than others. Likewise, breast cancer is higher for some groups of women than others. While the proximate causes may appear to be biological, research shows that environmental and socio-economic factors disproportionately place at greater risk members of socially subordinated racial and ethnic groups. For example, African Americans' and Hispanics' concentration in polluted and dangerous neighborhoods result in feelings of depression and powerlessness that, in turn, diminish the ability to improve these neighborhoods. Systematic investigation is necessary to uncover and distinguish what social forces, including race, contribute to disparate outcomes.

RESEARCH HIGHLIGHTS: RACE AND ETHNICITY AS FACTORS IN SOCIAL INSTITUTIONS

The following examples highlight significant research findings that illustrate the persistent role of race in primary social institutions in the United States, including the job market, neighborhoods, and the health care system. This scientific investigation would not have been possible without data on race.

Job Market

Sociological research shows that race is substantially related to workplace recruitment, hiring, firing, and promotions. Ostensibly neutral practices can advantage some racial groups and adversely affect others. For example, the majority of workers obtain their jobs through informal networks rather than through open recruitment and hiring practices. Business-as-usual recruitment and hiring practices include recruiting at predominantly White schools, advertising only in suburban newspapers, and employing relatives and friends of current workers. Young, White job seekers benefit from family connections, studies show. In contrast, a recent study revealed that word-of-mouth recruitment through family and friendship networks limited job opportunities for African Americans in the construction trades. Government downsizing provides another example of a "race neutral" practice with racially disparate consequences: Research shows that because African Americans have successfully established employment niches in the civil service, government workforce reductions displace disproportionate numbers of African American—and, increasingly, Hispanic—employees. These and other social processes, such as conscious and unconscious prejudices of those with power in the workplace, affecting the labor market largely explain the persistent two-to-one ratio of Black to White unemployment.

Neighborhood Segregation

For all of its racial diversity, the highly segregated residential racial composition is a defining characteristic of American cities and suburbs. Whites and African Americans tend to live in substantially homogenous communities, as do many Asians and Hispanics. The segregation rates of Blacks have declined slightly, while the rates of Asians and Hispanics have increased. Sociological research shows that the "hyper-segregation" between Blacks and Whites, for example, is a consequence of both public and private policies, as well as individual attitudes and group practices.

Sociological research has been key to understanding the interaction between these policies, attitudes, and practices. For example, according to attitude surveys, by the 1990s, a majority of Whites were willing to live next door to African Americans, but their comfort level fell as the proportion of African Americans in the neighborhood increased. Real estate and mortgage-industry practices also contribute to neighborhood segregation, as well as racially disparate homeownership rates (which, in turn, contribute to the enormous wealth gap between racial groups). Despite fair housing laws, audit studies show, industry practices continue to steer African American homebuyers away from White neighborhoods, deny African Americans information about available loans, and offer inferior property insurance.

Segregation profoundly affects quality of life. African American neighborhoods (even relatively affluent ones) are less likely than White neighborhoods to have high quality services, schools, transportation, medical care, a mix of retail establishments, and other amenities. Low capital investment, relative lack of political influence, and limited social networks contribute to these disparities.

Health

Research clearly documents significant, persistent differences in life expectancy, mortality, incidence of disease, and causes of death between racial groups. For example, African Americans have higher death rates than Whites for eight of the ten leading causes of death. While Asian–Pacific Islander babies have the lowest mortality rates of all broad racial categories, infant mortality for Native Hawaiians is nearly three times higher than for Japanese Americans. Genetics accounts for some health differences, but social and economic factors, uneven treatment, public health policy, and health and coping behaviors play a large role in these unequal health outcomes.

Socio-economic circumstances are the strongest predictors of both life span and freedom from disease and disability. Unequal life expectancy and mortality reflect racial disparities in income and incidence of poverty, education and, to some degree, marital status. Many studies have found that these characteristics and related environmental factors such as over-crowded housing, inaccessibility of medical care, poor sanitation, and pollution adversely impact life expectancy and both overall and cause-specific mortality for groups that have disproportionately high death rates.

Race differences in health insurance coverage largely reflect differences in key socio-economic characteristics. Hispanics are least likely to be employed in jobs that provide health insurance and relatively fewer Asian Americans are insured because they are more likely to be in small low-profit businesses that make it hard to pay for health insurance. Access to affordable medical care also affects health outcomes. Sociological research shows that highly segregated African American neighborhoods are less likely to have health care facilities such as hospitals and clinics, and have the highest ratio of patients to physicians. In addition, public policies such as privatization of medicine and lower Medicaid and Medicare funding have had unintended racial consequences; studies show a further reduction of medical services in African American neighborhoods as a result of these actions.

Even when health care services are available, members of different racial groups often do not receive comparable treatment. For example, African Americans are less likely to receive the most commonly performed diagnostic procedures, such as cardiovascular and orthopedic procedures. Institutional discrimination, including racial stereotyping by medical professionals, and systemic barriers, such as language difficulties for newer immigrants (the majority of whom are from Asia and Latin America), partly explain differential treatment patterns, stalling health improvements for some racial groups.

All of these factors interact to produce poorer health outcomes, indicating that racial stratification remains an important explanation for health disparities.

SUMMARY: THE IMPORTANCE OF SOCIOLOGICAL RESEARCH ON RACE

A central focus of sociological research is systematic attention to the causes and consequences of social inequalities. As long as Americans routinely sort each other into racial categories and act on the basis of those attributions, research on the role of race and race relations in the United States falls squarely within this scientific agenda. Racial profiling in law enforcement activities, "redlining" of predominantly minority neighborhoods in the mortgage and insurance industries, differential medical treatment, and tracking in schools, exemplify social practices that should be studied. Studying race as a social phenomenon makes for better science and more informed policy debate. As the United States becomes more diverse, the need for public agencies to continue to collect data on racial categories will become even more important. Sociologists are well qualified to study the impact of "race"—and all the ramifications of racial categorization—on people's lives and social institutions. The continuation of the collection and scholarly analysis of data serves both science and the public interest. For all of these reasons, the American Sociological Association supports collecting data and doing research on race.

DISCUSSION QUESTIONS

1. Why does the American Sociological Association think it is important to maintain race as a way of classifying humans?
2. What is some of the evidence of the significance of race in different social institutions?

Part I
Student Exercises

1. Observe your community and make note of its racial and ethnic makeup. What different groups do you find? Given the population of the United States, who is missing in the community? To find out, go to the U.S. Census Bureau home page (www.census.gov) and, using the American Fact Finder feature, select "Fact Sheet." Then fill in your own state and county and click on "Go." What does the resulting information reveal about the distribution of racial-ethnic groups in your area? How accurate were your own observations? Are these groups evenly distributed throughout the county or are they concentrated in particular areas?
2. For one week, keep a written log of every time the subject of race comes up in the conversations you hear around you. Make note of what people said and the tone in which they said it. You should also note, if possible, the age and race of the person making the comments. At the end of the week, review your log and answer the following questions:
   a. What evidence of racial attitudes did you find?
   b. How would you compare the attitudes of those belonging to different racial-ethnic groups?
   c. What do your observations reveal to you about the everyday reality of racism?

PART II

INTRODUCTION

The Social Construction of Race and Ethnicity

What does it mean to say that race is a social construction? The idea of race as a social construction is one of the most important things to learn in analyzing race in society. Most people tend to think of race in terms of skin color or other biological features that seemingly distinguish different groups, but is this a reasonable understanding of this concept? Race has been used historically to differentiate groups, but why not use eye color or height or some characteristic other than color to categorize people into so-called races? The answer lies in the historical treatment of different groups and the significance that society gives to features believed to mark different racial groups. Thus, to say that race is socially constructed is to say that what is important about race is not biological difference, but the different ways groups have been treated in society.

THE COMPLEXITY OF COMPLEXION

To begin with, race is not a matter of biology, although racist thinking claims this is so. Scientists in the Human Genome Project have concluded that there is no "race" gene. This project, which allows scientists to map people's genetic makeup, shows that there are far more similarities among people than there are differences, even when you take into account such things as eye color, skin color, and other physical characteristics. Scientists have concluded that genetic variation among human beings is indeed very small, and thus, as one of the biologists working in human genetics writes, "There is no biological basis for the separation of human beings into races" (Graves 2001: 1). What then is race?

The idea of race has developed within the context of the social institutions and practices in which groups socially defined as races have been exploited, controlled, and—in some cases—enslaved. Imagine this scenario: You have a social system in which some people are forced into slavery based on their presumed inferiority. The dominant group then has to create a belief system that supports this exploitation. Their solution is creating "race" as an idea they can use to justify the exploitation of groups perceived as different.

Sociologists define **race** as a group that is treated as distinct in society because of certain perceived characteristics that have been defined as signifying superiority or inferiority. Note that in this definition, perception, belief, and social treatment are the key elements defining race, not the actual characteristics of human groups. Furthermore, so-called "races" are created within a system of social dominance. *Race is thus a social construction, not an attribute of certain groups of people.*

Howard Taylor ("Defining Race") explores the complexity of defining race based on the multiple ways that race can be conceptualized. He helps us see that race is not a fixed personal attribute. Rather, it is fundamentally rooted in social definitions: how people see each other, how they define their own identity, and how they are situated within a social order—an order that has been structured along lines of inequality.

Ann Morning ("Race") furthers Taylor's analysis by looking at the different dimensions of how race is defined. She discusses how the census defines race and also writes about the increasing attention given to biological explanations of race. In Morning's discussion, assumptions about biological differences by race embed sociological processes. She is suggesting, as other sociologists would agree, that social processes are involved in the definition of race.

Abby Ferber ("Planting the Seed: The Invention of Race") takes us back to the establishment of race as a concept—a relatively recent development in modern history. She explains how race emerged through the work of quasi-scientists in the eighteenth century as White Europeans sought to explain and rationalize the exploitation of African people. She shows how the process of developing systems of racial classification is "intertwined with the history of racism." This explanation is important for understanding how racist thinking has emerged and how it is tied to the exploitation of people of color worldwide. Her work also shows how racism arose coterminously with the rise of science, and she will make you think twice about using the term *Caucasian*, commonly used to refer to White people, once you learn the term's racist origins.

Karen Brodkin ("How Did Jews Become White Folks?") illustrates the social construction of race in a different context. She writes about the long history of **anti-Semitism** (defined as the hatred of Jewish people) and tells us how at different times in world history anti-Semitic thinking constructed Jews, not as a religious ethnic group, but as a separate race. Thus, the Nazi regime of Germany in the 1930s and 1940s constructed Jewish people as an inferior race—and systematically murdered millions of them as a result. Brodkin's essay links anti-Semitism to other forms of racism in that anti-Semitism was supported by *scientific racism* (the use of quasi-science to support racism). As Brodkin shows, even in the United States, racism defined the only "real Americans" as native-born White people. American thinking, buttressed by exclusionary laws, defined Asian immigrants and many southern Europeans as outcasts, stripped them of rights, and excluded them from many social institutions. On the other hand, as Brodkin shows, Jewish Americans have been upwardly mobile. Why? Because over time they became seen as "White" and conditions in the United States after World War II opened up opportunities to those defined as "White."

Brodkin's essay also brings out another important point: that ethnic groups can become racialized. **Ethnic groups** are those that share a common culture and that have a shared identity. Ethnicity can thus stem from religion, national origin, or other shared characteristics. Important to this definition of ethnic group is not just the shared culture, but the sense of group belonging. Thus, Jewish people are an ethnic group, as are Italian Americans, Cajuns, and Irish American Catholics. Ethnicity can exist even within a so-called racial group (such as Jamaicans, Haitians, or Cape Verdeans among African Americans). Some ethnic groups have historically been "racialized," as Brodkin's essay about Jews shows. This has typically occurred during periods of high rates of immigration when anti-immigrant sentiment swelled and groups like the Irish were defined by dominant groups as racially inferior. The fact that we do not now consider Jews or the Irish to be a "race" shows how powerful the social construction of race can be.

That concepts of race shift and change does not mean that race is not real. Race is real—but real in a social and historical sense. By recognizing the socially constructed character of race, you are not denying that it exists but are recognizing its social reality. Sociologist Joe Feagin argues that race is systemic. The term **systemic racism** refers to the "recurring and unequal relations between groups and individuals" (2000: 19) that are organized along lines of race. In other words, racism is not just *in*, but is *of* society. Race and racism have been built into the institutions of society right from the founding of the nation.

As the result of systemic racism, dominant groups (namely, White elites) accumulate resources over time, providing them with what Feagin identifies as *unjust enrichment*. At the same time, those disadvantaged by a racist system experience *unjust impoverishment*. Within such a system, individuals—regardless of their own attitudes and beliefs—can benefit or be deprived of resources because of the systemic way that racism works. Thus, even the well-intentioned anti-racist White person can unintentionally benefit from what has been four centuries of racial hierarchy, because racism, as we will see further in the following part, extends beyond individual good and bad will.

Michael Omi and Howard Winant ("Racial Formation") introduce the concept of **racial formation** to refer to the social and historical process by which racial categories are created. Historically in the United States, one's racial membership was determined by law—though, interestingly enough, the meaning of race varied from state to state. In Louisiana, for example, you were defined as Black if you had one Black ancestor out of thirty-two; in Virginia, it was one in sixteen; in Alabama, you were Black by law if you had *any* Black ancestry. By just crossing state borders, a person would legally change his or her race!

Racial formation means that social structures, not biology, define race. Because a group's perceived racial membership has been the basis for group oppression, this understanding of race shows how systems of authority and governance construct concepts of race. There are, of course, sometimes observable differences between individuals, but it is what these differences have come to mean in history and society that matters.

In a society where certain groups are exploited—robbed of their lands, their labor appropriated for the benefit of others, their communities starved of resources—it does matter whether you are defined as Black, White, or some other so-called race. Historically, White, elite men determined the only groups who were counted as "White"—and, therefore, given full rights of citizenship. Who counts as a White person is a process of law and social judgments. Consider the case of Takao Ozawa, originally from Japan. Ozawa arrived in California in 1894, graduated from high school, and attended the University of California, Berkeley. After that, he worked for an American company, living with his family in the U.S. territory of Honolulu. Because he was not legally classified as White, he could not become a U.S. citizen, though he very much wanted to and filed for citizenship in 1914. Ozawa argued that he was a "true American," a person of good character who neither drank, smoked, nor gambled. His family went to an American church; his children went to American schools; he spoke only English at home and raised his children as Americans. In 1922, the U.S. Supreme Court denied his eligibility for citizenship based on the claim that he was not "Caucasian" (Takaki 1989; *Ozawa v. United States* 260 U.S. 178, 1922).

Why would our nation have such laws? Such laws emerge when social resources are distributed based on membership in groups being judged as superior or inferior, "alien," or "colored." During the history of racial formation in America, many groups have been denied the rights of freedom and full citizenship, as many of the articles in this book show.

The authors in this part explore the social construction of race in diverse ways. Michael Omi and Howard Winant ("Racial Formation") establish the framework of racial formation as a way of locating race in the social and historical processes that create racial categories. They also develop the concept of **racial projects** to explain organized efforts by any group to redistribute social and economic resources along racial lines. The authors note that resources are both material—such as money, housing, schooling, and so forth—as well as *representational*—that is, how groups are defined and depicted within a society and culture. Omi and Winant's perspective shows us that racism is not just a matter of individual attitudes and prejudices, although those surely are important, as we will see in the next part.

REFERENCES

Faegin, Joe R. 2000. *Racist America: Roots, Current Realities, and Future Reparations*. New York: Routledge.

Graves, Joseph L., Jr. 2001. The *Emperor's New Clothes: Biological Theories of Race at the Millennium*. New Brunswick, NJ: Rutgers University Press.

Takaki, Ronald. 1989. *Strangers from a Different Shore: A History of Asian Americans*. NewYork: Penguin.

HOWARD TAYLOR (DEFINING RACE)

ANN MORNING (RACE)

ABBY Ferber (Planting the seed: THE INVENTION OF RACE)

KAREN BRODKIN

Michael Omi + Howard Winant

(RACIAL FORMATION)

6

Defining Race

HOWARD F. TAYLOR

This article shows the complex and multiple dimensions around which the concept of race is socially constructed. Taylor shows that race is a social status, one that takes on meaning within the context of social identities and social relationships.

Race is with us every minute of every day. Despite our constant protestations against it and its realities, it, like gender, permeates every fiber of our very existence. Race is causally and intimately related to hundreds of forces and has hundreds of consequences. Our race determines, beyond chance, how long we will live. In general, racial minorities in America have a lower life expectancy than Whites. American minorities have less access to medical care; as a consequence, minorities are more burdened with chronic and life-threatening illnesses. American minorities, particularly Blacks, Hispanics, and American Indians, have considerably lower annual incomes relative to Whites, even Whites who have the same level of education as their minority counter-part. Thus Blacks, Hispanics, American Indians, and also Asians, have greater odds of being poor than Whites. Blacks, Hispanics and American Indians are on average promoted less often in the workplace than are Whites with the same or even less education. All three racial minorities routinely suffer housing discrimination and are less likely to be offered low-interest housing mortgages in an urban area than are Whites who live in the same urban area. Surveys find that members of all three minority groups feel more alienated from the institutions of society than do Whites of the same socioeconomic status. Finally, Blacks and Hispanics are more likely to be arrested for crime, more likely to be held without bail, more likely to be sentenced, and more likely to receive longer sentences than are Whites from the same part of the city or town, of the same or similar social class, and who have exactly the same record of prior arrests.

So: race matters. It matters a lot. What, then, is "race?" What are the definitions of race that are used in American society? How can one tell who is of which race? If the definitions of race in American society are inexact, as they are, then can one define race in some way so as to be able to research its effects? How can we actually measure one's race?

SOURCE: From Howard F. Taylor, "Defining Race." Reprinted by permission of the author.

Race is multiply defined in this society; that is, there is no one single definition, but several definitions. All these definitions apply simultaneously, and no one definition takes precedence over another. These definitions do however have one thing in common: They are all creations of society. They have been put into place by humans and by their social interactions and their societal institutions. Let us have a look at these definitions.

RACE AS A BIOLOGICAL CATEGORY

Most people grew up thinking that the real definition of "race" is that it is a strictly biological category. This is only partly correct. We were taught to place people into "race" categories such as Caucasian (White), Negroid (Black; Negro; African American), Mongolian (Asian), and so on, on the basis of secondary physical characteristics such as skin color, hair texture, lip form, nose form, and so on. This definition and classification, put forth in the late nineteenth and early twentieth centuries by some physical anthropologists and now considered vastly outdated, has had considerable influence on how people think about the matter of race right up to the present time. It gave rise to the idea that racial classification was "scientific" since it was based on physical traits and upon thinking at the time in the field of physical anthropology. Even some sociology texts as recently as the 1940s defined race in this manner, and furthermore even defined race as a "subspecies" of humankind (Young, 1942)!

It is now generally agreed upon by researchers in both the physical as well as the social and behavioral sciences that only a small part of the definition of race in this society is based on these secondary physical characteristics. Such characteristics do play a part in how people themselves define race, to be sure, but only a part. This is because the human variability between what are regarded in society as "racial groups" occurs mostly within racial groups rather than between them. Thus, the skin color of African Americans varies from extremely light/white (even blond and blue-eyed) all the way to very dark brown. There are White people who have dark skin and very curly hair and full lips yet they are still White, and are so regarded in their immediate communities. Similarly, there are African Americans with white skin, thin lips, and straight hair. So skin color (and lip form and hair texture) is not a very good indicator of race.

The population geneticist Richard Lewontin (1996) notes that even for strictly physical traits involving body chemistry, blood type, and other strictly physical traits, almost all the variability on such traits is within racial categories; almost no variability in such traits exists between races. Lewontin estimates that the overlap in physical traits between any two groups designated as "races" in American society is more than 99 percent. That applies as well to genetics: Any two racial groups compared are 99 percent similar genetically. Clearly, then, in physical characteristics, the races are far more similar than they are different. It is in this sense that one hears that race is "not a true scientific concept." In the sense of thousands of physical traits and characteristics, this is certainly true.

Does this mean that race is no longer important, that race no longer matters? Certainly not. We have already noted that race is a fundamental and firmly ingrained part of human existence, not only in America, but in most other societies as well. How then is it that a concept with virtually no "scientific" validity (at least, in the physical science sense) has come to be so important and intrusive in human existence? The answer lies in the realization that the definition of race in America, and in most other societies as well, is largely social.

RACE AS A SOCIAL CONSTRUCTION

To say that race is a social construction means that it and its definition grow out of the process of human interaction. This means that race is what interacting humans define it to be. In this sense, how you are perceived in a community of peers in part defines your race: If your friends and associates see you as African American, then in that one respect you are indeed African American.

As another example of social construction, if race-like divisions among people in this society were made only on the basis of who had red hair and who did not, then people would come to think of races as being defined by hair color—a physical characteristic. Over time, this definition would come to be upheld by society's institutions—by the courts, by the educational system, by the federal government, and so on. It might well come to be thought of as a truly "scientific" classification, since it is based on a physical characteristic (hair color instead of skin color), and as everyone knows, redheads are different from everyone else—they have fiery tempers and argue a lot, don't they? Thus, racial stereotypes would soon be applied to redheads!

Social construction means that people learn, through socialization and interaction processes, to attribute certain characteristics to people who are classified into a racial category. These are just what racial stereotypes are: attributions that are for the most part not true and yet stubbornly persist over time. These stereotypes are social constructions. They are generally based on only a small truth (if on any truth at all) and are then thought of in society as applying to all members, and to "typical" members, of some racial category.

Stereotypes are generally negative. We have all heard the common stereotypes: Blacks are inherently musical, possess "natural" rhythm, are loud, and crime-prone; Asians are sneaky and overly conforming; Hispanics are naturally violence-prone and carry knives; American Indians are quiet, subservient, and underachievers. These traits are seen by society as inherent to any member of the particular group. In fact they are seen as essential to the group identified by the stereotype. Sociologists call this process essentialization: Such negative stereotypical traits are regarded by society as essential (inherent) to the character of any person identified by the stereotype. Negative stereotypes, thus negative essentializations, are applied far more to minorities than to Whites, and they thus help define what it means to be minority in society, even though Whites are sometimes ridiculed as having a few negatively stereotyped traits ("White boys can't jump," "blondes are dumb," and so forth).

RACE AS AN ETHNIC GROUP

A group is an ethnic group if its members are united by a common culture. Culture includes language, religion, tools, music, habits, socialization practices, and many other elements. Racial groups are ethnic groups. They generally share a common culture—not perfectly, of course, but to an extent great enough so as to be able to identify some set of cultural elements held in common. Thus African Americans are not only regarded as a racial group but they are also an ethnic group. They possess many elements of a common culture (music, art, linguistic similarities, a sense of "we" feeling and identity, and so on). There are even multiple ethnic groups among African Americans such as Cape Verdeans, Haitian Americans, Gullah Islanders, and so forth. Jews are also an ethnic group—a group sharing a common or nearly common religion (religious culture), but Jews are not regarded as a race.

Sometimes an ethnic group in a society comes to be thought of in that society as a race. The best example of this is Hitler's definition of Jews as a race in Nazi Germany in the 1930s and early 1940s—the "Jewish race." This led to heavy negative stereotyping of Jews in Germany and other countries and eventually to the killing of millions of Jews by means of starvation and gas chambers during the Holocaust. What had been an ethnic group came to be defined in Nazi society as a "race," though we now know that Jewish people are not a race.

When any group—whether an ethnic group or a social class—comes to be thought of as a race and then is actually defined by society as a race, this means the group has become racialized. Thus Hitler racialized Jews. By so doing, it was easier to then stereotype Jews and to regard them as a separate and inferior category of people. It was the process of racialization that allowed Hitler to do this and to convince many German citizens that Jews were bad and needed to be totally eliminated. He almost succeeded. This underscores the point that races tend to be defined in a society by social processes, less so than by physical characteristics. This shows again that race is a social construction.

RACE AS A SOCIAL CLASS OR PRESTIGE RANK

In parts of Brazil, such as the province of Bahia, the higher one's social class status and wealth, the more likely one is to be formally listed in the country's census as "White" (Surratt and Inciardi 1998; Patterson 1982). This is true even if the person in question is dark-skinned and clearly of largely African ancestry. Thus while skin color is certainly important to defining who is of what race ("color") in Brazil, wealth and social standing are just as important. A favorite expression in Brazil is: "Money whitens" (o dinheiro embranquence)! People with mixed physical characteristics may be labeled "White" if they appear well-dressed and occupy a prestigious professional occupation. Similarly, a poor, light-skinned person may be labeled "Black" to indicate low social standing.

The use of color as a label for people in Brazil is far more complex than is the case in the United States. The simple distinctions of "Black," "White,"

"Yellow," and "Red" used in the United States are seen in Brazil as utterly ridiculous. There, a large number of labels are used to identify different colors of individuals—reddish brown; reddish dark brown; reddish brown with a hint of yellow; and so on. A fairly recent survey in Brazil revealed 143 color labels used for the population, including gray, pink, dirty white, and cinnamon (Ellison 1995). These race-like labels for individuals are used in conjunction with their wealth and social standing. This shows that how your color is perceived and thus labeled within a society can be significantly determined by your social class or prestige rank within that society.

RACE AS A "RACIAL FORMATION" OF SOCIETY'S INSTITUTIONS

What a race is, and who is of what race, can be defined by one or more of society's institutions, such as the government, the educational institution, the state and federal legal system, and the criminal justice system. An insightful theory in this regard is Omi and Winant's (1994) racial formation theory. This theory argues that over time, society's powerful institutions define what race is and who is to be classified within what racial category.

Such has been the case in the United States. During slavery, and after the Emancipation Proclamation of 1863, Black people were defined in the law as "Negro" if they had only a small fraction of Black ancestry ("Black blood"), even so small an amount as one great-great-great grandparent who was Black. This is the origin of the one-drop rule, namely, one was "Negro" if he or she had but one drop of Negro blood. The amount of ancestry ("blood") required varied somewhat from state to state, but it was a small amount. The negative stigma of Blackness needed to be only minuscule for one to be totally labeled. (In some states, one Black great-great-great-great grandparent was sufficient to render one a "total" Negro!) The point is that by law it was the government, via the law, that determined how race was to be defined. The individual was not allowed to define her or his own race; it was up to the government, and the government carried the strong sanction of the law. This is what is meant by racial formation. The "one-drop rule" exists even to this day in the public mind.

A clear illustration of how the government determines race is how one is designated as American Indian (Native American Indian) in the United States. According to the U.S. government, one cannot simply decide to call oneself "Indian." Through what is called the federal acknowledgement process, the U.S. government declares one to be an American Indian, but one must go through an elaborate bureaucratic process, involving paper forms and legal documents, and then be certified as a member of some Indian tribe. The U.S. government maintains a list of Indian tribes that it considers to be "legitimate," and the individual must be able to prove membership in one of those pre-designated tribes. Moreover, the tribes themselves must be certified as legitimate—and experts estimate that about 569 American Indian groups have been so designated. This means that the government, through this type of racial formation, has more say in whether or not American Indian people are American Indians than do the people themselves! Obviously, not all Native American Indians agree with this procedure.

RACE AS SELF-DEFINED

Finally, a person's race may be defined in part simply by what that person calls herself or himself. When asked "What race are you?" (or even simply "What are you"?), what then does the person say? A person with one Black parent and one White parent may well say "I am Black" (an implicit use of the one-drop rule); or, they may say "I am White" (which may conflict with how they appear to others); or, they may say "I am bi-racial" or "I am of mixed race." Or, still further, the person may refuse to be classified by race, arguing that race is a false classification in the first place. (If such a person has light skin and as a result chooses to live as a White person—severing all ties with Black relatives and also the Black community—that person by tradition could be accused by the Black community of "passing," that is, passing for White. Designating someone as "passing" is less common now than it was a decade or two ago, but it is still done.) On some present-day college campuses, a light-skinned Black person who appears to be passing for White and who denies being Black is (only somewhat) jokingly called an "incogNegro" (or simply, "incog")—a pun on "incognito!"

A person is, in principle, free to designate himself or herself in any way they might wish. The problem is that all the other five definitions that we have given also, simultaneously, come into play. What if the person with one Black parent and one White parent has brown skin and looks to everyone else like a Black person, that is, the person is dark enough of skin in order to be called "Black" among her or his peers? This would show intervention of both the biological as well as the social constructionist definitions, and perhaps the racial formation definition as well. (Golf star Tiger Woods is of African American and Asian parentage and has racially designated himself—though somewhat jokingly—as Cablinasian. What would you call him?)

The point is that there is more to race than one's personal definition even as applied to one's own self. It may sound odd to quarrel with a person about what they choose to call themselves racially. But such is the reality of race in the United States—and most other countries as well: It is not totally up to the person alone!

We are reminded in this context of the story of a dark-skinned Black man in the southern U.S. who in the early 1950s boarded a bus and proceeded to sit in the front row. The laws and norms of segregation in much of the South [at that time] prevented Black people from sitting up front on a bus. Consequently, the bus driver turned to the man and said: "Boy, you can't sit there!" The man replied, "Well, why not?" "Because," said the bus driver, "you are a Negro, and Negroes cannot sit up here in the front of the bus!" "Oh, well in that case," said the man, "I can stay right here. I have resigned from the Negro race!"

One is left to contemplate what race would be like in this country if one could easily "resign" from a race!

CONCLUSION

"Race" in the United States is not defined by one single definition, but simultaneously by several definitions. Six definitions of race were explored here. One's race is defined by a combination of the following: by one's physical appearance, such as by skin color (the biological definition); by social construction (any definition arising out of the process of human interaction, such as how those around you define you); as an ethnic group (Hitler defined the Jews as a race); as a social class rank (as in Brazil); as racial formation (such as the U.S. government's definition of who is American Indian); and finally, by one's own self-definition. No one definition is dominant over another in U.S. society. Each definition shows the significance of society in defining race.

REFERENCES

Ellison, K. 1995. "Brazil's Blacks, Building Pride, Invoke the Legend of Rebel Warrior Zumbi." *Miami Herald* (November 19): 1A.

Lewontin, Richard. 1966. *Human Diversity*. New York: W. H. Freeman.

Omi, Michael, and Howard Winant. 1994. *Racial Formation in the United States*, 2nd ed. New York: Routledge.

Patterson, Orlando. 1982. *Slavery and Social Death: A Comparative Study*. Cambridge: Harvard University Press.

Surratt, Hilary L., and James A. Inciardi. 1998. "Unraveling the Concept of Race in Brazil: Issues for the Rio de Janeiro Cooperative Agreement Site." *Journal of Psychoactive Drugs* 30 (July–September): 255–260.

Young, Kimball. 1942. *Sociology: A Study of Society and Culture*. New York: American Book Co.

DISCUSSION QUESTIONS

1. What are the different ways race can be defined in society? How does this challenge the understanding of race as simply a fixed, biological category?
2. How do social beliefs shape our understanding of race?
3. How does Taylor's essay indicate that race matters, even if it is a social construction?

7

Race

ANN MORNING

Morning discusses the difficulty of defining race. Her review of the various ways that race is defined shows how it is socially constructed. Morning also shows how recent claims to a biological definition of race can be sociologically interpreted.

Race is part of everyday life in the United States. We're asked for our race when we fill out forms at school or work, when we visit a government agency or doctor's office. We read or hear about race in the daily news, and it comes up in informal discussions in our neighborhoods and social circles. Most of us can apply—to ourselves and the people around us—labels like white or Asian. Yet for all its familiarity, race is strangely difficult to define. When I've asked people to explain what race is, many have trouble answering.

Two uncertainties are widespread. First, there is confusion about the relationship between *race and ethnicity.* Are these different concepts? How often have you come across descriptions of someone's "ethnicity" that use terms like *white* and *black?* The public, the media, politicians, and scientists often use *race* and *ethnicity* interchangeably. Both terms have something to do with our ancestral origins, or "background," and we often find both linked to ideas about "culture." As historian David Hollinger points out, we often use the term *multicultural* to refer to racial diversity. In doing so, we presume that racial groups have different cultural beliefs or practices, even though the way we classify people by race has little to do with their behavior, norms, or values.

Defining race is also a challenge because we are unsure how it is related to biology. Are racial categories based on surface physical characteristics? Do they reflect unobserved patterns of genetic difference? If race is a kind of biological taxonomy, we are uncertain about exactly which traits anchor it.

Clear-cut definitions of race are surprisingly elusive. *The New Oxford American Dictionary*, for example, equates *race* with *ethnic group*, and links it to a wide range of possible traits: "physical characteristics," "culture," "history," and "language." The U.S. Census Bureau is another place to look for an authoritative definition. In contrast to the dictionary definition, the federal government rejects both culture and biology as relevant to race. This is apparent in its approach to racial enumeration and how it explains its definition. The U.S. Census makes the most

SOURCE: From *Contexts*, Vol. 4, No. 4, 2005 pp. 44-46. Copyright © American Sociological Association. Reprinted by permission of The University of California Press.

visible use of the official racial categories that the Office of Management and Budget (OMB) first promulgated in 1977 and revised in 1997. These standards require all federal agencies to use the following classifications in their data collection and analysis:

1. American Indian and Alaska Native
2. Asian
3. Black or African American
4. Native Hawaiian and Other Pacific Islander
5. White

The OMB deliberately refrains from naming Hispanics as a race, instead identifying them as an "ethnic group" distinguished by culture (specifically, "Spanish culture or origin, regardless of race"). The growing tendency among journalists, researchers, and the public is to treat Latinos as a *de facto* racial group distinct from whites, blacks, and others, but the government view is that cultural differences do not determine racial boundaries.

Biological differences are also declared irrelevant to the official standards. The Census Bureau maintains that its categories "do not conform to any biological, anthropological or genetic criteria." Instead, the bureau says that its classification system reflects "a social definition of race recognized in this country"—but it does not elaborate further on that "social definition." The Census Bureau and OMB see themselves as technical producers of race-based statistics largely for the purpose of enforcing civil rights laws, not as the arbiters of the meaning of race.

We may take the Census Bureau's reference to a "social definition" as a version of the social-scientific understanding of race as a "social construct." In other words, race is whatever we as a society say it is. The American Sociological Association took this view in its 2002 "Statement on the Importance of Collecting Data and Doing Social Scientific Research on Race," where it defined race as "a social invention that changes as political, economic, and historical contexts change." The association also noted that concepts of race usually involve valuations of "physical, intellectual, moral, or spiritual superiority or inferiority." Both are crucial observations about the type of idea that race is: It arises at particular moments and in particular places, and has long served to perpetuate deep social fissures. However, the constructivist position does not necessarily define the actual content of racial beliefs. Many kinds of classification schemes are socially constructed and serve as the basis for class systems. So what distinguishes racial categories from other taxonomies?

Sociologist Max Weber (1864–1920) offers a useful starting point for seeing the elements of ancestry, culture, and biology in terms of socially shaped belief. However, it is his definition of "ethnic groups"—not races—that provides the template. Like most scientists of his time, Weber felt that races stemmed from "common inherited and inheritable traits that actually derive from common descent." In his definition of *ethnicity*, however, he introduced the notion of "believed" rather than actual commonality, describing ethnic groups as "those human groups that entertain a subjective belief in their common descent because of similarities of physical type or of customs or both, or because of memories of colonization and migration; this belief must be important for the propagation of group formation; conversely, it does not matter whether or not an objective blood relationship exists." Substitute "races" for "ethnic groups," and strike "customs" but retain "physical type," and we have the basic ingredients for a comprehensive definition of race.

An emphasis on *belief* in common descent, as well as *perception* of similarity and difference, is crucial for a useful definition of race. Without them, we could not account for the traditional American "one-drop" system of racial classification, for example. According to this logic, a person with one black great-grandparent and seven white great-grandparents is a black person, because their "drop of black blood" means they have more in common with blacks than with whites. This shows how we base racial classifications on socially contingent perceptions of sameness and difference, not on some kind of "natural" calculus.

Finally, regardless of our personal views on the biological basis of race, we must recognize that physical characteristics—Weber's "inherited and inheritable traits"—are central to the concept. As historian George Fredrickson points out, the word race, deriving from the late 15th-century Spanish designation for Jewish and Muslim origins, came freighted with Christians' belief that such people embodied an innate, permanent, and negative essence. Although Spaniards had previously believed that infidels could become Christians through religious conversion, the suspicion that they could never truly make the transition took root after 1492, when many Jews and Muslims chose conversion over expulsion. This early notion of inherent and unchangeable difference gave rise to the understandings of race we share today. We believe, for example, that a black person can "act white" or that a white person can "act black," but that no behavior, shared ideas, or values can actually determine the race that one truly is.

With these elements in place, we can improvise on Weber to define races as *groupings of people believed to share common descent, based on perceived innate physical similarities*. This formulation addresses the relation of culture and biology to race. Culture is absent here as an explicit basis of racial membership, leaving a clear distinction between race and ethnicity. Though both refer to beliefs about shared origins, ethnicity is grounded in the discourse of cultural similarity, and race in that of biological commonality. In addition, this definition emphasizes the constructivist observation that although racial categories are ostensibly based on physical difference, they need not be so in reality. Even if we disagree about whether or not races are biological entities, we can agree that they are based on *claims* about biological commonality. As a result, this definition gives us a shared starting point for the most contentious debate about the nature of race today, namely, whether advances in genetic and biomedical research have proven the essentialist claim that races are identifiable, biologically distinct groups that exist independently of our perceptions and preconceptions.

Sociological literature often suggests that the general acceptance by scholars of the idea of race as a "social construct" gives it the status of "conventional wisdom." Yet resurgent claims about the biological nature of race are increasingly difficult to ignore. After something of a postwar hiatus in "race science" following international condemnation of the Nazis, the question of the biological basis of race has received renewed scientific attention in recent years. In March 2005, for example, the *New York Times* published a geneticist's essay asserting that "scientists should admit that there is such a thing as race." A variety of professional scientific journals and popular science magazines have taken up the question of whether "there is such a thing as race." In December 2003, the cover of *Scientific American* inquired "Does Race Exist?" *Nature Genetics* devoted most of a November 2004 supplement to the same question. *Science*, the *New England Journal of Medicine, Genome Biology*, and the *International Journal of Epidemiology*, among others, have also addressed the issue. In fact, the relationship between race and human biology is far from settled in scientific circles.

The argument that race is socially constructed rests on two simultaneous claims, although we sociologists have tended to focus on just one. The constructionist idea implies that race is a product of particular historical circumstances and also that it is *not* rooted in biological difference—it only claims to be. By working harder to demonstrate the former—for example, the historical variability of racial categories, their roots in particular social institutions, their divergence from one society to the next—we have turned away from investigating why racial boundaries do not correspond to physical differences. In our teaching and writing, we have not tried to explain why race may not be rooted in biology even though Americans are accustomed to being able to see race; they see, for example, who is Asian or who is white. Yet a comprehensive constructionist account must explain why we consider only some of the many kinds of differences between geographic groupings of human beings to be racial differences. We see racial differences between Norwegians and Koreans, for example, but we do not consider the differences between Norwegians and Portuguese to be racial.

Although sociologists may be reluctant to evaluate geneticists' and medical practitioners' pronouncements on human differences, feeling that this is not our turf, the arguments made today about race and biology lend themselves to sociological analysis. As the recent *New York Times* essay demonstrated, the current claims about distinct racial genetic profiles involve assumptions about group membership that social scientists are generally accustomed to questioning. We are used to investigating problems of sample construction and potential bias, exploring how assumptions about boundary lines affect our results. If we find, for example, that African Americans and European Americans have different probabilities of having a particular gene variant, does that prove the existence of black and white "races" to which they belong? If Latinos have yet another probability of having that gene variant, have we proved that they too constitute a racial group? How about Ashkenazi Jews? Does it matter if only one in a thousand genes displays such a pattern of variation? Does it matter how many people provided the DNA samples, or how they were located? Should we be suspicious that the red/white/yellow/black racial classification that scientists and census-takers use today is at heart the same framework that Linnaeus established in the 18th century without the aid of genome sequencing? In short, despite the complexity of the human genome and the tools that we now have to study it, the debate about the nature of race revolves around broader questions of logic and reasoning—which makes it that much more important to establish a comprehensive but flexible definition of race.

DISCUSSION QUESTIONS

1. What role do social judgments play in the definition of race? How are these related to how groups have been treated?
2. What does Morning mean that arguments made about race and biology lend themselves to sociological interpretation?

8

Planting the Seed

The Invention of Race

ABBY L. FERBER

The history of the concept of race is deeply linked to racist thinking, particularly as it developed in some of the quasi-scientific notions developed in the sixteenth and seventeenth centuries. Ferber shows how racist thought is a relatively recent historical development, one that stems from the exploitation of human groups by others. In this history of racist thought, many have tried to use science (illegitimately) to try to justify such exploitation.

My students are always surprised to learn that race is a relatively recent invention. In their minds, race and racial antagonisms have taken on a universal character; they have always existed, and probably always will, in some form or another. Yet this fatalism belies the reality—that race is indeed a modern concept and, as such, does not have to be a life sentence.

Winthrop Jordan has suggested that ideas of racial inferiority, specifically that blacks were savage and primitive, played an essential role in rationalizing slavery.[1] There was no conception of race as a physical category until the eighteenth century.[2] There was, however, a strong association between blackness and evil, sin, and death, long grounded in European thought. The term "race" is believed to have originated in the Middle Ages in the romance languages, first used to refer to the breeding of animals. Race did not appear in the English language until the sixteenth century and was used as a technical term to define human groups in the seventeenth century. By the end of the eighteenth century, as emphasis upon the observation and classification of human differences grew, "race" became the most commonly employed concept for differentiating human groups according to Northern European standards. Audrey Smedley argues that because "race" has its roots in the breeding of animal stock, unlike other terms used to categorize humans, it came to imply an innate or inbred quality, believed to be permanent and unchanging.[3]

Until the nineteenth century, the Bible was consulted and depended upon for explanations of human variation, and two schools of thought emerged. The first asserted that there was a single creation of humanity, monogenesis, while the

SOURCE: From Abby L. Ferber, *White Man Falling: Race, Gender, and White Supremacy* (Lanham, MD: Rowman & Littlefield, 1998), pp. 27–43. Reprinted by permission of Rowman & Littlefield Publishers, Inc.

second asserted that various human groups were created separately, polygenesis. Polygenesis and ideas about racial inferiority, however, gained few believers, even in the late 1700s when the slave trade was under attack, because few were willing to support doctrines that conflicted with the Bible.[4]

While European Americans remained dedicated to a biblical view of race, the rise of scientific racism in the middle of the eighteenth century shaped debate about the nature and origins of races.[5] The Enlightenment emphasized the scientific practices of observing, collecting evidence, measuring bodies, and developing classificatory schemata. In the early stages of science, the most prevalent activity was the collection, examination, and arrangement of data into categories. Carolus Linnaeus, a prominent naturalist in the eighteenth century, developed the first authoritative racial division of humans in his *Natural System*, published in 1735.[6] Considered the founder of scientific taxonomy, he attempted to classify all living things, plant and animal, positioning humans within the matrix of the natural world. As Cornel West demonstrates, from the very beginning, racial classification has always involved hierarchy and the linkage of physical features with character and cultural traits.[7] For example, in the descriptions of his racial classifications, Linnaeus defines Europeans as "gentle, acute, inventive ... governed by customs," while Africans are "crafty, indolent, negligent ... governed by caprice."[8] Like most scientists of his time, however, Linnaeus considered all humans part of the same species, the product of a single creation.

Linnaeus was followed by Georges Louis Leclerc, Comte de Buffon, who is credited with introducing the term "race" into the scientific lexicon. Buffon also believed in monogenesis and in his 1749 publication *Natural History*, suggested that human variations were the result of differences in environment and climate. Whiteness, of course, was assumed to be the real color of humanity. Buffon suggested that blacks became dark-skinned because of the hot tropical sun and that if they moved to Europe, their skin would eventually lighten over time. Buffon cited interfertility as proof that human races were not separate species, establishing this as the criterion for distinguishing a species.

Buffon and Johann Friedrich Blumenbach are considered early founders of modern anthropology. Blumenbach advanced his own systematic racial classification in his 1775 study *On the Natural Varieties of Mankind*, designating five human races: Caucasian, Mongolian, Ethiopian, American, and Malay. While he still considered races to be the product of one creation, he ranked them on a scale according to their distance from the "civilized" Europeans.[9] He introduced the term "Caucasian," chosen because he believed that the Caucasus region in Russia produced the world's most beautiful women. This assertion typifies the widespread reliance upon aesthetic judgments in ranking races....

The science of racial classifications relied upon ideals of Greek beauty, as well as culture, as a standard by which to measure races. Race became central to the definition of Western culture, which became synonomous with "civilization."[10]...

The history of racial categorizations is intertwined with the history of racism. Science sought to justify a priori racist assumptions and consequently rationalized and greatly expanded the arsenal of racist ideology. Since the eighteenth century, racist beliefs have been built upon scientific racial categorizations and the linking of social and cultural traits to supposed genetic racial differences. While some social critics have suggested that contemporary racism has replaced biology with a concept of culture, the [1994] publication of *The Bell Curve*[11] attests to the staying power of these genetic notions of race. Today, as in the past, racism weaves together notions of biology and culture, and culture is assumed to be determined by some racial essence.

Science defined race as a concept believed to be hereditary and unalterable. The authority of science contributed to the quick and widespread acceptance of these ideas and prevented their interrogation. Equally important, the study of race and the production of racist theory also helped establish scientific authority and aided discipline building. While the history of the scientific concept of race argues that race is an inherent essence, it reveals, on the contrary, that race is a social construct. Young points out that "the different Victorian scientific accounts of race each in their turn quickly became deeply problematic; but what was much more consistent, more powerful and long-lived, was the cultural construction of race."[12]

Because race is not grounded in genetics or nature, the project of defining races always involves drawing and maintaining boundaries between those races. This was no easy task. It is important to pay attention to the construction of those borders: how was it decided, in actual policy, who was considered white and who was considered black? What about those who did not easily fit into either of those categories? What were the dangers of mixing? How could these dangers be avoided? These issues preoccupied policy makers, popular culture, and the public at large....

Throughout the second half of the nineteenth century, discussion of race and racial purity grew increasingly popular in both academic and mainstream circles as Americans developed distinctive beliefs and theories about race for the first time. As scientific beliefs about race were increasingly accepted by the general public, support for the one-drop rule became increasingly universal. Popular opinion grew to support the belief that no matter how white one appeared, if one had a single drop of black blood, no matter how distant, one was black....

Throughout the history of racial classification in the West, miscegenation and interracial sexuality have occupied a place of central importance. The science of racial differences has always displayed a preoccupation with the risks of interracial sexuality. Popular and legal discourses on race have been preoccupied with maintaining racial boundaries, frequently with great violence. This [essay] suggests that racial classification, the maintenance of racial boundaries, and racism are inexorably linked. The construction of biological races and the belief in maintaining the hierarchy and separation of races has led to widespread fears of integration and interracial sexuality....

The history of racial classification, and beliefs about race and interracial sexuality, can be characterized as inherently white supremacist. White supremacy has been the law and prevailing worldview throughout U.S. history, and the ideology of what is today labeled the white supremacist movement is firmly rooted in this tradition. Accounts that label the contemporary white supremacist movement as fringe and extremist often have the consequence of rendering this history invisible. Understanding this history, however, is essential to understanding and combating both contemporary white supremacist and mainstream racism.

REFERENCES

1. Jordan, Winthrop. 1969. *White over black.* Chapel Hill: University of North Carolina Press.
2. Banton, Michael, and Jonathan Harwood. 1975. *The race concept.* New York: Praeger; Mencke, John G. 1979. *Mulattoes and race mixture: American attitudes and images, 1865–1918.* Ann Arbor, Mich.: University Microfilms Research Press.
3. Smedley, Audrey. 1993. *Race in North America: Origin and evolution of a worldview.* Boulder, Colo.: Westview Press.
4. Banton and Harwood 1975, 19.
5. Banton and Harwood 1975, 24.
6. West, Cornel. 1982. *Prophesy deliverance! An Afro-American revolutionary Christianity.* Philadelphia: Westminster Press.
7. West 1982.
8. West 1982, 56.
9. Smedley 1993, 166.
10. Young, Robert J. C. 1995. *Colonial desire: Hybridity in theory, culture and race.* New York: Routledge.
11. Hernstein, Richard J., and Charles Murray. 1994. *The bell curve: Intelligence and class structure in American life.* New York: Free Press.
12. Young. 1995, p. 94.

DISCUSSION QUESTIONS

1. What does Ferber mean when she writes that "the history of racial categorizations is intertwined with the history of racism"?
2. What role have science and religion played in the social construction of racism?

9

How Did Jews Become White Folks?

KAREN BRODKIN

Brodkin's discussion of how Jewish people have come to be defined as "White" shows how race can be socially constructed in particular social and historical contexts. As Jewish immigrants to this United States became upwardly mobile, racial ideologies changed to redefine Jewish people in different terms than those with which they were first met in America.

> The American nation was founded and developed by the Nordic race, but if a few more million members of the Alpine, Mediterranean and Semitic races are poured among us, the result must inevitably be a hybrid race of people as worthless and futile as the good-for-nothing mongrels of Central America and Southeastern Europe.
>
> KENNETH ROBERTS,
> "WHY EUROPE LEAVES HOME"

The late nineteenth century and early decades of the twentieth saw a steady stream of warnings by scientists, policymakers, and the popular press that "mongrelization" of the Nordic or Anglo-Saxon race—the real Americans—by inferior European races (as well as by inferior non-European ones) was destroying the fabric of the nation.

I continue to be surprised when I read books that indicate that America once regarded its immigrant European workers as something other than white, as biologically different. My parents are not surprised; they expect anti-Semitism to be part of the fabric of daily life, much as I expect racism to be part of it. They came of age in the Jewish world of the 1920s and 1930s, at the peak of anti-Semitism in America. They are rightly proud of their upward mobility and think of themselves as pulling themselves up by their own bootstraps. I grew up during the 1950s in the Euro-ethnic New York suburb of Valley Stream, where Jews were simply one kind of white folks and where ethnicity meant little more to my generation than food and family heritage. Part of my ethnic heritage was the belief that Jews were smart and that our success was due to our own efforts and

SOURCE: From Karen Brodkin. 1999. *How Jews Became White Folks and What That Says About Race in America* (Piscataway, NJ: Rutgers University Press.). Copyright © 1999 by Karen Brodkin. Reprinted by permission of the author.

abilities, reinforced by a culture that valued sticking together, hard work, education, and deferred gratification.

I am willing to affirm all those abilities and ideals and their contribution to Jews' upward mobility, but I also argue that they were still far from sufficient to account for Jewish success. I say this because the belief in a Jewish version of Horatio Alger has become a point of entry for some mainstream Jewish organizations to adopt a racist attitude against African Americans especially and to oppose affirmative action for people of color. Instead I want to suggest that Jewish success is a product not only of ability but also of the removal of powerful social barriers to its realization.

It is certainly true that the United States has a history of anti-Semitism and of beliefs that Jews are members of an inferior race. But Jews were hardly alone. American anti-Semitism was part of a broader pattern of late-nineteenth-century racism against all southern and eastern European immigrants, as well as against Asian immigrants, not to mention African Americans, Native Americans, and Mexicans. These views justified all sorts of discriminatory treatment, including closing the doors, between 1882 and 1927, to immigration from Europe and Asia. This picture changed radically after World War II. Suddenly, the same folks who had promoted nativism and xenophobia were eager to believe that the Euro-origin people whom they had deported, reviled as members of inferior races, and prevented from immigrating only a few years earlier, were now model middle-class white suburban citizens.

It was not educational epiphany that made those in power change their hearts, their minds, and our race. Instead, it was the biggest and best affirmative action program in the history of our nation, and it was for Euromales. That is not how it was billed, but it is the way it worked out in practice. I tell this story to show the institutional nature of racism and the centrality of state policies to creating and changing races. Here, those policies reconfigured the category of whiteness to include European immigrants. There are similarities and differences in the ways each of the European immigrant groups became "whitened." I tell the story in a way that links anti-Semitism to other varieties of anti-European racism because this highlights what Jews shared with other Euro-immigrants.

EURORACES

The U.S. "discovery" that Europe was divided into inferior and superior races began with the racialization of the Irish in the mid-nineteenth century and flowered in response to the great waves of immigration from southern and eastern Europe that began in the late nineteenth century. Before that time, European immigrants—including Jews—had been largely assimilated into the white population. However, the 23 million European immigrants who came to work in U.S. cities in the waves of migration after 1880 were too many and too concentrated to absorb. Since immigrants and their children made up more than 70 percent of the population of most of the country's largest cities, by the 1890s urban American had taken on a distinctly southern and eastern European immigrant flavor. Like the Irish in Boston and New York, their urban concentrations in dilapidated neighborhoods put them cheek by jowl next to the rising elites and the middle class with whom they shared public space and to whom their working-class ethnic communities were particularly visible.

The Red Scare of 1919 clearly linked anti-immigrant with anti-working-class sentiment—to the extent that the Seattle general strike by largely native-born workers was blamed on foreign agitators. The Red Scare was fueled by an economic depression, a massive postwar wave of strikes, the Russian Revolution, and another influx of postwar immigration. Strikers in the steel and garment industries in New York and New England were mainly new immigrants....

Not surprisingly, the belief in European races took root most deeply among the wealthy, U.S.-born Protestant elite, who feared a hostile and seemingly inassimilable working class. By the end of the nineteenth century, Senator Henry Cabot Lodge pressed Congress to cut off immigration to the United States; Theodore Roosevelt raised the alarm of "race suicide" and took Anglo-Saxon women to task for allowing "native" stock to be outbred by inferior immigrants. In the early twentieth century, these fears gained a great deal of social legitimacy thanks to the efforts of an influential network of aristocrats and scientists who developed theories of eugenics—breeding for a "better" humanity—and scientific racism.

Key to these efforts was Madison Grant's influential *The Passing of the Great Race*, published in 1916. Grant popularized notions developed by William Z. Ripley and Daniel Brinton that there existed three or four major European races, ranging from the superior Nordics of northwestern Europe to the inferior southern and eastern races of the Alpines, Mediterraneans, and worst of all, Jews, who seemed to be everywhere in his native New York City. Grant's nightmare was race-mixing among Europeans. For him, "the cross between any of the three European races and a Jew is a Jew." He didn't have good things to say about Alpine or Mediterranean "races" either. For Grant, race and class were interwoven: the upper class was racially pure Nordic; the lower classes came from the lower races.

Far from being on the fringe, Grant's views were well within the popular mainstream. Here is the *New York Times* describing the Jewish Lower East Side of a century ago:

> The neighborhood where these people live is absolutely impassable for wheeled vehicles other than their pushcarts. If a truck driver tries to get through where their pushcarts are standing they apply to him all kinds of vile and indecent epithets. The driver is fortunate if he gets out of the street without being hit with a stone or having a putrid fish or piece of meat thrown in his face. This neighborhood, peopled almost entirely by the people who claim to have been driven from Poland and Russia, is the eyesore of New York and perhaps the filthiest place on the western continent. It is impossible for a Christian to live there because he will be driven out, either by blows or the dirt and stench. Cleanliness is an

> unknown quantity to these people. They cannot be lifted up to a higher plane because they do not want to be. If the cholera should ever get among these people, they would scatter its germs as a sower does grain.[1]

Such views were well within the mainstream of the early twentieth-century scientific community....

By the 1920s, scientific racism sanctified the notion that real Americans were white and that real whites came from northwest Europe. Racism by white workers in the West fueled laws excluding and expelling the Chinese in 1882. Widespread racism led to closing the immigration door to virtually all Asians and most Europeans between 1924 and 1927, and to deportation of Mexicans during the Great Depression.

Racism in general, and anti-Semitism in particular, flourished in higher education. Jews were the first of the Euro-immigrant groups to enter college in significant numbers, so it was not surprising that they faced the brunt of discrimination there. The Protestant elite complained that Jews were unwashed, uncouth, unrefined, loud, and pushy. Harvard University President A. Lawrence Lowell, who was also a vice president of the Immigration Restriction League, was open about his opposition to Jews at Harvard. The Seven Sister schools had a reputation for "flagrant discrimination." M. Carey Thomas, Bryn Mawr president, may have been some kind of feminist, but she was also an admirer of scientific racism and an advocate of immigration restriction....

Jews are justifiably proud of the academic skills that gained them access to the most elite schools of the nation despite the prejudices of their gatekeepers. However, it is well to remember that they had no serious competition from their Protestant classmates. This is because college was not about academic pursuits. It was about social connection—through its clubs, sports and other activities, as well as in the friendships one was expected to forge with other children of elites. From this, the real purpose of the college experience, Jews remained largely excluded.

This elite social mission had begun to come under fire and was challenged by a newer professional training mission at about the time Jews began entering college. Pressures for change were beginning to transform the curriculum and to reorient college from a gentleman's bastion to a training ground for the middle-class professionals needed by an industrial economy.... Occupational training was precisely what had drawn Jews to college. In a setting where disparagement of intellectual pursuits and the gentleman C were badges of distinction, it certainly wasn't hard for Jews to excel. Jews took seriously what their affluent Protestant classmates disparaged, and, from the perspective of nativist elites, took unfair advantage of a loophole to get where they were not wanted.

Patterns set by these elite schools to close those "loopholes" influenced the standards of other schools, made anti-Semitism acceptable, and "made the aura of exclusivity a desirable commodity for the college-seeking clientele."[2] Fear that colleges "might soon be overrun by Jews" were publicly expressed at a 1918 meeting of the Association of New England Deans. In 1919 Columbia University took steps to decrease the number of its Jewish students by a set of practices that soon came to be widely adopted. They developed a psychological test based on the World War I army intelligence tests to measure "innate ability—and middle-class home environment"; and they redesigned the admission application to ask for religion, father's name and birth place, a photo, and personal interview. Other techniques for excluding Jews, like a fixed class size, a chapel requirement, and preference for children of alumni, were less obvious....

Columbia's quota against Jews was well known in my parents' community. My father is very proud of having beaten it and been admitted to Columbia Dental School on the basis of his skill at carving a soap ball. Although he became a teacher instead because the tuition was too high, he took me to the dentist every week of my childhood and prolonged the agony by discussing the finer points of tooth-filling and dental care. My father also almost failed the speech test required for his teaching license because he didn't speak "standard," i.e., nonimmigrant, nonaccented English. For my parents and most of their friends. English was the language they had learned when they went to school, since their home and neighborhood language was Yiddish. They saw the speech test as designed to keep all ethnics, not just Jews, out of teaching.

There is an ironic twist to this story. My mother always urged me to speak well, like her friend Ruth Saronson, who was a speech teacher. Ruth remained my model for perfect diction until I went away to college. When I talked to her on one of my visits home, I heard the New York accent of my version of "standard English," compared to the Boston academic version.

My parents believe that Jewish success, like their own, was due to hard work and a high value placed on education. They attended Brooklyn College during the Depression. My mother worked days and went to school at night; my father went during the day. Both their families encouraged them. More accurately, their families expected it. Everyone they knew was in the same boat, and their world was made up of Jews who were advancing just as they were. The picture for New York—where most Jews lived—seems to back them up. In 1920, Jews made up 80 percent of the students at New York's City College, 90 percent of Hunter College, and before World War I, 40 percent of private Columbia University. By 1934, Jews made up almost 24 percent of all law students nationally and 56 percent of those in New York City. Still, more Jews became public school teachers, like my parents and their friends, than doctors or lawyers....

How we interpret Jewish social mobility in this milieu depends on whom we compare them to. Compared with other immigrants, Jews were upwardly mobile. But compared with nonimmigrant whites, that mobility was very limited and circumscribed. The existence of anti-immigrant, racist, and anti-Semitic barriers kept the Jewish middle class confined to a small number of occupations. Jews were excluded from mainstream corporate management and corporately employed professions, except in the garment and movie industries, in which they were pioneers. Jews were almost totally excluded from university faculties (the few who made it had powerful patrons). Eastern European Jews were concentrated in small businesses, and in professions where they served a largely Jewish clientele....

Although Jews, as the Euro-ethnic vanguard in college, became well established in public school teaching—as well as visible in law, medicine, pharmacy, and librarianship before the postwar boom—these professions should be understood in the context of their times. In the 1930s they lacked the corporate context they have today, and Jews in these professions were certainly not corporation-based. Most lawyers, doctors, dentists, and pharmacists were solo practitioners, depended upon other Jews for their clientele, and were considerably less affluent than their counterparts today.

Compared to Jewish progress after World War II, Jews' pre-war mobility was also very limited. It was the children of Jewish businessmen, but not those of Jewish workers, who flocked to college. Indeed, in 1905 New York, the children of Jewish workers had as little schooling as the children of other immigrant workers. My family was quite the model in this respect. My grandparents did not go to college, but they did have a modicum of small business success. My father's family owned a pharmacy. Although my mother's father was a skilled garment worker, her mother's family was large and always had one or another grocery or deli in which my grandmother participated. It was the relatively privileged children of upwardly mobile Jewish immigrants like my grandparents who began to push on the doors to higher education even before my parents were born.

Especially in New York City—which had almost one and a quarter million Jews by 1910 and retained the highest concentration of the nation's 4 million Jews in 1924—Jews built a small-business-based middle class and began to develop a second-generation professional class in the interwar years. Still, despite the high percentages of Jews in eastern colleges, most Jews were not middle class, and fewer than 3 percent were professionals—compared to somewhere between two-thirds and three-quarters in the postwar generation.

My parents' generation believed that Jews overcame anti-Semitic barriers because Jews are special. My answer is that the Jews who were upwardly mobile were special among Jews (and were also well placed to write the story). My generation might well respond to our parents' story of pulling themselves up by their own bootstraps with "But think what you might have been without the racism and with some affirmative action!" And that is precisely what the post-World War II boom, that decline of systematic, public, anti-Euro racism and anti-Semitism, and governmental affirmative action extended to white males let us see.

WHITENING EURO-ETHNICS

By the time I was an adolescent, Jews were just as white as the next white person. Until I was eight, I was a Jew in a world of Jews. Everyone on Avenue Z in Sheepshead Bay was Jewish. I spent my days playing and going to school on three blocks of Avenue Z, and visiting my grandparents in the nearby Jewish neighborhoods of Brighton Beach and Coney Island. There were plenty of Italians in my neighborhood, but they lived around the corner. They were a kind of Jew, but on the margins of my social horizons. Portuguese were even more distant, at the end of the bus ride, at Sheepshead Bay. The *shul*, or temple, was on Avenue Z, and I begged my father to take me like all the other fathers took their kids, but religion wasn't part of my family's Judaism. Just how Jewish my neighborhood was hit me in first grade, when I was one of two kids to go to school on Rosh Hashanah. My teacher was shocked—she was Jewish too—and I was embarrassed to tears when she sent me home. I was never again sent to school on Jewish holidays. We left that world in 1949 when we moved to Valley Stream, Long Island, which was Protestant and Republican and even had farms until Irish, Italian, and Jewish ex-urbanites like us gave it a more suburban and Democratic flavor.

Neither religion nor ethnicity separated us at school or in the neighborhood. Except temporarily. During my elementary school years, I remember a fair number of dirt-bomb (a good suburban weapon) wars on the block. Periodically, one of the Catholic boys would accuse me or my brother of killing his god, to which we'd reply, "Did not," and start lobbing dirt bombs. Sometimes he'd get his friends from Catholic school and I'd get mine from public school kids on the block, some of whom were Catholic. Hostilities didn't last for more than a couple of hours and punctuated an otherwise friendly relationship. They ended by our junior high years, when other things became more important. Jews, Catholics and Protestants, Italians, Irish, Poles, "English" (I don't remember hearing WASP as a kid), were mixed up on the block and in school. We thought of ourselves as middle class and very enlightened because our ethnic backgrounds seemed so irrelevant to high school culture. We didn't see race (we thought), and racism was not part of our peer consciousness. Nor were the immigrant or working-class histories of our families.

As with most chicken-and-egg problems, it is hard to know which came first. Did Jews and other Euro-ethnics become white because they became middle-class? That is, did money whiten? Or did being incorporated into an expanded version of whiteness open up the economic doors to middle-class status? Clearly, both tendencies were at work.

Some of the changes set in motion during the war against fascism led to a more inclusive version of whiteness. Anti-Semitism and anti-European racism lost respectability. The 1940 Census no longer distinguished native whites of native parentage from those, like my parents, of immigrant parentage, so Euro-immigrants and their children were more securely white by submersion in an expanded notion of whiteness.

Theories of nurture and culture replaced theories of nature and biology. Instead of dirty and dangerous races that would destroy American democracy, immigrants became ethnic groups whose children had successfully assimilated into the mainstream and risen to the middle class. In this new myth, Euro-ethnic suburbs like mine became the measure of American democracy's victory over racism.... Jewish mobility became a new Horatio Alger story. In time and with hard work, every ethnic group would get a piece of the pie, and the United States would be a nation with equal opportunity for all its people to become part of a prosperous middle-class majority. And it seemed that Euro-ethnic immigrants and their children were delighted to join middle America.

REFERENCES

1. *The New York Times*, July 30, 1893; cited in Allon Schoener, 1967. *Portal to America: The Lower East Side 1870–1925*. New York: Holt, Rinehart, and Winston, pp. 57–58.
2. Synott, Marcia Graham. 1986. "Anti-Semitism and American Universities: Did Quotas Follow the Jews?" In *Anti-Semitism in American History*, edited by David A. Gerber. Urbana, IL: University of Illinois Press, p. 250.

DISCUSSION QUESTIONS

1. How does Brodkin see anti-Semitism as linked to other forms of racism? How is this revealed by historical events?
2. What does Brodkin mean by arguing the "Jews have become white folks?" How does her own family history reveal this process?

10

Racial Formation

MICHAEL OMI AND HOWARD WINANT

The concept of racial formation, *first developed by Omi and Winant, has become central to the sociological study of race. It refers to the social and historical processes by which groups come to be defined in racial terms and it specifically locates those processes in state-based institutions, such as the law.*

In 1982–83, Susie Guillory Phipps unsuccessfully sued the Louisiana Bureau of Vital Records to change her racial classification from black to white. The descendant of an 18th-century white planter and a black slave, Phipps was designated "black" in her birth certificate in accordance with a 1970 state law which declared anyone with at least 1/32nd "Negro blood" to be black.

The Phipps case raised intriguing questions about the concept of race, its meaning in contemporary society, and its use (and abuse) in public policy. Assistant Attorney General Ron Davis defended the law by pointing out that some type of racial classification was necessary to comply with federal record-keeping requirements and to facilitate programs for the prevention of genetic diseases. Phipps's attorney, Brian Begue, argued that the assignment of racial categories on birth certificates was unconstitutional and that the 1/32nd designation was inaccurate. He called on a retired Tulane University professor who cited research indicating that most Louisiana whites have at least 1/20th "Negro" ancestry.

In the end, Phipps lost. The court upheld the state's right to classify and quantify racial identity....

Phipps's problematic racial identity, and her effort to resolve it through state action, is in many ways a parable of America's unsolved racial dilemma. It illustrates the difficulties of defining race and assigning individuals or groups to racial categories. It shows how the racial legacies of the past—slavery and bigotry—continue to shape the present. It reveals both the deep involvement of the state in the organization and interpretation of race, and the inadequacy of state institutions to carry out these functions. It demonstrates how deeply Americans both as individuals and as a civilization are shaped, and indeed haunted, by race.

Having lived her whole life thinking that she was white, Phipps suddenly discovers that by legal definition she is not. In U.S. society, such an event is indeed

SOURCE: From Michel Omia and Howard Winant, *Racial Formation in the United States: From the 1960s to the 1990s*, 2nd ed. (New York: Routledge, 1994), pp. 53–61. Reprinted by permission of Routledge/Taylor & Francis Books, Inc.

catastrophic. But if she is not white, of what race is she? The state claims that she is black, based on its rules of classification ... and another state agency, the court, upholds this judgment. But despite these classificatory standards which have imposed an either-or logic on racial identity, Phipps will not in fact "change color." Unlike what would have happened during slavery times if one's claim to whiteness was successfully challenged, we can assume that despite the outcome of her legal challenge, Phipps will remain in most of the social relationships she had occupied before the trial. Her socialization, her familial and friendship networks, her cultural orientation, will not change. She will simply have to wrestle with her newly acquired "hybridized" condition. She will have to confront the "Other" within.

The designation of racial categories and the determination of racial identity is no simple task. For centuries, this question has precipitated intense debates and conflicts, particularly in the U.S.—disputes over natural and legal rights, over the distribution of resources, and indeed, over who shall live and who shall die.

A crucial dimension of the Phipps case is that it illustrates the inadequacy of claims that race is a mere matter of variations in human physiognomy, that it is simply a matter of skin color. But if race cannot be understood in this manner, how can it be understood? We cannot fully hope to address this topic—no less than the meaning of race, its role in society, and the forces which shape it—in one [article], nor indeed in one book. Our goal in this [article], however, is far from modest: we wish to offer at least the outlines of a theory of race and racism.

WHAT IS RACE?

There is a continuous temptation to think of race as an essence, as something fixed, concrete, and objective. And there is also an opposite temptation: to imagine race as a mere illusion, a purely ideological construct which some ideal nonracist social order would eliminate. It is necessary to challenge both these positions, to disrupt and reframe the rigid and bipolar manner in which they are posed and debated, and to transcend the presumably irreconcilable relationship between them.

The effort must be made to understand race as an unstable and "decentered" complex of social meanings constantly being transformed by political struggle. With this in mind, let us propose a definition: race is a concept which signifies and symbolizes social conflicts and interests by referring to different types of human bodies. Although the concept of race invokes biologically based human characteristics (so-called "phenotypes"), selection of these particular human features for purposes of racial signification is always and necessarily a social and historical process. In contrast to the other major distinction of this type, that of gender, there is no biological basis for distinguishing among human groups along the lines of race.... Indeed, the categories employed to differentiate among human groups along racial lines reveal themselves, upon serious examination, to be at best imprecise, and at worst completely arbitrary.

If the concept of race is so nebulous, can we not dispense with it? Can we not "do without" race: at least in the "enlightened" present? This question has been posed often, and with greater frequency in recent years.... An affirmative answer would of course present obvious practical difficulties: it is rather difficult to jettison widely held beliefs, beliefs which moreover are central to everyone's identity and understanding of the social world. So the attempt to banish the concept as an archaism is at best counterintuitive. But a deeper difficulty, we believe, is inherent in the very formulation of this schema, in its way of posing race as a *problem*, a misconception left over from the past, and suitable now only for the dustbin of history.

A more effective starting point is the recognition that, despite its uncertainties and contradictions, the concept of race continues to play a fundamental role in structuring and representing the social world. The task for theory is to explain this situation. It is to avoid both the utopian framework which sees race as an illusion we can somehow "get beyond," and also the essentialist formulation which sees race as something objective and fixed, a biological datum. Thus we should think of race as an element of social structure rather than as an irregularity within it; we should see race as a dimension of human representation rather than an illusion. These perspectives inform the theoretical approach we call racial formation.

Racial Formation

We define *racial formation* as the sociohistorical process by which racial categories are created, inhabited, transformed, and destroyed. Our attempt to elaborate a theory of racial formation will proceed in two steps. First, we argue that racial formation is a process of historically situated projects in which human bodies and social structures are represented and organized. Next we link racial formation to the evolution of hegemony, the way in which society is organized and ruled. Such an approach, we believe, can facilitate understanding of a whole range of contemporary controversies and dilemmas involving race, including the nature of racism, the relationship of race to other forms of differences, inequalities, and oppression such as sexism and nationalism, and the dilemmas of racial identity today.

From a racial formation perspective, race is a matter of both social structure and cultural representation. Too often, the attempt is made to understand race simply or primarily in terms of only one of these two analytical dimensions.... For example, efforts to explain racial inequality as a purely social structural phenomenon are unable to account for the origins, patterning, and transformation of racial difference.

Conversely, many examinations of racial difference—understood as a matter of cultural attributes à la ethnicity theory, or as a society-wide signification system, à la some poststructuralist accounts—cannot comprehend such structural phenomena as racial stratification in the labor market or patterns of residential segregation.

An alternative approach is to think of racial formation processes as occurring through a linkage between structure and representation. Racial *projects do the ideological "work" of making these links. A racial project is simultaneously an interpretation, representation, or explanation of racial dynamics, and an effort to recognize and redistribute resources along particular racial lines.* Racial projects connect what race means in a particular discursive practice and the ways in which both social structures and everyday experiences are racially *organized*, based upon that meaning. Let us consider this proposition, first in terms of large-scale or macro-level social processes, and then in terms of other dimensions of the racial formation process.

Racial Formation as a Macro-Level Social Process *To interpret the meaning of race is to frame it social structurally.* Consider for example, this statement by Charles Murray on welfare reform:

> My proposal for dealing with the racial issue in social welfare is to repeal every bit of legislation and reverse every court decision that in any way requires, recommends, or awards differential treatment according to race, and thereby put us back onto the track that we left in 1965. We may argue about the appropriate limits of government intervention in trying to enforce the ideal, but at least it should be possible to identify the ideal: Race is not a morally admissible reason for treating one person differently from another. Period....

Here there is a partial but significant analysis of the meaning of race: it is not a morally valid basis upon which to treat people "differently from one another." We may notice someone's race, but we cannot act upon that awareness. We must act in a "color-blind" fashion. This analysis of the meaning of race is immediately linked to a specific conception of the role of race in the social structure: it can play no part in government action, save in "the enforcement of the ideal." No state policy can legitimately require, recommend, or award different status according to race. This example can be classified as a particular type of racial project in the present-day U.S.—a "neoconservative" one.

Conversely, *to recognize the racial dimension in social structure is to interpret the meaning of race.* Consider the following statement by the late Supreme Court Justice Thurgood Marshall on minority "set-aside" programs:

> A profound difference separates governmental actions that themselves are racist, and governmental actions that seek to remedy the effects of prior racism or to prevent neutral government activity from perpetuating the effects of such racism....

Here the focus is on the racial dimensions of social structure—in this case of state activity and policy. The argument is that state actions in the past and present have treated people in very different ways according to their race, and thus the government cannot retreat from its policy responsibilities in this area. It cannot suddenly declare itself "color-blind" without in fact perpetuating the same type of differential, racist treatment.... Thus, race continues to signify difference and structure inequality. Here, racialized social structure is immediately linked to an interpretation of the meaning of race. This example too can be classified as a particular type of racial project in the present-day U.S.—a "liberal" one.

To be sure, such political labels as "neoconservative" or "liberal" cannot fully capture the complexity of racial projects, for these are always multiply determined, politically contested, and deeply shaped by their historical context. Thus encapsulated within the neoconservative example cited here are certain egalitarian commitments which derive from a previous historical context in which they played a very different role, and which are rearticulated in neoconservative racial discourse precisely to oppose a more open-ended, more capacious conception of the meaning of equality. Similarly, in the liberal example, Justice Marshall recognizes that the contemporary state, which was formerly the architect of segregation and the chief enforcer of racial difference, has a tendency to reproduce those patterns of inequality in a new guise. Thus he admonishes it (in dissent, significantly) to fulfill its responsibilities to uphold a robust conception of equality. These particular instances, then, demonstrate how racial projects are always concretely framed, and thus are always contested and unstable. The social structures they uphold or attack, and the representations of race they articulate, are never invented out of the air, but exist in a definite historical context, having descended from previous conflicts. This contestation appears to be permanent in respect to race.

These two examples of contemporary racial projects are drawn from mainstream political debate; they may be characterized as center-right and center-left expressions of contemporary racial politics.... We can, however, expand the discussion of racial formation processes far beyond these familiar examples. In fact, we can identify racial projects in at least three other analytical dimensions: first, the political spectrum can be broadened to include radical projects, on both the left and right, as well as along other political axes. Second, analysis of racial projects can take place not only at the macro-level of racial policy-making, state activity, and collective action, but also at the micro-level of everyday experience. Third, the concept of racial projects can be applied across historical time, to identify racial formation dynamics in the past.

DISCUSSION QUESTIONS

1. What do Omi and Winant mean by *racial formation*? What role does the law play in such a process? How is this shown in the history of the United States?
2. What difference does it make to conceptualize race as a property of social structures versus as a property (or attribute) of individuals?

Part II
Student Exercises

1. Using some random method of assignment, your instructor will divide your class (or some other grouping) into two groups, one of which is designated the Blues and the other the Greens. Over a period of a week, the Greens should serve the Blues in any way the Blues ask—such as carrying their books, running errands for them, delivering meals to places they designate, or any other job that the Blues design. (Since this is a course assignment, you should be reasonable in your demands).

   As the week progresses, observe how the Blues act among themselves and in front of the Greens. Also observe how the Greens act among themselves and in front of the Blues. What attitudes do the two groups develop toward each other and toward themselves? How do they talk publicly about the other group? Do classmates begin to generalize about the assumed characteristics of the two different groups?

   What does the experiment reveal about the *social construction of race*?

2. Have members of your class describe the ethnic background of family members. You can describe such things as when and how your family arrived in the United States, ethnic traditions that your families may observe, whether ethnic pride is a part of your family experience. After hearing from classmates belonging to different ethnic groups, list what you learned about ethnicity from listening to these different experiences. Is ethnicity more significant for some groups than others? Is ethnicity more important to some generations within a family than others? Why? What does this teach you about the *social construction of ethnicity* in society?

PART III

INTRODUCTION

Representations of Race and Group Beliefs: Prejudice and Racism

When you think of beliefs about race, the term *prejudice* most likely comes to mind. Prejudice is an attitude that tends to denigrate individuals and groups who are perceived to be somehow different and undesirable. The social scientific definition of prejudice dates back to the 1950s and the work of psychologist Gordon Allport. Allport defined **prejudice** as "a hostile attitude directed toward a person or group simply because the person is presumed to be a member of that group and is perceived as having the negative characteristics associated with the group" (Allport 1954: 7).

Prejudice can be directed at many groups. One can be prejudiced against women or gays or athletes or foreigners—anyone who is perceived as a member of an "out-group," that is, a group different from one's own. Although prejudice can be positive (as in thinking all women are nurturing), it is generally a negative attitude involving hostile or derogatory feelings as well as false generalizations about people in the out-groups. Prejudice can also be expressed by any group, whether dominant or subordinate; thus, racial minorities may be prejudiced against other racial minorities or against the dominant group, just as more powerful people may be prejudiced against less powerful people. In other words, prejudice is a prejudgment and is the basis for much racial intolerance.

Prejudice rests on social **stereotypes** that is, oversimplified beliefs about members of a particular social group (Andersen and Taylor 2004). Stereotypes categorize people based on false generalizations along a narrow range of presumed characteristics, such as the belief that all Jewish people are greedy or that all blondes are dumb. Although stereotypes are perpetuated in many ways in society (e.g., in families, where parents teach children about other groups), one of the most influential mediums is popular culture—music, magazines, films, and television, among others. Thus, because men of color are portrayed in the media as criminals, this is the most common way that they are stereotyped. Or Asian American women may be stereotyped as sexy and beguiling—an image repeatedly shown in magazines, videos, and other popular media.

Because stereotypes are not neutral depictions of people, they can be enormously harmful in that they influence how people define others and themselves. Research has found, for example, that the exposure of young Native American children to cartoonish depictions of Native Americans as school mascots actually depresses the children's self-esteem, that is, how they value themselves (Fryberg 2003). Stereotypes control how people come to define each other and, as such, are "controlling images" (Collins 1990) in U.S. culture.

The prejudice that grows out of social stereotypes is an *attitude*. This is distinct from discrimination, which is a behavior. **Discrimination** is the negative and unequal *treatment* of members of a social group based on their perceived membership in a particular group (Andersen and Taylor 2004: 320). Although the term *discrimination* sometimes has a positive connotation (as in "she has a discriminating attitude"), such behavior generally categorically deprives people of their civil rights.

Generally, prejudice and discrimination are thought of as related—prejudice causes discrimination—but things are not that simple. Over fifty years ago, sociologist Robert Merton (1949) developed a four-square typology showing different ways that prejudice and discrimination are and are not related. Look at the following:

	Prejudice:	**Positive (+)**	**Negative (–)**
Discrimination:	*Positive(+)*	Case 1: The bigot (++)	Case 2: Non-prejudiced discriminator (–+)
	Negative (–)	Case 3: Prejudiced non-discriminator (+–)	Case 4: The all-weather liberal (––)

In Case 1, someone may be both prejudiced and discriminate—the classic bigot. Both prejudice and discrimination are overt, intentional, and hostile. In Case 4, someone may be free of prejudice and not discriminate (the person Merton called the "all-weather liberal"). It is in Cases 2 and 3 that we see prejudice and discrimination may not have a causal relationship. In Case 2, one may not be prejudiced, but still discriminate, such as a homeowner who holds no racial prejudice but will only buy a home in an all-White neighborhood "to protect their property value." Such a person may say they hold no prejudice but most look out for their own interests. In Case 3, someone may be prejudiced, but not discriminate—for example, when the law prohibits discrimination. A landlord may, for example, rent to a Black tenant despite holding prejudice. The point is that both prejudice and discrimination occur in a larger context—that of society as a whole. This societal context predicts the occurrence of discrimination as much as people's individual attitudes.

Prejudice is not just a free-floating attitude, however. It is linked to group positions in society (Blumer 1958; Bobo and Hutchings 1996). It is a learned behavior and it can be "unlearned." It can be intentional or unintentional. Public opinion surveys show that although most people do not now express overt prejudice, they do blame minorities themselves for their socioeconomic standing and many are resistant to policy changes that would actually bring about more racial equality. Lawrence Bobo refers to this trend in "The Color Line, the Dilemma, and the Dream" as laissez-faire racism. In **laissez-faire** ("hands-off") **racism**, antiracist practices are abandoned and overt bigotry is minimal. Bobo shows how most people support equal rights in principle but are unwilling to support practices and policies that would move our society in that direction. The problem is institutional racism, not prejudice per se.

Racism is a principle of social domination in which a group that is seen as inferior or different because of presumed biological or cultural characteristics is oppressed, controlled, and exploited—socially, economically, culturally, politically, psychologically—by a dominant group (Wilson 1973). Note the key elements of this definition. First, racism is a *principle of domination*; that is, embedded in this definition is the thought that racism involves one or more groups' subordinate position within a system of racial inequality.

Second the emphasis is on "presumed." As we learned in Part II, race is not "real" in the biological or cultural sense, but it has real meaning in society and throughout history. Race takes on meaning in the context of power relations (Feagin 2000); thus, how people are perceived within a system of hierarchy and power is the key to understanding racism. Who gets the power to define different groups and what are the means by which they do it? Law? Media? Schools? All of the these? Those who shape how people are represented have enormous power to shape people's consciousness about race.

Third, racism involves domination on a number of fronts: social, economic, political, cultural, and psychological. Although its economic effects are more easily seen, racism also involves the suppression of some groups' cultures and can shape the psychology of both dominant and subordinate groups. Most people think of racism only at the individual level, but it is present in society (the social front), and, as such, shapes people's minds, their interactions with others, and the opportunity structures in which groups find themselves.

Institutional racism is the complex and cumulated pattern of racial advantage and disadvantage built into the structure of a society. Institutional racism, reflected in the prejudice and discrimination seen in a society, comprises more than an attitude or behavior. It is a system of power and privilege that advantages some groups over others. Thus you might say that prejudice is lodged in people's minds, but racism is lodged in society.

As several of the authors in Part III show, many people who benefit from institutional racism are often blind to the systemic advantage that it gives them. Thus, just being a White person may open some opportunities that might not be as readily available to others—independent of that White person's own attitudes and behavior.

The invisibility of racial privilege to dominant groups is reflected in the growth of so-called **color-blind racism**, the belief that race should be ignored and that race-conscious practices and policies only foster more racism. When dominant groups think that racism is no longer an issue, despite its ongoing reality, they are not likely to engage in practices or support policies that challenge racism (Brown et al. 2003; Bonilla-Silva 2003). To be color-blind in a society in which race still structures people's relationships, identities, and opportunities is to be blind to the continuing realities of race.

Charles Gallagher examines color-blind privilege in his essay "Color-Blind Privilege: The Social and Political Functions of Erasing the Color Line in Post-Race America." He points out that we live with the appearance of a multiracial society in which we "all get along." In this context, race has actually become a commodity, something that White people can buy and display, while at the same time not challenging the privilege that underlies a system of racial stratification. Products are mass-marketed using multiracial images and presumably sell across color lines, but such images legitimate color-blindness. Gallagher's research shows that while many White people believe themselves to be color-blind, once you scratch the surface of this belief, they are quick to defend the status quo.

In "Learning to be White through the Movies," Hernán Vera and Andrew M. Gordon show another dimension of racial representations in everyday life—popular culture. People generally go to the movies just to be entertained, but movies, like other visual and printed media, communicate powerful images of race, images that influence how we see ourselves and others. Vera and Gordon examine films critically, asking who do we see in films and how are they portrayed? They find that films implicitly celebrate and promote White people and simultaneously obscure White privilege by making it seem natural or "just normal." The next time you see a movie or a television show, or read a popular magazine, or watch the news, go beyond the surface. Rather than just taking what you see at face value, try to examine how White people and people of color are portrayed. Think about how these portrayals shape people's understandings of race in society.

Charles Springwood and C. Richard King ("Playing Indian": Why Native American Mascots Must End") take on the use of Native American icons in popular culture. These icons, mostly in the form of caricatures, are everywhere—as part of the names of sports teams, on products in the grocery store, and in school textbooks. These images distort understanding of the realities of Native American life and stereotype Indians as either noble warriors or cartoonish savages. Note, too, how gender is displayed in such images. Although some defenders claim that such portrayals are all in fun, you might ask: Would it seem so light-hearted if your own group were depicted in such narrow, repetitive, and joking ways?

Cornel West ("The Necessary Engagement with Youth Culture") puts the discussion of race and popular culture in a different light. He writes about hip-hop and rap music as emerging from the critical stance of urban, Black youth. Rap and hip-hop in the early years—and currently in some more underground music—expressed strong criticism of dominant social institutions. But, West argues, as hip-hop has become more commercialized and dominated by White capitalist interests (and purchased mostly by White youth), the messages in this popular music have changed, becoming not only more sexist but also less critical of current social injustices. West appeals to the young generation to be less consumed by materialistic values and more attuned to social and political justice.

Racial inequality in the United States has resulted in enormous misunderstanding of and misinformation about different groups, something that is true for both dominant and subordinate groups. Because many people, especially White people, do not interact regularly with people outside their own group, understanding across racial-ethnic lines is hindered. As you read the articles in this section, think about the racial representations you see around you and how they might be transformed or challenged.

REFERENCES

Allport, Gordon. 1954. *The Nature of Prejudice*. Reading, MA: Addison-Wesley.

Andersen, Margaret L., and Howard F. Taylor. 2004. *Sociology: Understanding a Diverse Society*. Belmont, CA: Wadsworth.

Blumer, Herbert. 1958. "Race Prejudice as a Sense of Group Position." *Pacific Sociological Review* 1 (Spring): 3–7.

Bobo, Lawrence, and Vincent L. Hutchings. 1996. "Perceptions of Racial Group Competition: Extending Blumer's Theory of Group Position to a Multiracial Social Context." *American Sociological Review* 25 (December): 951–972.

Bonilla-Silva, Eduardo. 2003. *Racism without Racists: Colorblind Racism and the Persistence of Racial Inequality in the United States.* Lanham, MD: Rowman & Littlefield.

Brown, Michael, Martin Carnoy, Elliott Currie, Troy Duster, David Oppenheimer, Marjorie M. Schultz, and David Wellman. 2003. *Whitewashing Race: The Myth of a Color-Blind Society.* New York: Oxford University Press.

Collins, Patricia Hill. 1990. *Black Feminist Thought: Knowledge, Consciousness, and the Politics of Empowerment.* Boston: Unwin Hyman.

Feagin, Joe. 2000. *Racist America: Roots, Realities, and Reparations.* New York: Routledge.

Fryberg, Stephanie. 2003. "Really, You Don't Look Like an American Indian: Social Representations and Social Group Identities." Ph.D. dissertation, Department of Psychology, Stanford University.

Merton, Robert. 1949. "Discrimination and the American Creed." *Discrimination and the National Welfare*, ed. Robert W. MacIver. New York: Harper and Brothers. pp. 99–126.

Wilson, William Julius. 1973. *Power, Racism, and Privilege: Race Relations in Theoretical and Sociohistorical Perspectives.* New York: Macmillan.

11

The Color Line, the Dilemma, and the Dream

Race Relations in America at the Close of the Twentieth Century

LAWRENCE D. BOBO

National studies of public opinion show a significant decline in overtly expressed prejudice, but most White Americans are nonetheless unwilling to support social policies that would actually reduce the persistence of racial inequality. Bobo refers to this new form of racism as laissez-faire racism, *meaning people's attitudes that are without overt prejudice but that do not support programs to address racial inequities.*

At the dawning of the twentieth century W. E. B. Du Bois forecast that the defining problem of the twentieth century would be "the color line."[1] His analysis was penetrating. He wrote at a time when most African Americans lived as disenfranchised, purposely miseducated, and brutally oppressed second-class citizens. He wrote at a time when popular conceptions of African Americans were overtly racist, even among the well-educated white elite. As we stand near the dawning of the twenty-first century, the problem to which Du Bois so presciently drew our attention appears no closer to a fundamental resolution in the United States or in much of the rest of the world than it did a century ago. The color line endures....

The review of information of changing racial attitudes discussed below will support three conclusions.... First, the available data suggest that the United States has experienced a genuine and tremendous positive transformation in racial attitudes. A once predominant ideology of Jim Crow racism has, over the past five decades, steadily receded from view. Nonetheless, ... the collapse of Jim Crow racism has not been followed by a full embrace of African Americans as true co-equals with whites, deserving of a complete measure of all the fruits of membership in the polity. Instead, a new configuration of negative racial attitudes has recently crystallized. This new cultural pattern of understanding the

SOURCE: From *Civil Rights and Social Wrongs: Black-White Relations Since World War II*, ed. by John Higham (University Park, PA: Pennsylvania State University Press, 1999), pp. 33–55. Reprinted by permission.

actual and normative position of blacks in American society is appropriately labeled "laissez-faire racism."

Second, racial discrimination remains a barrier to blacks' full economic, political, and social participation in American institutions. The problem of racial discrimination today is less extreme, absolute, and all-encompassing. Hence, I reject claims that only superficial change in the position of blacks has taken place. However, direct discrimination in jobs, in housing, and in myriad forms of interpersonal interaction continue to face African Americans almost irrespective of the social class background and achievements of the individual black person. In short, the significance of race in social life goes on.

This is an extremely nettlesome issue. Black and white Americans, the surveys show, could not be further apart in their thinking about the problem of discrimination. Blacks perceive it, experience it, and feel it acutely. They become frustrated when whites do not see the problem in the same way. Many whites see tremendous positive gains, cannot understand the steady litany of complaints, and have grown resentful and impatient. Miscommunication and mounting resentments accumulate on both sides as a result.

Third, the problem of social breakdown occurring in poor urban communities has a strong racial overlay and increasing political potence as a device to mobilize voters at the local, state, and national levels. The linkage in the minds of many white Americans between black culture and the problem of family dissolution, welfare dependency, crime, failing schools, and drug use may be setting the stage for a new period of deep retrenchment in civil rights and social welfare provision. All too often, a major subtext of campaigns about reducing welfare and fighting crime is a narrative about generally retaining white status privilege over blacks, and specifically about controlling and punishing poor black communities. This thinly veiled racial subtext of American politics is not lost on the black community. It feeds a growing suspicion and distrust among African Americans that white-dominated institutions may be moving toward overt hostility to the aspirations of African American communities....

THE DECLINE OF JIM CROW RACISM

The available survey data suggest that antiblack attitudes associated with Jim Crow racism were once widely accepted. The Jim Crow social order called for a society based on deliberate segregation by race. It gave positive sanction to antiblack discrimination in economics, education, and politics. It prohibited race-mixing, especially in the form of miscegenation or racial intermarriage. It involved an etiquette of interaction designed to reinforce the inferior status imposed on African Americans. All of this was expressly premised on the notion that blacks were the innate intellectual, cultural, and temperamental inferiors to whites.

Survey-based questions dealing with racial principles essentially asked about the degree of popular endorsement of these tenets of Jim Crow racism. The evidence from national sample surveys of the white population show that at one time the Jim Crow ideology was widely accepted, especially among residents of southern states but outside the South as well. Over the ensuing five decades since the first baseline surveys were conducted in the early 1940s, however, support for Jim Crow has steadily declined. It has been replaced by popular support for integration and equal treatment as principles that should guide relations between blacks and whites.

For example, whereas a solid majority, 68 percent, of white Americans in 1942 favored racially segregated schools, only 7 percent took such a position as early as 1985. Similarly, 55 percent of whites surveyed in 1944 believed whites should receive preference over blacks in access to jobs, as compared with only 3 percent who offered such an opinion as early as 1972. Indeed, so few people are willing to endorse a discriminatory response to either question that both have been dropped from ongoing social surveys. On both of these issues, once-pivotal features of the Jim Crow racist ideology—majority endorsements of the principles of segregation and discrimination—have given way to overwhelming support for the principles of integration and equal treatment.

This pattern of movement away from support for Jim Crow toward apparent support for racial egalitarianism holds with equal force for those questions dealing with issues of residential integration, access to public transportation and public accommodations, choice among qualified candidates for political office, and even racial intermarriage. It is important to note that the high absolute levels of support seen for the principles of school integration and equal access to jobs (both better than 90 percent nationwide) is not achieved for all racial principle–type questions. Despite improvement from extraordinarily low levels of support in the 1950s and 1960s, survey data continue to show substantial levels of white discomfort with the prospect of interracial dating and marriage.

Opinions among whites have never been uniform or monolithic. Both historical and sociological research have pointed to lines of cleavage and debate in the thinking of whites about the place of African Americans. The survey-based literature has shown that views on issues of racial principle vary greatly by region (at least along the traditional South versus non-South divide), level of education, age or generation, and other ideological factors. As might be expected, opinions in the South more lopsidedly favored segregation and discrimination at the time baseline surveys were conducted than was true outside the South. Patterns of change, save for a period of unusually rapid change in the South, have usually been parallel.

Level of education matters for racial attitudes. The highly educated are also typically found to express greater support for principles of racial equality and integration. Indeed, one can envision separating the white population into a multitiered reaction to issues of racial justice based on the interaction of level of education and region. At the more progressive and liberal end, one finds college-educated whites who live outside the South. At the bottom, one finds Southern whites with the least amount of schooling.

The degree of expressed support for racial integration and equality is also responsive to a person's age. Younger people are usually more racially tolerant than older people, where issues of principle are concerned. A small but meaningful part of the apparent generational differences in racial attitudes is attributable to the increasing levels of education obtained by younger cohorts. As the average level of education has risen, so too has the level of support for racial equality and integration....

The transformation of attitudes regarding the rules that should guide interaction between blacks and whites in public and impersonal spheres of life has been large, steady, and sweeping. Individuals living outside the South, the highly educated, and younger-age cohorts led the way to these changes. However, positive changes usually occurred across regions, age-groups, and education levels. At least with regard to racial principles, this change was so sweeping that Schuman, Steeh, and myself characterized it as a fundamental transformation of social norms with regard to race. Analysts of in-depth interview material reached similar conclusions. Bob Blauner's discussions with a small group of blacks and whites living in the San Francisco Bay area led him to conclude:

> The belief in a right to dignity and fair treatment is now so widespread and deeply rooted, so self-evident, that people of all colors would vigorously resist any effort to reinstate formalized discrimination. This consensus may be the most profound legacy of black militancy, one that has brought a truly radical transformation in relations between the races (1989).[2]

Some read this change as having far-reaching implications for transcending the color line and overcoming the American Dilemma. Based on their assessment of the available trend data, one group of survey researchers concluded: "Without ignoring real signs of enduring racism, it is still fair to conclude that America has been successfully struggling to resolve its Dilemma and that equality has been gaining ascendancy over racism."[3]

THE EMERGENCE OF LAISSEZ-FAIRE RACISM

The sanguine picture of change, however, changes substantially when attention shifts from principle to policy. Where issues of implementing the social changes needed to bring about greater integration of communities, schools, and workplaces are concerned, we find evidence of an important qualification on the extent of change in racial attitudes. Likewise, where issues of enforcing antidiscrimination laws and taking steps to improve the economic standing of African Americans are concerned, the survey data again point to significant bounds on the scope of positive change in racial attitudes.

At one level, it is not surprising that there are sharp differences in level of support between racial principles and policy implementation. Principles, when viewed in isolation, need not conflict with other principles, interests, or needs. At another, more concrete level, however, choices must often be made and priorities set. Particularly in the domain of racial attitudes and relations, there are large gaps between the principles most white Americans advocate and the practical steps they are willing to undertake to pursue those ends. For example, a 1964 survey showed that 64 percent of whites nationwide supported the principle of integrated schooling, but that only 38 percent believed that the federal government had a role to play in bringing about greater integration. The gap had actually grown larger by 1986. At that time, 93 percent of whites supported the principle of integrated schooling, but only 26 percent endorsed government efforts to increase the level of school integration. Analysis suggests that little of this disjuncture can be accounted for by distrust of government.

Similar patterns emerge in the areas of jobs and housing. Support for the principle of equal access to jobs stood at 97 percent in 1972, while support for federal efforts to prevent discrimination in jobs had only reached 39 percent. A 1976 national survey showed that 88 percent of whites supported the idea that blacks have a right to live wherever they can afford, that only 35 percent would vote in favor of a law prohibiting homeowners from discriminating on the basis of race when selling a home. While the sort of near-exact pairing of principle and implementation items that Schuman, Steeh, and myself were able to compare for the 1942–85 period is no longer possible with the available data, it seems likely that this disjuncture continues.

Implementation items not only typically exhibit lower absolute levels of support than principle items do, and less evidence of positive change, they also are less responsive to a number of other factors. There are smaller differences by age, education level, and region on questions of implementation. The lack of strong age-group differences, education effects, and regional differences implies that policy change in the area of race is not likely to witness the sort of great positive transformation seen for broad issues of principle.

Furthermore, the black-white divide on implementation questions is also sharp. Most of the implementation questions Schuman, Steeh, and myself analyzed showed majority—sometimes overwhelming majority—black support for government action to bring about integration, to fight discrimination, or to improve the economic conditions of the black community. Responses of whites were often a mirror opposite, with a clear majority opposing such a role for government.

The black-white division of opinion is often sharpest on questions of affirmative action. Bobo and Smith reported that 69 percent of white respondents in a major 1990 national survey opposed affirmative action in higher education for blacks, while only 26 percent of blacks opposed it. An equally large gap between black and white views emerged for the question of affirmative action in employment. Whereas 82 percent of whites opposed giving preference to blacks in hiring and promotion, only 37 percent of blacks adopted a similar position.

Bobo and Smith also attempted to determine whether the black-white gap in opinions could be explained by differences in social class background, political ideology and values, and racial attitudes. Their analyses of opinions on several race-specific social policies showed that virtually none of the black-white gap in opinion could be explained by social class background factors, that a small part reflected differences in political ideology and basic beliefs about economic inequality, and that a somewhat larger fraction was attributable to different racial outlooks. However, a substantial difference in views persisted, despite controls for a wide range of factors. This suggests, as myself and James Kluegel argue, that there is a set of collective interests that divide blacks and whites on policy questions of race.

Part of the reason for the opposition of whites to policy changes favorable to blacks may be found in the persistence of antiblack stereotypes. A 1990 national survey found evidence of widespread negative stereotypes of blacks, Hispanics, and Asian Americans. The study showed that whites tend to perceive blacks as more likely than whites to be unpatriotic, violence-prone, unintelligent, lazy, and to prefer to live off of welfare rather than being self-supporting. Fully 78 percent of whites in this national survey adopted this position. Such high levels of negative stereotyping were found because of a measurement procedure that did not force people to make categorical judgments, but rather allowed the expression of differing magnitudes of group difference. Whereas an overall 78 percent of whites saw blacks as more likely than white people to prefer to live off welfare, only a small number understood this to be a stark difference. Most whites expressed only a small difference between the races in tendencies on this and the other trait dimensions examined. The evidence of persistent negative stereotyping is not at odds with evidence of positive change in racial principles. These data merely qualify the scope and meaning of those positive changes.

Negative stereotypes help explain the gap between principle and implementation described above and shed light on current racial tensions. The more negative stereotypes a person holds about blacks, the less likely he or she is to support affirmative action policies. In addition, those with the highest negative stereotypes are strongly predisposed to maintain social distance between themselves and blacks. In sum, while Jim Crow racism has fallen into disrepute, images of blacks appear to have remained sufficiently negative to prompt many whites to reject affirmative action and to resist close association with blacks.

Another factor contributing to black-white polarization on the policy solutions to racial inequality is a substantial difference in their thinking about the problem of racial discrimination. The available survey data suggest that most blacks see racial discrimination as a more prevalent problem than do most whites, as having a stronger institutional base than do most whites, and as carrying responsibility for the generally disadvantaged position blacks occupy in American society than do most whites. The black-white gap in thinking about discrimination may be the core factor underlying modern misunderstanding and miscommunication across the color line.

In their 1992 study of the Detroit housing market, Farley and colleagues asked a general question about how much discrimination blacks faced in finding housing wherever they want. They found that 85 percent of blacks perceived "a lot" or at least "some" discrimination, compared with 80 percent of whites. While this difference is not large, they also found that blacks perceived discrimination as having changed little or gotten worse, whereas whites saw discrimination as declining. In addition, they wrote: "Blacks see much more institutionalized discrimination than do whites.... While 86 percent of blacks believe that blacks miss out on good housing because real estate agents discriminate, only 61 percent of whites believe this. When it comes to banks and lenders, 89 percent of blacks see discrimination, in contrast, to 56 percent of whites."[4]

Thus, although many whites acknowledge the existence of discrimination, especially in the housing market, they are much less likely than blacks to think of discrimination as having a systematic, institutionalized social basis....

According to Bobo and Smith, laissez-faire racism is a new form of racist ideology that has emerged in the post–Jim Crow, post–civil rights era. Although the full argument about laissez-faire racism cannot be developed here, its emergence reflects crucial changes in the economy and polity that at once undermined the structural basis for the Jim Crow social order of the American South and yet left in place the patterns of residential segregation, economic inequality, racialized identities, and antiblack outlooks that existed on a national basis. This new pattern of belief involves staunch rejection of an active role for government in undoing racial segregation and inequality, an acceptance of negative stereotypes of African Americans, a denial of discrimination as a current societal problem, and attribution of primary responsibility for black disadvantage to blacks themselves....

Does the color line still divide us? Does the dilemma still persist? Is the dream ever to be realized? There is little question that the United States is still sharply divided by race, that the struggle to fulfill national democratic ideals goes on, and that the dream remains far, very far, from realization. As pessimistic as these answers sound, they do not, however, amount to an acceptance of the idea that no meaningful change about American racism has or can take place. The "racial chasm" and "permanence of racism" to which [some authors]... have written are seriously overstated.

We have witnessed the substantial disappearance of one epochal form of racist ideology. Jim Crow racism once dominated the views white Americans had of black people. This ideology is no longer widely accepted or publicly espoused by any significant number of people. While segregationist notions, including explicitly biological racist ideas, have not completely vanished, they show no sign of making a quick or easy return to popular acceptance.

Nonetheless, a new epochal form of racism has emerged. Bobo and Smith label it "laissez-faire racism." It is a cultural pattern of belief that connects opposition to substantial policy change and activism with regard to improving the status of blacks with negative stereotyping of African Americans, a tendency to deny the potency of modern discrimination, and a view of blacks as largely responsible for their own disadvantaged circumstances. The emergence of this new form of American racism can be traced, in the first instance, to the historical erosion of the structural basis for the Jim Crow economic, political, and social institutions. In the second instance, it can be traced to the important but partial victories of the civil rights movement. The 1954 *Brown v. Board of Education* decision, the Civil Rights Act of 1964, and the Voting Rights Act of 1965 secured the basic citizenship rights of African Americans. As fundamental as these gains were, these accomplishments did not directly undo racial residential segregation or the individual and institutional actors that sustain it; these accomplishments did not directly undo tremendous disparities in earnings, prospects for employment, and wealth-holding that constituted the core structural bases of racial economic inequality; and these accomplishments did not wipe away racial identities and a long cultural heritage of disparaging views of African Americans. Thus, a new antiblack ideology has crystallized, an ideology that is appropriate to a historical epoch in which we have a formally race-neutral state and economy but a still racially divided social order and quality of life experience.

Laissez-faire racism has crystallized despite mounting evidence of discrimination against blacks with regard to housing, jobs, and access to many public spaces. The political climate makes the challenge of improving race relations that much greater. The very center of American politics is now infused with thinly veiled racial appeals that call for rolling back the welfare state, erecting more prisons, lengthening prison sentences, making it easier to convict the accused, and bringing greater certainty and swiftness to the execution of death sentences. While not based solely in efforts to harm the black community and its interests, or simply in the cynical pursuit of votes by opportunistic politicians, the current political climate has a powerful and undeniably racial subtext. The product of these trends is a potentially widening gulf of perception, of understanding, of feelings, and of interests across the color line.

Some, … look at these complex problems, see no real change, and see little hope for progress in the future, but that prognosis is difficult to accept. Racial categories and identities are social constructs. They are not given in nature, and they are subject to enormous variation over historical time and space. To be sure, the black-white divide in the United States is deeply entrenched structurally, culturally, and psychologically. But simply as a matter of logic and historical experience, the color line is not unmodifiable.

If historical figures like Thurgood Marshall, Fannie Lou Hamer, Rosa Parks, and Martin Luther King Jr. had simply accepted Jim Crow racism rather than challenging it head on, we almost surely would not have had a Brown decision, the Civil Rights Act, or the Voting Rights Act. We would not have witnessed the emergence of a black middle class that is at least modestly better residentially integrated and as large and accomplished as we have now. We would just as surely not have as many appointed and elected black public officials.

We stand at a moment of great ambiguity, uncertainty, and potentially momentous change in race relations. Lack of clarity about the future, however, does not warrant pessimistic certainty. The present is a time of deeply contradictory trends, not one of unequivocal backlash and polarization. Positive changes in racial attitudes and relations do not simply happen. They are made, in both intended and unintended ways. Laissez-faire racism, modern discrimination, and the current subtext of race in American politics are harder to confront directly than the obvious racism of the Jim Crow era. But the only path to transcending the color line is a continuous struggle to resolve the … dilemma of race and to steadfastly pursue Dr. King's dream.

REFERENCES

1. W. E. B. Du Bois, *The Souls of Black Folk* (Greenwich, Conn.: Fawcett, 1903).
2. Bob Blauner, *Black Lives, White Lives: Three Decades of Race Relations in America* (Berkeley and Los Angeles: University of California Press, 1989), 317.
3. Richard G. Niemi, John Mueller, and Tom W. Smith, *Trends in Public Opinion: A Compendium of Survey Data* (New York: Greenwood Press, 1989), 168.
4. Farley, Reynolds, Charlotte Steeh, Tara Jackson, Maria Kryan, and Heith Reeves. 1993. "Continued Racial Residential Segregation in Detroit; (Chocolate City, Vanilla Suburb) Revisited." *Journal of Housing Research* 4: 1–38.

DISCUSSION QUESTIONS

1. What does Bobo mean by *laissez-faire racism*? How does it contrast with Jim Crow racism?
2. What evidence does Bobo present to support his argument that there is a gap in the principles and policies that most White Americans are willing to support with regard to race in the United States?

12

Color-Blind Privilege

The Social and Political Functions of Erasing the Color Line in Post Race America

CHARLES A. GALLAGHER

Popular culture is full of images that portray a multiracial society, but such representations encourage a new form of thinking—color-blind racism. *Color-blind racism refers to the dominant belief that race no longer matters in shaping people's experiences—a belief that is contradicted by the reality of race in America.*

The young white male sporting a FUBU (African-American owned apparel company "For Us By Us") shirt and his white friend with the tightly set, perfectly braided cornrows blended seamlessly into the festivities at an all white bar mitzvah celebration. A black model dressed in yachting attire peddles a New England, yuppie boating look in Nautica advertisements. It is quite unremarkable to observe white, Asian or African-Americans with dyed purple, blond or red hair. White, black and Asian students decorate their bodies with tattoos of Chinese characters and symbols. In cities and suburbs young adults across the color line wear hip-hop clothing and listen to white rapper Eminem and black rapper 50-cent. It went almost unnoticed when a north Georgia branch of the NAACP installed a white biology professor as its president. Subversive musical talents like Jimi Hendrix, Bob Marley and The Who are now used to sell Apple Computers, designer shoes and SUVs. Du-Rag kits, complete with bandana headscarf and elastic headband, are on sale for $2.95 at hip-hop clothing stores and family centered theme parks like Six Flags. Salsa has replaced ketchup as the best selling condiment in the United States. Companies as diverse as Polo, McDonalds, Tommy Hilfiger, Walt Disney World, Master Card, Skechers sneakers, IBM, Giorgio Armani and Neosporin antibiotic ointment have each crafted advertisements that show an integrated, multiracial cast of characters interacting and consuming their products in post-race, color-blind world.

Americans are constantly bombarded by depictions of race relations in the media which suggests that discriminatory racial barriers have been dismantled. Social and cultural indicators suggest that America is on the verge, or has already

SOURCE: From *Race, Gender & Class* 10 (2003): 22–37. Reprinted by permission.

become, a truly color-blind nation. National polling data indicate that a majority of whites now believe discrimination against racial minorities no longer exists. A majority of whites believe that blacks have "as good a chance as whites" in procuring housing and employment or achieving middle class status while a 1995 survey of white adults found that a majority of whites (58%) believed that African Americans were "better off" finding jobs than whites (Gallup, 1997; Shipler, 1998). Much of white America now see a level playing field, while a majority of black Americans sees a field which is still quite uneven.... The color-blind or race neutral perspective holds that in an environment where institutional racism and discrimination have been replaced by equal opportunity, one's qualifications, not one's color or ethnicity, should be the mechanism by which upward mobility is achieved. Color as a cultural style may be expressed and consumed through music, dress, or vernacular but race as a system which confers privileges and shapes life chances is viewed as an atavistic and inaccurate accounting of U.S. race relations.

Not surprisingly, this view of society blind to color is not equally shared. Whites and blacks differ significantly, however, on their support for affirmative action, the perceived fairness of the criminal justice system, the ability to acquire the "American Dream," and the extent to which whites have benefited from past discrimination (Moore, 1995; Moore & Saad, 1995; Kaiser, 1995). This article examines the social and political functions colorblindness serves for whites in the United States. Drawing on interviews and focus groups with whites from around the country I argue that colorblind depictions of U.S. race relations serves to maintain white privilege by negating racial inequality. Embracing a colorblind perspective reinforces whites' belief that being white or black or brown has no bearing on an individual's or a group's relative place in the socioeconomic hierarchy.

DATA AND METHOD

I use data from seventeen focus groups and thirty-individual interviews with whites from around the country. Thirteen of the seventeen focus groups were conducted in a college or university setting, five in a liberal arts college in the Rocky Mountains and the remaining eight at a large urban university in the Northeast. Respondents in these focus groups were selected randomly from the student population. Each focus group averaged six respondents ... equally divided between males and females. An overwhelming majority of these respondents were between the ages of eighteen and twenty-two years of age. The remaining four focus groups took place in two rural counties in Georgia and were obtained through contacts from educational and social service providers in each county. One county was almost entirely white (99.54%) and in the other county whites constituted a racial minority. These four focus groups allowed me to tap rural attitudes about race relations in environments where whites had little or consistent contact with racial minorities....

COLORBLINDNESS AS NORMATIVE IDEOLOGY

The perception among a majority of white Americans that the socio-economic playing field is now level, along with whites' belief that they have purged themselves of overt racist attitudes and behaviors, has made colorblindness the dominant lens through which whites understand contemporary race relations. Colorblindness allows whites to believe that segregation and discrimination are no longer an issue because it is now illegal for individuals to be denied access to housing, public accommodations or jobs because of their race. Indeed, lawsuits alleging institutional racism against companies like Texaco, Denny's, Coke, and Cracker Barrel validate what many whites know at a visceral level is true: firms which deviate from the color blind norms embedded in classic liberalism will be punished. As a political ideology, the commodification and mass marketing of products that signify color but are intended for consumption across the color line further legitimate colorblindness. Almost every household in the United States has a television that, according to the U.S. Census, is on for seven hours every day (Nielsen 1997). Individuals from any racial background can wear hip-hop clothing, listen to rap music (both purchased at Wal-Mart) and root for their favorite, majority black, professional sports team. Within the context of racial symbols that are bought and sold in the market, colorblindness means that one's race has no bearing on who can purchase a Jaguar, live in an exclusive neighborhood, attend private schools or own a Rolex.

The passive interaction whites have with people of color through the media creates the impression that little, if any, socio-economic difference exists between the races....

Highly visible and successful racial minorities like [former] Secretary of State Colin Powell and ... [Secretary of State] Condelleeza Rice are further proof to white America that the state's efforts to enforce and promote racial equality have been accomplished.

The new color-blind ideology does not, however, ignore race; it acknowledges race while disregarding racial hierarchy by taking racially coded styles and products and reducing these symbols to commodities or experiences that whites and racial minorities can purchase and share. It is through such acts of shared consumption that race becomes nothing more than an innocuous cultural signifier. Large corporations have made American culture more homogenous through the ubiquitousness of fast food, television, and shopping malls but this trend has also created the illusion that we are all the same through consumption. Most adults eat at national fast food chains like McDonalds, shop at mall anchor stores like Sears and J.C. Penney's and watch major league sports, situation comedies or television drama. Defining race only as cultural symbols that are for sale allows whites to experience and view race as nothing more than a benign cultural marker that has been stripped of all forms of institutional, discriminatory or coercive power. The post-race, color-blind perspective allows whites to imagine that depictions of racial minorities working in high status jobs and consuming the same products, or at least appearing in commercials for products whites desire or consume, is the same as living in a society where color is no longer used to allocate resources or shape group outcomes. By constructing a picture of society where racial harmony is the norm, the color-blind perspective functions to make white privilege invisible while removing from public discussion the need to maintain any social programs that are race-based.

How then, is colorblindness linked to privilege? Starting with the deeply held belief that America is now a meritocracy, whites are able to imagine that the socio-economic success they enjoy relative to racial minorities is a function of individual hard work, determination, thrift and investments in education. The color-blind perspective removes from personal thought and public discussion any taint or suggestion of white supremacy or white guilt while legitimating the existing social, political and economic arrangements which privilege whites. This perspective insinuates that class and culture, and not institutional racism, are responsible for social inequality. Colorblindness allows whites to define themselves as politically and racially tolerant as they proclaim their adherence to a belief system that does not see or judge individuals by the "color of their skin." This perspective ignores, as Ruth Frankenberg puts it, how whiteness is a "location of structural advantage societies structured in racial dominance" (2001 p. 76).... Colorblindness hides white privilege behind a mask of assumed meritocracy while rendering invisible the institutional arrangements that perpetuate racial inequality. The veneer of equality implied in colorblindness allows whites to present their place in the racialized social structure as one that was earned.

OPPORTUNITY HAS NO COLOR

Given this norm of colorblindness it was not surprising that respondents in this study believed that using race to promote group interests was a form of (reverse) racism....

Believing and acting as if America is now color-blind allows whites to imagine a society where institutional racism no longer exists and racial barriers to upward mobility have been removed. The use of group identity to challenge the existing racial order by making demands for the amelioration of racial inequities is viewed as racist because such claims violate the belief that we are a nation that recognizes the rights of individuals not rights demanded by groups....

The logic inherent in the colorblind approach is circular; since race no longer shapes life chances in a color-blind world there is no need to take race into account when discussing differences in outcomes between racial groups. This approach erases America's racial hierarchy by implying that social, economic and political power and mobility is equally shared among all racial groups. Ignoring the extent or ways in which race shapes life chances validates whites' social location in the existing racial hierarchy while legitimating the political and economic arrangements that perpetuate and reproduce racial inequality and privilege.

REFERENCES

Frankenberg, R. (2001). The mirage of an unmarked whiteness. In B. B. Rasmussen, E. Klineberg, I. J. Nexica & M. Wray (eds.) *The making and unmaking of whiteness.* Durham: Duke University Press.

Gallup Organization. (1997). Black/white relations in the U.S. June 10, pp. 1–5.

Kaiser Foundation. (1995). *The four Americas: Government and social policy through the eyes of America's multi-racial and multi-ethnic society*. Menlo Park, CA: Kaiser Family Foundation.

Moore, D. (1995). "Americans" most important sources of information: Local news." *The Gallup Poll Monthly*, September, pp. 2–5.

Moore, D. & Saad, L. (1995). No immediate signs that Simpson trial intensified racial animosity. *The Gallup Poll Monthly*, October, pp. 2–5.

Nielsen, A. C. (1997). *Information please almanac* (Boston: Houghton Mifflin).

Shipler, D. (1998). *A country of strangers: Blacks and whites in America.* New York: Vintage Books.

DISCUSSION QUESTIONS

1. How does Gallagher see color-blind racism as resulting from White people's privilege? How does privilege influence what White people can understand about racism?
2. In what ways does color-blind racism support the traditional American ideal that any individual can succeed if they only try hard enough?

13

Learning to Be White through the Movies

HERNÁN VERA AND ANDREW M. GORDON

Watching a film may seem like a purely innocent leisure activity, but Vera and Gordon show how films, like other forms of popular culture, produce cultural narratives with specific, even if implicit, racialized themes, thus shaping the social construction of race and racism.

Why does Gone with the Wind *touch such deep chords inside me? Maybe because it put those chords there in the first place. This is the movie that taught me and three generations how to be Southerners. It doesn't move us because we are Southern; we are Southern because we have taken this movie to heart.*
—*Susan Stewart*[1]

I was the only Negro in the theater, and when Butterfly McQueen went into her act, I felt like crawling under the rug.—*Malcolm X*[2]

We need to study movies because ordinarily we do not want to think about the influence that they have on us and on our society. We tend to dismiss the cinema as mere entertainment; yet it has profound effects, shaping our thinking and our behavior....Movies can teach us who we are: what our identity is and what it should be. "Radio, television, film and the other products of the culture industries," Douglas Kellner argues, "provide the models of what it means to be male or female, successful or a failure, powerful or powerless. Media culture also provides the materials out of which many people construct their sense of class, of ethnicity and race, of nationality, of sexuality, of "us" and "them." Movies manufacture the way we see, think of, feel, and act towards others."[3]

We need to study movies not only because of what they tell us about the world we live in but also, and most importantly, because movies are a crucial part of that world. In the simulations of the moving pictures we learn who has

SOURCE: From Hernán Vera and Andrew M. Gordon, *Screen Saviors: Hollywood Fictions of Whiteness* (Lanham, MD: Rowman & Littlefield, 2003), pp. 8–13. Reprinted by permission of Rowman & Littlefield Publishers, Inc.

the power and who is powerless, who is good and who is evil. "Media spectacles," Kellner writes, "dramatize and legitimate the power of the forces that be and demonstrate to the powerless that if they fail to conform, they risk incarceration or death."[4]

We live in a cinematic society, one that presents and represents itself through movie and television screens. By 1930 the movies had become a weekly pastime for a majority of Americans. After 1950, with the advent of television, watching moving pictures became a daily activity, even an addiction, in the United States and other countries in the industrialized world. One report projected that for the year 2001, the average American spent 1577 hours in front of the TV set, 13 hours in movie theaters, and 55 hours watching prerecorded videos at home.[5] This represents 28 percent of our waking time. It is also five times the number of hours the average American spent in 2001 reading books, newspapers, and magazines.

In the same way that literate societies are dramatically different from illiterate societies, the social organization of cinematic societies is dramatically different from that of noncinematic ones. Without taking into account the impact that the moving pictures in television and cinema screens have on the people of a country, we could no more understand contemporary society than we could understand it without realizing the impact of literacy. The daily rhythms of our lives, what we know and what we ignore, are set by the rhythm of and the information contained in the screens of cinema and television. Countries without a film industry can be considered colonies for foreign filmmakers. Within countries, one can, of course, speak of diversely cinematized segments of the population because the time, energy, and money spent on media consumption vary greatly by age, class, religion, income, race, geography, and other such sociodemographic variables.

The Hollywood film industry does not portray all the segments of society and the world populations equally, with the same frequency, accuracy, or with the same respect. Consider that Latinos, who according to the 2000 U.S. Census constitute one of the largest U.S. minority groups, have seldom been represented as the protagonist of Hollywood films. The *Video Hound Golden Movie Retriever* Index, for example, lists only 17 films under the category of "Hispanic America," roughly half of which are Hollywood main releases. In contrast, the same index lists 69 films in the category "Ireland," 151 under "Judaism," 45 under "British Royalty," and even 119 under "Zombies"![6]

Allan G. Johnson notes that of the films that have won the Academy Award in the category Best Picture from 1965 through 1999, "none set in the United States places people of color at the center of the story without their having to share it with white characters of equal importance" (e.g., *In the Heat of the Night* [1967] or *Driving Miss Daisy* [1989]). "Anglo, heterosexual males, even though they are less than twenty percent of the U.S. population," he proposes, "represent ninety percent of the characters in the most important movies ever made."[7] Until recently, most minority characters in Hollywood movies have usually been caricatured and portrayed with disrespect.

Much of what we know about people we consider to be "others" we learn through the movies. The moving pictures allow access to private spaces, scenes that would normally be out of the reach of our eyes. Through the film media we learn what life supposedly is like or used to be like and what it is in distant lands and in private places.

Films also represent us, the spectators, who find enjoyment and solace in them. The streets we walk; the landmarks in our cities we go by; the appliances, furniture, and gadgets we use every day; the cars and buses we ride; and the music we listen to all appear in the movies. The social roles we play—as children, parents, workers, and lovers—are also recognizable in films. The words and the jargon the characters use in the movies are part of the language we speak. In this sense, we, the audience, watch ourselves. Much of the attraction and power of film, its ability to make us laugh or cry and to teach us about the world and about ourselves, rests on our being, simultaneously, spectators and subjects being gazed on.

The cinematic viewing experience, in our opinion, is one of recognizing and mis-recognizing ourselves in the moving pictures. Watching a movie is the experience of sharing—or sometimes, of resisting—the way of seeing, the ideology, and the values of the filmmakers, their gaze, and their imagination. Through their technology and their language, films implement ways of looking at class, gender, and race differences. Filmmakers can make us see these differences, but they can also hide them from our sight by creating pleasing fictions. This way of seeing carries the individual and social biases of the filmmakers but also the biases and standpoints of the culture of the people for which films are produced, the culture to which the film belongs....

DIALECTICS OF RACE IN FILM

... We regard film in a dialectical fashion, considering them in two opposed ways. First, we consider them as a means to celebrate whiteness, to teach what it is like to be white and to enjoy the privilege of being white. Second, we consider movies as social therapeutic devices to help us cope with the unjust racial divide by denying or obscuring white privilege and the practices on which it depends.... We believe that unless we capture the tension and contradictions between these two intentions and the central need of Hollywood to entertain and to be profitable, we would miss critical elements of the role films play in the production and reproduction of racism in the United States and around the world.

Consider that in Susan Stewart's earlier remark about *Gone with the Wind*, although she recognizes that the film taught her how to be "Southern," she fails to recognize that it only taught her how to be a white Southerner: white becomes so normative and universal a category that she does not even need to mention it. She does not notice the ways the film forces African Americans into the background and occludes their story. Recall Malcolm X's humiliation, when he was the sole black patron in a white theater, at seeing how blacks were portrayed in *Gone with the Wind*. Lorraine Hansberry, the black playwright, confirms Malcolm X's response when she writes that *Gone with the Wind* did not teach blacks to be blacks. The fact that in the United States white "goes without saying" in statements such as Susan Stewart's is an important trait of what … we call "the white self." …

The concept of self, of the white self—the portrayal of which we will be examining—is used by scholars to designate who and what we are. The self is the human person, the place in which all experience—our memories, our pain and pleasure, our emotions—is organized. By self, we mean the sense of being a person, the experience of existing as an individual contained in the space of a body over time. As universal as this notion might be, it is highly culture specific. In the United States, the fundamental entries on our birth certificates are name, birth date, gender, and race. These constitute our legal sense of self. Race is also crucial to our psychological sense of self. Without it we would be fatally disoriented, like Joe Christmas in *Faulkner's Light in August*, who goes mad and is destroyed because he never knows whether he is white or black.

WHITENESS

The key element to understanding racial thinking in the United State and in much of the world today is white supremacy. The modern concept of race and the notion of whiteness were invented during the period of European colonization of the Americas and Africa. The stock of knowledge we call racism has been developed in the past five hundred years precisely to establish the superiority of whites and to contribute a veneer of legitimacy to colonial domination, exploitation, or extermination of people of color, both domestically and internationally, by whites.

One difficulty in studying "whiteness" is that, until recently, it was an empty or invisible category, not perceived as a distinctive racial identity. Richard Dyer writes, "As long as race is something only applied to non-white peoples, as long as white people are not racially seen and named, they/we function as a human norm. Other people are raced, we are just people."[8] Thus, most white Americans either do not think of their "whiteness" or think of it as neutral. The power of whiteness rests in its apparent universality and invisibility, in the way it has gone unexamined. Nevertheless, the images of film, especially of films in which whites interact with persons of another color, offer a way to study white self-representation across the twentieth century. As has often been said, "Whites don't have a color until a person of color enters the room."

Until recently, sociologists and culture critics concentrated on prejudice, that is, on the distorted images that we construct of others we perceive as different. For example, Bogle (1997), Cripps (1997), Snead (1994), and Guerrero (1993), among others, have studied the prejudicial images of African Americans in American films.[9] We want to shift the focus to the representation of the white self-concept.

Whiteness as we understand it today in the United States is a construct, a public fiction that has evolved throughout American history in response to changing political and economic needs and conditions. Whiteness has always been a shifting category used to police class and sexual privilege. At the beginning, only "free white persons" could become American citizens. White privilege depended on the exclusion of "others," but the definition of who was non-white constantly changed. Thus, previous historic categories such as Celt, Slav, Alpine, Hebrew, Iberic, Anglo-Saxon, and Nordic have been incorporated into the contemporary concept of "white" or "Caucasian." "Caucasians are made and not born."[10] We argue that the notion of whiteness has become so integral to the American identity that it is embedded in the national unconscious.

RACE

In practice, the term "race" designates one or more biological traits (e.g., skin color) from which a sociopolitical hierarchy is derived and the assumption that some races are superior and therefore deserve to be more powerful than others. In spite of the concentrated efforts by scientists over the past one hundred years, the concept of "race" has become progressively more elusive, to the extent that today we can say that race is an illusion, a fiction that no longer leads to a meaningful classification of humans in the biological or social sciences. It is not an objective or fixed category. Race, according to Omi and Winant, is "an unstable and 'decentered' complex of social meanings constantly being transformed by political struggle." Although the concept of race may be a fiction, we cannot simply jettison it because it "continues to be central to everyone's identity and understanding of the social world."[11]

Today the vast majority of humans across the globe still think, feel, and act as if "race" were real, as if it pointed to true, useful differences among people. Furthermore, no one alive today has lived in a world in which race did not matter. In the United States, race matters in the chance each of us had of being born alive and healthy. Race matters in the neighborhoods where we grow up, the quality of the education we obtain, the persons we choose as friends, spouses, or lovers, the careers we pursue, the health and opportunities of the children we are going to have, and the churches we attend. Race matters in the length of our life span. Finally, race matters even in the cemetery where we lie after death.

At the beginning of the twenty-first century, the memories of the horrors in which race was the operative concept are still fresh. Among others, the horror of racial segregation and lynching in the United States, of the Nazi Holocaust in Europe, of apartheid in South Africa, and of the "ethnic cleansing" in Bosnia and Kosovo cannot be ignored. One can also not deny that members of oppressed groups find identity, self-expression, and solidarity in racial and ethnic categories. Historically and today, race and the violent or subtle practices we call racism shape both the structures of our societies and the daily rhythms of our lives.

We cannot begin to explain the contradiction between the scientific uselessness of the concept of race and the real consequences the application of racial categories bring about. At the interpersonal level, the biological trait or set of traits thought to reveal "race" are used as assumptions about other physical, intellectual, emotional, or spiritual traits of persons with those characteristics. In the United States, for example, those who are not considered white are often automatically assumed to be smelly, "greasy," less intelligent, lazy, dirty, not in control of their emotions, unreliable, and so on. The category of race, however fictional, is taken for real and is real in its consequences.

Today, white supremacy still dominates America. Consider that the Constitution of the United States of 1789, the fundamental document of the first democratic society, accepted the slavery of Africans and African Americans within its borders and gave Congress the authority to suppress slave insurrections. For tax distribution purposes, a slave was counted as three-fifths of a person. In 1861, both houses of the U.S Congress passed a bill that would have made slavery a permanent feature of the American legal system. Today, decades after the Civil Rights revolution of the 1960s, the enforcement of civil rights laws is very weak, at best. Film production is one of the resources through which power is wielded by the classes that benefit from the racial status quo.

REFERENCES

1. Stewart, Susan. 2000. "Lessons Learned: The Enduring Truths of *Gone with the Wind*," *TV Guide*, December 23–29, p. 26.
2. Malcolm X, with Alex Haley. 1965. *The Autobiography of Malcolm X*. New York: Grove, p. 42.
3. Kellner, Douglas. 1995. *Media Culture: Cultural Studies, Identity, and Politics between the Modern and the Postmodern*. New York: Routledge, p. 1.
4. Ibid, p. 2.
5. Veronis, Suhler & Associates. 1999. "Table 920. Media Usage and Consumer Spending: 1992 to 2002: Communications Industry Report." New York.
6. Craddock, Jim, ed. 2001. *Video Hound's Golden Movie Retriever 2001: The Complete Guide to Movies on Videocassette, DVD, and Laserdisc*. Detroit, Mich.: Visible Ink.
7. Johnson, Alan G. 2001. *Privilege, Power and Difference*. Mountain View, Calif.: Mayfield, p. 108.
8. Dyer, Richard. 1997. *White*. New York: Routledge, p. 1.
9. Bogle, Donald. 1997. *Toms, Coons, Mulattoes, Mammies, and Bucks*. New York: Continuum; Cripps, Thomas. 1997. *Slow Fade to Black: The Negro in American Film, 1900–1942*. New York: Oxford University Press; Guerrero, Ed. 1993. *Framing Blackness: The African American Image in Film*. Philadelphia: Temple University Press; Snead, James A., Colin MacCabe, and Cornel West. 1994. *White Screens, Black Images: Hollywood from the Dark Side*. New York: Routledge.
10. Jacobson, Mathew Frye. 1998. *Whiteness of a Different Color: European Immigrants and the Alchemy of Race*. Cambridge, Mass.: Harvard University Press, p. 4. See also

Roediger, David. 1991. *The Wages of Whiteness: Race and the Making of the American Working Class*. London: Verso; Allen, Theodore. 1994. *The Invention of the White Race*. London: Verso; Ignatiev, Noel. 1995. *How the Irish Became White*. New York: Routledge; Brodkin, Karen. 1998. *How Jews Became White Folks & What that Says About Race in America*. New Brunswick, N.J.: Rutgers University Press.

11. Omi, Michael, and Howard Winant. 1994. *Racial Formation in the United States: From the 1960s to the 1990s*. 2d ed. New York: Routledge, p. 55.

DISCUSSION QUESTIONS

1. Think of the last three movies that you saw (and, if you can, view them again). How are White people depicted in these movies? How are other groups depicted? How do your observations relate to Vera and Gordon's argument that movies are one of the lenses through which we come to understand race?
2. What other forms of popular culture do you think contribute to the "celebration of White privilege" that Vera and Gordon identify? Do you agree with Vera and Gordon that movies (and other forms of popular culture) have the ability to define who is powerful and who is powerless?

14

"Playing Indian"

Why Native American Mascots Must End

CHARLES FRUEHLING SPRINGWOOD AND C. RICHARD KING

Springwood and King show how the popular use of Native American mascots leads to disrespect and stereotyping of Native American people. Such mascots typically identify Native Americans as fierce warriors, sometimes with comiclike features. Research has also shown that exposure to such mascots is damaging to the self-esteem of young Native Americans.

American Indian icons have long been controversial, but 80 colleges still use them, according to the National Coalition on Racism in Sports and Media. Recently, the struggles over such mascots have intensified, as fans and foes across the country have become increasingly outspoken.

At the University of Illinois at Urbana-Champaign, for example, more than 800 faculty members ... signed a petition against retaining Chief Illiniwek as the university's mascot.[1] Students at Indiana University of Pennsylvania have criticized the athletics teams' name, the Indians. The University of North Dakota has experienced rising hostilities on campus against its Fighting Sioux. Meanwhile, other students, faculty members, and administrators have vehemently defended those mascots.

Why, nearly 30 years after Dartmouth College and Stanford University retired their American Indian mascots, do similar mascots persist at many other institutions? And why do they evoke such passionate allegiance and strident criticism?

American Indian mascots are important as symbols because they are intimately linked to deeply embedded values and worldviews. To supporters, they honor indigenous people, embody institutional tradition, foster shared identity, and intensify the pleasures of college athletics. To those who oppose them, however, the mascots give life to racial stereotypes as well as revivify historical patterns of appropriation and oppression. They often foster discomfort, pain, and even terror among many American Indian people.

SOURCE: "Playing Indian: Why Native American Mascots Must End" by Charles F. Springwood and C. Richard King in *Chronicle of Higher Education*, Dec. 9, 2001. Reprinted by permission of the authors.

1. Note that since the writing of this article, the University of Illinois retired its mascot, Chief Illiniwek, in 2007.

The December 1999 cover of *The Orange and Blue Observer*, a conservative student newspaper at Urbana-Champaign, graphically depicts the multilayered and value-laden images that American Indian mascots evoke. Beneath the publication's masthead, a white gunslinger gazes at the viewer knowingly while pointing a drawn pistol at an Indian dancer in full regalia. A caption in large letters spells out the meaning of the scene: "Manifest Destiny: Go! Fight! Win!" Although arguably extreme, the cover, when placed alongside what occurs at college athletic events—fans dressing in paint and feathers, half-time mascot dances, crowds cheering "the Sioux suck"—reminds us that race relations, power, and violence are inescapable aspects of mascots.

We began to study these mascots while we were graduate students in anthropology at the University of Illinois in the early 1990s. American Indian students and their allies were endeavoring to retire Chief Illiniwek back then, as well, and the campus was the scene of intense debates. Witnessing such events inspired us to move beyond the competing arguments and try to understand the social forces and historical conditions that give life to American Indian mascots—as well as to the passionate support of, and opposition to, them. We wanted to understand the origins of mascots: how and why they have changed over time: how arguments about mascots fit into a broader racial context: and what they might tell us about the changing shape of society.

Over the past decade, we have developed case studies on the role that mascots have played at the halftime ceremonies of the University of Illinois. Marquette University, Florida State University, and various other higher-education institutions. Recently, we published an anthology, *Team Spirits: The Native American Mascots Controversy*, in which both American Indian and European American academics explored "Indian-ness," "whiteness," and American Indian activism. They also suggested strategies for change—in a variety of contexts that included Syracuse University and Central Michigan University, the Los Angeles public schools, and the Washington Redskins. Our scholarship and that of others have confirmed our belief that mascots matter, and that higher-education institutions must retire these hurtful symbols.

The tradition of using the signs and symbols of American Indian tribes to identify an athletic team is part of a much broader European American habit of "playing Indian," a metaphor that Philip Joseph Deloria explores in his book of that title (Yale University Press, 1998). In his historical analysis, Deloria enumerates how white people have appropriated American Indian cultures and symbols in order to continually refashion North American identities. Mimicking the indigenous, colonized "other" through imaginary play—as well as in literature, in television, and throughout other media—has stereotyped American Indian people as bellicose, wild, brave, pristine, and even animalistic.

Educators in particular should realize that such images, by flattening conceptions of American Indians into mythological terms, obscure the complex histories and misrepresent the identities of indigenous people. Moreover, they literally erase from public memory the regnant terror that so clearly marked the encounter between indigenous Americans and the colonists from Europe.

That higher-education institutions continue to support such icons and ensure their presence at athletics games and other campus events—even in the face of protest by the very people who are ostensibly memorialized by them—suggests not only an insensitivity to another race and culture, but also an urge for domination. Power in colonial and postcolonial regimes has often been manifested as the power to name, to appropriate, to represent, and to speak—and to use such powers over others. American Indian mascots are expressive practices of precisely those forms of power.

Consider, for example, the use of dance to feature American Indian mascots. Frequently, the mascot, adorned in feathers and paint, stages a highly caricatured "Indian dance" in the middle of the field or court during halftime. At Urbana-Champaign, Chief Illiniwek sports an Oglala war bonnet to inspire the team: at Florida State University, Chief Osceola rides across the football field, feathered spear held aloft.

Throughout U.S. history, dance has been a controversial form of expression. Puritans considered it sinful; when performed by indigenous people, the federal government feared it as a transgressive, wild, and potentially dangerous form of expression. As a result, for much of the latter half of the 19th century, government agents, with the support of conservative clergy, attempted to outlaw native dance and ritual. In 1883, for example, the Department of the Interior established rules for Courts of Indian Offenses. Henry Teller, the secretary of the department, anticipated the purpose of such tribunals in a letter that he wrote to the Bureau of Indian Affairs stating that they would end the "heathenish practices" that hindered the assimilation of American Indian people. As recently as the 1920s, representatives of the federal government criticized American Indian dance, fearing the "immoral" meanings animated by such performances.

The majority of Indian mascots were invented in the first three decades of the 20th century, on the heels of such formal attempts to proscribe native dance and religion, and in the wake of the massive forced relocation that marked the 19th-century American Indian experience. European Americans so detested and feared native dance and culture that they criminalized those "pagan" practices. Yet at the same time they exhibited a passionate desire for certain Indian practices and characteristics—evidenced in part by the proliferation of American Indian mascots.

Although unintentional perhaps, the mascots' overtones of racial stereotype and political oppression have routinely transformed intercollegiate-athletic events into tinderboxes. Some 10 years ago at Urbana-Champaign, several Fighting Illini boosters responded to American Indian students who were protesting Chief Illiniwek by erecting a sign that read "Kill the Indians, Save the Chief." And, in the wake of the North Dakota controversy, faculty members who challenged the Fighting Sioux name have reported to us that supporters of the institution's symbol have repeatedly threatened those who oppose it.

Although many supporters of such mascots have argued that they promote respect and understanding of American Indian people, such symbols and the spectacles associated with them are often used in insensitive and demeaning ways that further shape how many people perceive and engage American Indians. Boosters of teams employing American Indians have enshrined largely romanticized stereotypes—noble warriors—to represent themselves. Meanwhile, those who support competitive teams routinely have invoked images of the frontier, Manifest Destiny, ignoble savages, and buffoonish natives to capture the sprit of impending athletics contests and their participants. In our studies, we find countless instances of such mockery on the covers of athletics programs, as motifs for homecoming floats, in fan cheers, and in press coverage.

For example, in 1999, *The Knoxville News-Sentinel* published a cartoon in a special section commemorating the appearance of the University of Tennessee at the Fiesta Bowl. At the center of the cartoon, a train driven by a team member in a coonskin cap plows into a buffoonish caricature of a generic Indian, representing the team's opponent, the Florida State Seminoles. As he flies through the air, the Seminole exclaims. "Paleface speak with forked tongue! This land is ours as long as grass grows and river flows. Oof!"

The Tennessee player retorts. "I got news, pal. This is a desert. And we're painting it orange!" Below them, parodying the genocide associated with the conquest of North America Smokey, a canine mascot of the University of Tennessee, and a busty Tennessee fan speed down Interstate 10, dubbed "The New and Improved Trail of Tears." What effect can such a cartoon have on people whose ancestors were victims of the actual Trail of Tears?

The tradition of the Florida State Seminoles bears its share of responsibility for inviting that brand of ostensibly playful, yet clearly demeaning, discourse. For, at FSU, the image of the American Indian as warlike and violent is promoted without hesitation. Indeed, the Seminoles' football coach, Bobby Bowden, is known to scribble "Scalp 'em" underneath his autograph.

Such images and performances not only deter cross-cultural understanding and handicap social relations, they also harm individuals because they deform indigenous traditions, question identities, and subject both American Indians and European Americans to threatening experiences. For example, according to a *Tampa Tribune* article, a Florida resident and Kiowa tribe member, Joe Quetone, took his son to a Florida State football game during the mid-1990s. As students ran through the stands carrying tomahawks and sporting war paint, loincloths, and feathers, Quetone and his son overheard a man sitting nearby turn to a little boy and say, "Those are real Indians down there. You'd better be good, or they'll come up and scalp you!"

Environmental historian Richard White has suggested that "[White Americans] are pious toward Indian peoples, but we don't take them seriously; we don't credit them with the capacity to make changes. Whites readily grant certain nonwhites a 'spiritual' or 'traditional' knowledge that is timeless. It is not something gained through work or labor; it is not contingent knowledge in a contingent world." The omnipresence of American Indian mascots serves only to advance the inability to accept American Indians as indeed contingent, complicated, diverse, and genuine Americans.

Ultimately, American Indian mascots cannot be separated from their origins in colonial conditions of exploitation. Because the problem with such mascots is one of context, they can never be anything more than a white man's Indian.

Based on our research and observations, we cannot imagine a middle ground for colleges with Indian mascots to take—one that respects indigenous people, upholds the ideals of higher education, or promotes cross-cultural understanding. For instance, requiring students to take courses focusing on American Indian heritage, as some have suggested, reveals a troubling vision of the fit between curriculum, historic inequities, and social reform. Would we excuse colleges with active women's studies curricula if their policies and practices created a hostile environment for women?

Others have argued that colleges with American Indian mascots can actively manage them, promoting positive images and restricting negative uses. Many institutions have already exerted greater control over the symbols through design and licensing agreements. But they can't control the actions of boosters at their institutions or competitors at others. For example, the University of North Dakota would probably not prefer fans at North Dakota State University to make placards and T-shirts proclaiming that the "Sioux suck." Such events across the nation remind us that mascots are useful and meaningful because of their openness and flexibility—the way that they allow individuals without institutional consent or endorsement to make interpretations of self and society.

American Indian mascots directly contradict the ideals that most higher-education institutions seek—those of transcending racial and cultural boundaries and encouraging respectful relations among all people who live and work on their campuses. Colleges and universities bear a moral responsibility to relegate the unreal and unseemly parade of "team spirits" to history.

DISCUSSION QUESTIONS

1. What arguments do different groups make in favor of eliminating the use of Native American mascots? What arguments are there to keep them?
2. What would happen if similar caricatures of other groups were used as school mascots?

15

The Necessary Engagement with Youth Culture

CORNEL WEST

West's article argues that as rap and hip-hop have become more co-opted by commercial interests, the earlier forms of social protest that were part of this urban, black youth movement have become blunted. At the same time, his article is an appeal to the need for young people to mobilize around issues of racial and other forms of social justice, even when their consumerist-based culture trains them to be oblivious to the need for social change.

In past moments of national division, young people have played a disproportionate role in deepening the American democratic experiment. The black freedom struggle and the antiwar movement in the 1960s were largely sustained owing to their vision and courage. As older folk become jaded, disillusioned, and weary, the lively moral energy of reflective and compassionate young people can play a vital role in pushing democratic momentum. Yet one of the most effective strategies of corporate marketeers has been to target the youth market with distractive amusement and saturate them with pleasurable sedatives that steer them away from engagement with issues of peace and justice. The incessant media bombardment of images (of salacious bodies and mindless violence) on TV and in movies and music convinces many young people that the culture of gratification—a quest for insatiable pleasure, endless titillation, and sexual stimulation—is the only way of being human. Hedonistic values and narcissistic identities produce emotionally stunted young people unable to grow up and unwilling to be responsible democratic citizens. The market-driven media lead many young people to think that life is basically about material toys and social status. Democratic ideas of making the world more just, or striving to be a decent and compassionate person, are easily lost or overlooked.

This media bombardment not only robs young people of their right to struggle for maturity—by glamorizing possessive individualism at the expense of democratic individuality—but also leaves them ill equipped to deal with the spiritual malnutrition that awaits them after their endless pursuit of pleasure. This

SOURCE: From *Democracy Matters* by Cornel West, copyright © 2004 by Cornel West. Used by permission of The Penguin Press, a division of Penguin Group (USA) Inc.

sense of emptiness of the soul holds for wealthy kids in the vanilla suburbs and poor kids in the chocolate cities. Neither the possession of commodities nor the fetishizing of commodities satisfies young people's need for love and self-confidence. Instead we witness personal depression, psychic pain, and individual loneliness fueling media-influenced modes of escapism. These include the high use of drugs like cocaine and Ecstasy; the growing popularity of performing sex acts at incredibly young ages, such as middle-school-age girls giving boys blow jobs because it will make them "cool"; and the way in which so many kids have become addicted to going online and instant messaging or creating Weblogs in which they assume an alternate personality. This disgraceful numbing of the senses, dulling of the mind, and confining of life to an eternal present—with a lack of connection to the past and no vision for a different future—is an insidious form of soul murder. And we wonder why depression escalates and suicides increase among our precious children.

The most dangerous mode of dealing with this bombardment is addiction—to drugs, alcohol, sex, or narrow forms of popularity or success. These addictions leave little room or time for democratic efforts to become mature, concerned about others, or politically engaged in social change. The popular way of escaping from the pain and emptiness is self-medication—the first step toward self-violation and self-destruction. This is why so many—too many—of the youth of America are drifting, rootless, deracinated, and denuded. They have hardly a sense of their history, little grasp of what shapes them, and no vital vision of their human potential. Many have been reduced to a bundle of desires targeted by corporate America for consumption. Their armor of life is often too feeble to enable them to withstand the emotional trauma generated, in part, by the fast-paced capitalist culture of consumption that confronts them. In short, many lack the necessary navigational skills to cope with the challenges and crises in life—disappointment, disease, death. This is why so many are enacting the nihilism of meaninglessness and hopelessness in their lives that mirrors the nihilism of the adult world—often they are so disillusioned in large part because they can see that the adult world itself is so bereft of morality.

Yet some young folk do persevere and prevail: those who are dissatisfied with mere material toys and illusions of security. They hunger for something more, thirst for something deeper. They want caring attention, wise guidance, and compassionate counsel. They desire democratic individuality, community, and society. They know something is wrong with America and something is missing in their lives. They long for energizing visions worthy of pursuit and sacrifice that will situate their emaciated souls in a story bigger than themselves and locate their inflated egos (that only conceal deep insecurities and anxieties) in a narrative grander than themselves. Their emaciated souls contain a rage that often strikes out at the world; their inflated egos yield a cocky pose and posture that defies authority, whether legitimate or illegitimate. A grand story and a large narrative—especially democratic ones—can channel their longings into mature efforts to contribute in a meaningful way to making the world a better place. This longing is the raw stuff of democracy matters.

Like every younger generation, our kids today see clearly the hypocrisies and mendacities of our society, and as they grow up they begin to question in a fundamental way some of the lies that they've received from society. They also begin to see that their education has been distorted and sugarcoated and has sidestepped so many uncomfortable truths. This often leads to an ardent disappointment, and even anger, about the failures of our society to consistently uphold the democratic and humanitarian values that can be born in youths in this phase of their life. This new sense of conscience in young people is a profound force that adult society should take much more seriously. In fact, we should understand the expressions of this moral outrage as having a profound kind of wisdom, even as we must also help to channel that outrage into a more productive sense of commitment to find a positive way forward.

In the political sphere, the most significant expression today of this mix of anger, disappointment, and yet a tough-edged longing is the democratic globalization movement here and abroad. Although still in the early stages, this movement to establish democratic accountability of the American empire and its global corporate behemoths is disproportionately led by the youth culture. The historical day of protest—February 15, 2003—in which millions of people in over six hundred cities (including nearly two hundred U.S. cities) protested the likelihood of a U.S. violation of international law in its invasion of Iraq exemplifies the deep democratic energy and moral fervor that youth can bring to bear. Other protests in Seattle, Prague, Washington, Rome, and Davos, Switzerland—driven largely by young people—focused international attention on the antidemocratic character of global world power centers that reinforce American imperial rule.

The central thrust of this movement is criticism of the dogma of free-market fundamentalism and the increasing wealth inequality all around the world that the slavish devotion to the dogma has produced. The movement also targets the aggressive militarism of the U.S. government and the escalating authoritarianism here and around the world. The impressive efforts to create lasting institutions out of the energy of these protests—such as the public-interest groups Move On and Global Citizens Campaign—exemplify democratic commitment in action. Much of the support for and enthusiasm generated by these organizations is owing to youth culture. One of the tasks to which I am devoted—as a democratic intellectual of middle age!—is to help make this movement more multiracial by linking it to black youth culture. One way I've worked at doing this is by engaging with the profound power and energy of hip-hop culture and rap music, by taking it as seriously as it should be taken.

Although hip-hop culture has become tainted by the very excesses and amorality it was born in rage against, the best of rap music and hip-hop culture still expresses stronger and more clearly than any cultural expression in the past generation a profound indictment of the moral decadence of our dominant society. An unprecedented cultural breakthrough created by talented poor black youths in the hoods of the empire's chocolate cities, hip-hop has by now transformed the entertainment industry and culture here and around the world. The fundamental irony of hip-hop is that it has become viewed as a nihilistic, macho, violent, and bling-bling phenomenon when in fact its originating impulse was a fierce disgust with the hypocrisies of adult culture—disgust with the selfishness, capitalist callousness, and xenophobia of the culture of adults, both within the hood and in the society at large.

... The first stages of hip-hop were hot. Coming from the margins of society, the lyrics and rhythms of Grandmaster Flash and the Furious Five, Kool Here, Rakim, Paris, the Poor Righteous Teachers, Afrikaa Bambaataa, and, above all, KRS-ONE and Public Enemy (led by Chuck D) unleashed incredible democratic energies. Their truth telling about black suffering and resistance in America was powerful. The political giants of hip-hop all expressed and continue to express the underground outlook of Outkast: righteous indignation at the dogmas and nihilism of imperial America. Yet hip-hop was soon incorporated into the young American mainstream and diluted of its prophetic fervor.

With the advent of the giants of the next phase—Tupac Shakur, Ice-T, Ice Cube, Biggie Smalls, and Snoop Dogg—linguistic genius and gangster sentiments began to be intertwined. Ironically, their artistic honesty revealed subversive energy and street prowess in their work and life. As the entertainment industry began to mainstream the music, that street prowess became dominant—with the racist stereotypes of black men as hypercriminal and hypersexual and black women as willing objects of their conquests. The companies perceived that white kids were much more interested in the more violence-ridden, misogynist mode than in the critical, prophetic mode. This packaging for eager rebellious youth in vanilla suburbs—now 72 percent of those who buy hip-hop CDs and even more who illegally download them—led to an economic boom for the industry, until its recent downturn.

... It is important not to confuse prophetic hip-hop with Constantinian hip-hop. Prophetic hip-hop remains true to the righteous indignation and political resistance of deep democratic energies. Constantinian hip-hop defers to the dogmas and nihilisms of imperial America. As DA Smart says in "Where Ya At?":

> What you trying to pull eatin' us like cannibals
> Whatever happened to that forty acres and that animal
> Now you tryin to use integration just to fool us
> Like Malcolm said we been hoodwinked and bamboozled.

That such powerful poetry and insightful social critiques could be created by youths who have been flagrantly disregarded, demeaned, and demonized by the dominant market-driven culture—targeted as cannon fodder by a racist criminal-justice system and a growing prison-industrial complex, in disgraceful schools and shattered families (including too many irresponsible, unemployed fathers) and violent environments—is a remarkable testament to the vital perspective and energy that can be injected into our democracy by the young, who have not made their compromises yet with the corrupted system.

What a horrible irony it is that this poetry and critique could be co-opted by the consumer preferences of suburban white youths—white youths who long for rebellious energy and exotic amusement in their own hollow bourgeois world. But the black voices from the hood were the most genuine, authentic voices from outside the flaccid mainstream market culture that they could find. So the recording and fashion industries seized on this market opportunity.

... Like the forms of black music in the past, hip-hop seized the imaginations of young people across the globe. Prophetic hip-hop has told painful truths about their internal struggles and how the decrepit schools, inadequate health care, unemployment, and drug markets of the urban centers of the American empire have wounded their souls. Yet Constantinian hip-hop revels in the fetishism of commodities, celebrates the materialism, hedonism, and narcissism of the culture (the bling! bling!) and promotes a degrading of women, gays, lesbians, and gangster enemies. In short, hip-hop is a full-scale mirror of the best and worst, the virtuous and vicious, aspects of our society and world.

Hip-hop culture and rap music are, in many ways, an indictment of the old generation even as they imitate and emulate us in a raw and coarse manner. The defiant and insightful voices of this new generation lyrically proclaim that they have been relatively unloved, uncared for, and unattended to by adults too self-indulgent, too self-interested, and too self-medicated to give them the necessary love, care, and attention to flower and flourish. Only their beloved mothers—often overworked, underpaid, and wrestling with a paucity of genuine intimacy—are spared. They also indict the American empire for its mendacity and hypocrisy—not in a direct anti-imperialist language but in a poetic rendering of emotional deficits and educational defects resulting from the unequal institutional arrangements of the empire.

It is important that all democrats engage and encourage prophetic voices in hip-hop—voices that challenge youth to be self-critical rather than self-indulgent, Socratic rather than hedonistic. This is why I strongly support and participate in the efforts of Russell Simmons and Benjamin Chavis to organize hip-hop into a political force that accents the plight of youth. I also support the vision of KRS-ONE and others behind the Hip Hop Temple, which teaches youth the prophetic aims of underground hip-hop. There is also the organization of L. Londell Mcmillan—the Empowerment Arts Collective—which protects prophetic artists from abuse by the industry; and there are annual gatherings of the great musical genius Prince at Paisley Park, which bring the older generation together with the young artists in order to wrestle with political issues and enjoy performances. Prophetic hip-hop is precious soil in which the seeds of democratic individuality, community, and society can sprout.

DISCUSSION QUESTIONS

1. How has the commercialization of hip-hop blunted the social criticism regarding race that West argues was part of the inception of hip-hop music?
2. How does West depict the role of youth culture—and, specifically, Black youth culture—in preserving democracy?

Part III
Student Exercises

1. Identify a particular form of media that interests you—film, television, magazines, or books, for example—and design a research plan that will examine some aspect of the images you find of a racial-ethnic group. Narrow your topic so it won't be overly general. For example, if you choose films, pick only those nominated for Best Film in a given year, or if you choose television, look only at prime-time situation comedies. Alternatively, you could examine images of women of color in top fashion magazines or watch Saturday morning children's cartoons to see how people of color are portrayed. Once you have narrowed your topic, design a systematic way to catalog your observations, such as counting the number of times people of color are represented in the medium you select, listing the type of characters portrayed by Asian men, or comparing the portrayal of White men and men of color in women's fashion magazines.

   What do your observations tell you about the representation of race in the form you chose? If you were to design your project to study such images as seen now and in the past, what might you expect to find? What impact do you think the images you found have on the beliefs of different racial-ethnic groups?

2. The readings in Part III identify *color-blind racism* as a new form of racism in which dominant group members (and some subordinate group members) think that race no longer matters and that to recognize race is to be racist. Some of the authors claim that when you delve under the surface of these beliefs, you will find that people still harbor stereotypical ideas about racial minorities. Design a series of interview questions, perhaps modeled on some of the research reported, and then interview a small sample of people. Do you find evidence of color-blind racism among those you interviewed? How does the race and ethnicity of those you interviewed influence your findings?

PART IV

INTRODUCTION

Race and Identity

This book shows how race and ethnicity are part of social structure. As such, you might think they are "out there"—but they are also in us and in our relationships with other people and groups. As Peter Berger (1963) once wrote about the sociological perspective: "Yes, people live in society but society also lives in people." Similarly, in the United States people live within a system of race and ethnic relations, but race and ethnic relations also live within us. How society has organized race and ethnicity is reflected in our identities and in our relationships with others. And, as society becomes more racially and ethnically diverse, so do people's identities become more diverse, and the possibilities for multiracial identities and relationships increase.

Identity means the self-definition of a person or group, but it is not free-floating. Identity is anchored in a social context: We define ourselves in relationship to the social structures that surround us. Moreover, identities are multidimensional and thus include many of the social spaces we occupy. At any given time, some identities may be more salient than others—age as you grow older, sexuality if you are questioning your sexual orientation, race as you confront the realities of a racially stratified society, and so forth.

Racial identity is learned early in life, although those in the dominant group may take it for granted. People of color likely learn explicit lessons about racial identity early on, as parents prepare them for living in a society in which their racial status will make them vulnerable to harm. As Beverly Tatum (1997) shows, forming a positive racial identity—for both dominant and subordinate groups—means having to grapple with the realities of race. For people of color, this can mean surrounding themselves with others of their group, even though they may then be blamed for "self-segregating" by Whites who do not understand or appreciate the support this affiliation can provide.

The formation of racial identity is especially complex when multiple races are involved. The children of biracial couples may define themselves as being of two or more races. Thus, a child born to a White parent and a Black parent may identify as Black, but appear White to others, and then identify herself or himself as "biracial." As the society becomes more diverse, multiracial identities are becoming increasingly common. Racial and ethnic identities can also be complex because we have so many immigrants from nations in which race and ethnicity may be "constructed" differently than in the United States. Such complex identities hold out the possibility that the rigid thinking about race that has prevailed for so long might break down.

The articles in this part each explore different dimensions of racial identity. We open with the views of psychologist Beverly Tatum ("Why Are the Black Kids Sitting Together?," interviewed by John O'Neil) whose work is well known for explaining the formation of racial identity. Tatum explains how racial identity emerges within the context of racial inequality. She challenges us to think about what it means when people of color, especially young people, choose to be among people like them. Although they then get accused of "self-segregating" (even while White people who do the same thing are rarely so accused), Tatum shows the affirmation that such behavior provides.

In a different light, Heather Dalmage ("Tripping on the Color Line") shows how multiracial people have to negotiate the boundaries that racist societies produce. Multiracial (or biracial) people cross racial borders, negotiate relationships across such borders, and shape identities in the context of complex race relations. The burden of having to negotiate these barriers, Dalmage argues, makes its harder to form interracial relationships. Min Zhou's article ("Are Asian Americans Becoming 'White'?") shows how racial identities can be redefined in different historical circumstances. Stereotyped as the "model minority," Asian Americans, though still marked as a minority group, are perceived as more successful than others. Zhou criticizes the "myth of the model minority" by pointing to the diverse experiences among and within Asian American groups, but she also shows how racial identities can fluctuate—thus affirming their socially constructed character.

Finally, racial identity is not just acquired by people of color. White people are generally not considered to have a racial identity, because they are not socially "marked" as are people of color. Because they are the dominant group, their identity has been considered transparent, taken-for-granted, not marked as are the identities of racial and ethnic minorities. As Mark Chesler and his colleagues ("Blinded by Whiteness") point out, White people actually do have a racial identity, but it is often not salient until they encounter experiences wherein that identity is brought to light. Thus, White college students may confront their own racial experiences for the first time when they interact with students of color on campus. Chesler and his colleagues analyze the different phases that constitute the development of White identity and, in so doing, show how identity can develop to produce a commitment to a more racially just society.

Together, the articles in Part IV challenge fixed definitions of race and ethnicity and show how racial and ethnic identities are emerging in an increasingly diverse society.

REFERENCES

Berger, Peter L. 1963. *Invitation to Sociology: A Humanist Perspective*. New York: Doubleday-Anchor.

Tatum, Beverly Daniel. 1997. *Why Are All the Black Kids Sitting Together in the Cafeteria? And Other Conversations about Race*. New York: Basic.

16

Why Are the Black Kids Sitting Together?

A Conversation with Beverly Daniel Tatum

BEVERLY TATUM AND JOHN O'NEIL

Beverly Tatum, well known educational leader, tackles the often-asked question here of why Black students (and, by implication, other students of color) tend to stick together when they are in predominantly White settings, such as a high school or college campus. Rather than blaming the students, as many do, for supposed "self-segregating," she interprets this behavior as a matter of social support and racial identity formation. Her work has been very influential in antiracism education.

Educators and students themselves need to explore racial stereotypes, beliefs, and perspectives if classrooms are to become places where equity is valued, author Beverly Daniel Tatum says.

The question is one I'm asked over and over again when I do a workshop on racial identity at a racially mixed school. Educators notice that kids often group themselves with others of the same race, for reasons I'll explain later. But people have other concerns as well. How do we talk to young children about race? How do we address these issues with our colleagues? How do you even engage in conversations about such hot topics as affirmative action without alienating one another?

Why do we have such problems discussing racial issues? Is it because we don't really understand one another's experiences?

I think that's part of it. It's interesting to watch people's reactions when they are really forced to experience being in the minority. One of the exercises that I ask white students and educators to engage in is to create a situation in which they will be in the minority, for a short period of time. A common choice, for example, is to attend a black church on a Sunday morning. Another is to go to a place where you know there's going to be a large Spanish-speaking population.

Usually, whites are very nervous and anxious about doing this. Some are even unwilling to do it alone, so they find a partner to go with, which is fine.

SOURCE: *Educational Leadership* 55 (December 1997): 12–15.

But it's just interesting to me how fearful people are about this kind of experience. When they come back, they often report how welcomed they felt, what a positive experience it was. But I do point out to them how worried they were about their own discomfort. And I hope that they develop a greater sense of understanding of how a person of color might feel in an environment that is predominantly white.

Some people suggest that race relations among kids are much improved, compared to our generation or our parents'. What do you see?

Young children do interact across racial lines fairly comfortably at the elementary school grade level. If you visit racially mixed schools at the elementary level, you will see kids interacting in the lunch room and on the playground. To the extent that neighborhoods are segregated, their interracial friendships might be limited. But you see much more cross-racial interaction at the elementary level than you do at the junior high or high school level.

Why?

I think the answer has to do with the child's transition into adolescence. Adolescents are searching for identity; they're asking questions like: Who am I in the world? How does the world see me? How do I see others? What will I be in the future? and Who will love me? All those questions of identity are percolating during that time period. And particularly for adolescents of color, these questions cannot be answered without also asking: Who am I ethnically? Who am I racially? and What meaning does this have for how people view me and interact with me?

Can you give an example of how these issues might emerge in students of color?

Sure. Imagine a 7-year-old black boy who everybody thinks is cute. So he's used to the world responding to him in a certain way: Look at that cute kid!

Now imagine that same kid at 15. He's six feet tall, and people don't think he's cute anymore. They think he's dangerous or a potential criminal; maybe people are now crossing the street to avoid him. So the way he sees himself perceived by others is very different at 15 than at seven. And that 15-year-old has to start figuring out what this means. Why are people crossing the street when I walk down it? Why am I being followed around by the security guards at the mall? And as that young person is trying to make sense of his experience, in all likelihood he is going to seek out and try to connect with other people who are having similar experiences.

As a result, even the young person who has grown up in a multiracial community and had a racially mixed group of friends tends to start to pull away from his non-black friends, his white friends in particular. This happens, in part, because their white friends are not having the same experiences; they're not having the same encounters with racism. And, unfortunately, many white youth don't have an understanding of how racism operates in our society, so they're not able to respond in ways that would be helpful.

The example I used was of a black boy, but a similar process unfolds among black girls or children of another race or ethnic heritage.

What about white students? What are they experiencing during this time of self-identification?

They can be confused and hurt by some of the changes. For example, it's not uncommon for a white student in my college class on the psychology of racism to say: "You know, I had a really close black or Latino friend in elementary school, and when we went to junior high she didn't want to hang around with me anymore." The student reporting the story usually is quite confused about that; it's often a very hurtful experience.

The observation I make is that, again, many white students are oblivious to the power of racism and the way that it's operating in society. And so when their friends are starting to have encounters with racism, they don't necessarily know how to respond. An example from my book is when a teacher makes a racist remark to a young black female. Afterward, her white friend comes up to her and says, "You seem upset, what's the problem?" So she explains what upset her to her friend, and the friend says, quite innocently, "Gee, Mr. Smith is such a nice guy, I can't imagine he would intentionally hurt you. He's not a racist, you know."

So white students might discount it because they can't identify with it?

Exactly. They can't identify, and also in many situations people who try to comfort often end up invalidating the person's feelings, by saying things like, "Oh, come on, it wasn't that bad." What happens is that you withdraw from the conversation. The feeling is "Well, you don't get it, so I'll find somebody who does." It would have been a very different response, however, if this young white student understood stereotypes and the reality of racism in society and told her friend, "You know, that was a really offensive thing he said."

Should teachers or principals be concerned when students self segregate? Should they actively seek to integrate the groups?

During "downtime" like lunch or recess, students should be able to relax with their friends, regardless of whether or not those friendship groups are of the same race or ethnic groups. However, it is important to create opportunities for young people to have positive interactions across group lines in school. So structuring racially mixed work groups—for example, by using cooperative learning strategies in the classroom—can be a very positive thing to do.

Similarly, intentionally working to recruit diverse members of the student body to participate in extracurricular events is worthwhile. We need to take advantage of every opportunity we have to bring young people together where they can work cooperatively as equals toward a common goal. Sports teams are a good example of the kind of mutually cooperative environment where young people often develop strong connections across racial lines, and we should look to create more such opportunities in schools. Unfortunately, school policies like tracking (which tends to sort kids along racial lines) impede rather than facilitate such opportunities.

You've talked with students of color who attend integrated schools but find themselves isolated in honors or advanced classes. What are those students experiencing in terms of their identity?

Even in racially mixed schools, it is very common for young people of color, particularly black and Latino students, to find that the upper-level courses have very few students of color in them. And, of course, honors chemistry is only offered during a certain time period in the day, which means you might also be taking English and other courses with the same kids. And so those students in honors chemistry or advanced algebra may find that their black or Latino peers accuse them of "trying to be white" because they're hanging around with all white kids. So to the extent that you're frequently in the company of white students, and your black classmates who are in the lower tracks see you as somehow separating yourself from them—it's a hard place to be.

What can educators do to support the healthy development of kids as they work through these issues?

I think students of color really need to see themselves reflected in positive ways in the curriculum. And that probably sounds very obvious, but the fact is that too often they don't see themselves reflected in the curriculum.

When and how they see themselves reflected in the curriculum is so important, though. To use African Americans as an example: Most schools teach about slavery, and for many black students that's a point of real discomfort. Their experience of that is that the teacher's talking about slavery and all the white kids in the class are looking at us, to see what our reaction is. I'm certainly not suggesting we shouldn't teach about slavery, but I think it's important to teach it in an empowering way. Teachers need to focus on resistance to victimization. Students of color need to see themselves represented not just as victims but as agents of their own empowerment. And there are lots of ways to do that. You can talk about Sojourner Truth, you can talk about Harriet Tubman, Frederick Douglass, and so on.

At the same time, I think white children need to be helped to understand how racism operates. Inevitably, when you talk about racism in a predominantly white society, you generate feelings of discomfort and often guilt among white people because they might feel that you're saying that white people are bad. What do we do? In these discussions, we need to include examples of white people as agents of change. Teach students about the abolitionists. Teach students about Virginia Foster Durr, who was so active during the Civil Rights movement. All children need to learn about those white folks who worked against oppression. Unfortunately, many white students don't have that information.

Some people have suggested that the school curriculum be heavily focused on cultural heritage; that black students need an Afrocentric curriculum, and so on. What's your perspective?

It's important to have as diverse a curriculum as possible because all students need to be able to view things through multiple perspectives.

A high school teacher told me recently that the young white men in her English classes were reluctant to read about somebody's experience other than their own. For example, she had the class read *House on Mango Street* (Cisneros 1994), a book about a young Chicana adolescent coming of age in Chicago. These young men were complaining: "What does this have to do with me? I can't identify with this experience." But, at the same time, they never wonder why the Latino students in the class have to read Ernest Hemingway. We need to help them develop that understanding. All of us need to develop a sense of multiple perspectives, regardless of the composition of our classrooms.

The teaching ranks are predominantly white, even though the student population is becoming increasingly diverse. What does this mean for efforts to increase racial understanding?

It makes it harder, but it's not impossible. We should be working very hard to increase the diversity of the teaching pool, and many teacher education programs are trying to do that. Still, we need to recognize that it's going to be a long time before the teaching population reflects the classroom population. So it's really important for white teachers to recognize that it is possible for them to become culturally sensitive and to be proactive in an antiracist way. Many white educators have grown up in predominantly white communities, attended mostly white schools, and may have had limited experiences with people of color, and that is a potential barrier. But what that means is that people need to expand their experiences.

Dreamkeepers: Successful Teachers of African-American Children, by Gloria Ladson-Billings (1994) is a great resource. I often encourage educators I work with to read that because she profiles several teachers, some of them white. Those are teachers who probably didn't grow up in neighborhoods or communities where they had a lot of interaction with people of color, but, one way or another, they have really been able to establish great teaching relationships with kids of color. So it certainly can be done.

You train teachers to work on issues involving race in their schools. What kinds of things do they learn?

For a number of years, I've taught a professional development course called *Effective Anti-Racist Classroom Practices for All Students*. It's basically designed to help teachers recognize what racism is, how it operates in schools, and what the impact of that is personally and professionally. So the focus is not just the impact of racism on the racial identity development of the students but also on the teachers. I've found that teachers who have not reflected on their own racial identity development find it very difficult to understand why young people are reflecting on theirs. So it's important to engage in self-reflection even as we're trying to better understand our students.

The course also looks at stereotypes, omissions, distortions, how those have been communicated in our culture and in our classroom, and then, what that means in terms of how we think about ourselves either as people of color or as whites. And, finally, what we can do about it. I talk in my book about racism as a sort of "smog." People who aren't aware of it can unwittingly perpetuate a cycle of oppression. If you breathe that smog too long, you internalize these messages. We can't really interrupt that process until it becomes visible to us. That's the first step—making the process visible. And once it is visible, we can start to strategize about how we're going to interrupt it.

Many teachers have been caught short by a racist incident or comment in their classroom. It often happens suddenly, and the teacher may be at a loss for how to respond. How have you handled it?

Well, I've been teaching a course on the psychology of racism since 1980, so I feel like I've probably heard it all.

It is a difficult situation, because you want the classroom to be a safe place, where students can say anything, knowing that only by opening up will they get feedback about their comments and learn another perspective. At the same time, you want the classroom to be a safe place for someone who may be victimized by a comment.

One time, a white student in my class made a very offensive remark about Puerto Ricans being responsible for crimes. Well, one of the things I've learned is that it really helps to validate somebody's comment initially, even if it is outrageous. So I said something like: "You know, I'm sure there are many people who feel that way, and if you've been victimized by a crime, that's a very difficult experience. At the same time, I think it's important to say that not all Puerto Ricans are car thieves." From there, you can move into how making such statements can reinforce stereotypes.

It must be hard to make it a teachable moment for everyone in the class.

Absolutely. One time I was observing somebody else's teaching when there was a similar kind of incident—a student made comments that the teacher thought were inappropriate, but she didn't know how to respond. So she didn't respond to them. After the class, we talked, and she said she felt terrible—she knew she should have done something, but she didn't know what to do. And we talked about what the choices might have been.

Even though she felt badly about how she handled it, those moments can be revisited. So in this particular case, the teacher opened her next class by saying: "You know, in our last class something happened that really bothered me, and I didn't say anything. I didn't say anything because I wasn't sure what to say, but in my silence I colluded with what was being said. So I would really like to talk about it now." And she brought the class back to the incident, and it was not an easy conversation. But I think it really deepened the students' understanding—both of the teacher herself and of how racism operates, because it showed how even well-intentioned people may unwittingly contribute to perpetuating the problem.

Although integrated schools have been a goal for decades, current statistics show a growth in schools that are nearly all-black or all-Hispanic. What do you see as the likely impact of the trend toward even more racially identifiable schools?

It's a very difficult issue from a number of perspectives. The continuing pattern of white flight is one of the main reasons that schools resegregate. A lot of money is put into a busing plan, and then white families leave the school. So now many people are asking questions about whether it's a good idea to spend all that money transporting kids instead of just using it to improve the neighborhood school, regardless of who attends it.

Many parents of color experience this as a double-edged sword. They're offended by the notion that children of color can only learn when they are in classes with white kids. They know there is nothing magical about sitting next to a white kid in class. On the other hand, the reality of school funding is that schools with more white students receive more financial support.

And so the question that I hear people asking now is: Can separate ever be equal? That's one I don't have the answer to!

... Real progress is being made in starting conversations at the local community level. Many grass-roots organizations are encouraging this kind of dialogue.

For example, an organization in Connecticut has a program of Study Circles.[1] They actually have a guidebook for facilitating conversations about race. Using the guides, people come together and begin to discuss the questions together to improve their understanding. Also, many houses of worship encourage cross-group dialogue, whether it means interfaith dialogue or cross-racial dialogue.

It's sometimes frustrating for people who have been doing this work for years, because it may seem like there's talk, talk, talk and it doesn't go anywhere. However, I do think that when you engage in open and honest dialogue, you start to recognize the other person's point of view, and that helps you see where your action might be needed most. So if people engage in dialogue with the understanding that dialogue is supposed to lead somewhere, it can be a very useful thing to do.

We can't afford to forget the institutional nature of racism. And so it's not just about personal prejudices, though obviously we want to examine those. We can't just aspire to be prejudice-free. We need to examine how racism persists in our institutions so we don't perpetuate it.

REFERENCES

Cisneros, S. (1994). *House on Mango Street.* New York: Random House.

Ladson Billings, G. (1994). *Dreamkeepers: Successful Teachers of African-American Children.* San Francisco, Calif.: Jossey-Bass.

DISCUSSION QUESTIONS

1. How does Tatum's analysis link the everyday behavior of minority students to the social structures in which they live?
2. What lessons are there in Tatum's discussion for reducing racial prejudice?
3. What are the implications of Tatum's argument for educating teachers about race?

1. For more Information contact Study Circles Resource Center, P. O. Box 203, 697 Pomfret St., Pomfret. CT 06258: tel: 203-928-2616, fax: 20392N 3713.

17

Tripping on the Color Line

HEATHER M. DALMAGE

Dalmage develops the concept of "border patrolling" to refer to the many ways that dominant groups develop and maintain boundaries between racial groups. In this essay, she discusses specifically how such border patrolling affects multiracial (or biracial) people.

To live near the color line, in the space Gloria Anzaldúa calls the borderlands, means to contend constantly with what I call borderism.[1] Borderism is a unique form of discrimination faced by those who cross the color line, do not stick with their own, or attempt to claim membership (or are placed by others) in more than one racial group. Like racism, borderism is central to American society. It is a product of a racist system yet comes from both sides. The manner in which people react to individuals who cross the color line highlights the investment, the sense of solidarity, and perhaps the comfort these observers have with existing categories. Perhaps most important, the reaction shows the wide acceptance of racial essentialism as the explanation for the color line. When the color line is crossed, the idea of immutable, biologically based racial categories is threatened. The individual who has crossed the line must be explained away or punished so that essentialist categories can remain in place. Ironically, if race were natural and essential, individuals would not have to engage in borderism. The act of borderism is one of the many ways in which individuals construct or "do" race.

Multiracial family members contend with borderism in many aspects of life. It is both part of the workings of larger institutions and the outcome of individual actions. When families are unable to find accepting places of worship and comfortable neighborhoods, they contend with examples of institutional borderism. A nefarious individual is not responsible for creating these situations. Rather, this borderism has developed in the context of a deeply racist and segregated society. Some borderism, however, does play out on an individual level and is meant to be hurtful. A family disowns a child for not sticking with his own. Peers tell a multiracial child that she is not black enough. An interracial couple is physically accosted in the street. Such borderism may stem from hostility, hatred, or feelings of betrayal and is grounded in ideas about how people ought to act. But it may also reveal concern. For example, before returning to

SOURCE: From Heather M. Dalmage, *Tripping on the Color Line: Black-White Multiracial Families in a Racially Divided World* (Piscataway, NJ: Rutgers University Press, 2000), pp. 41–63. Copyright © 2000. Reprinted by permission of Rutgers University Press.

graduate school I was working for a corporation that sent me to the Baltimore–Washington, D.C., area for a summer. It was a racially tense season. The Klan was active, hanging notices all over one town declaring it a "nigger-free zone." In another town white men killed a black man because he was walking with a white woman in a white neighborhood. When my husband, Philip, came to visit, we were cautious. One weekend went without incident until I dropped him off at the train station. We were talking when an elderly black woman walked up to Philip and scolded, "Get away from her; she's going to get you killed. You need to stick with your own." I believe she meant to be helpful to Philip. After all, another black man had just been killed. Nonetheless, these moments can be paralyzing. In such situations there is no way for me to say, "Please, my skin color belies my politics. I am likewise outraged and live with the fear of violence." For me to speak at these moments is to belittle the history of racial oppression and the fact that I do represent, in my physical being, the object and the symbol that has been used to justify so much oppression. Protecting white women's bodies has long been the justification for abusing and lynching thousands of black men and women.[2]

All members of multiracial families face borderism, although as individuals we face specific forms of discrimination based on our race, physical features, gender, and other socially significant markers....

BORDER PATROLLING

The belief that people ought to stick with their own is the driving force behind efforts to force individuals to follow prescribed racial rules. Border patrollers often think (without much critical analysis) that they can easily differentiate between insiders and outsiders. Once the patroller has determined a person's appropriate category, he or she will attempt to coerce that person into following the category's racial scripts. In *Race, Nation, Class: Ambiguous Identities*, Etienne Balibar and Immanuel Wallerstein observe that "people shoot each other every day over the question of labels. And yet, the very people who do so tend to deny that the issue is complex or puzzling or indeed anything but self-evident."[3] Border patrollers tend to take race and racial categories for granted. Whether grounding themselves in essentialist thinking or hoping to strengthen socially constructed racial categories, they believe they have the right and the need to patrol. Some people, especially whites, do not recognize the centrality and problems of the color line, as evinced in color-blind claims that "there is only one race: the human race" or "race doesn't really matter any more." Such thinking dismisses the terror and power of race in society. These individuals may patrol without being aware of doing so. In contrast, blacks generally see patrolling the border as both problematic and necessary.

While border patrolling from either side may be scary, hurtful, or annoying, we must recognize that blacks and whites are situated differently. The color line was imposed by whites, who now have institutional means for maintaining their power; in contrast, blacks must consciously and actively struggle for liberation....

White Border Patrolling

Despite the institutional mechanisms in place to safeguard whiteness, many whites feel both the right and the obligation to act out against interracial couples. If a white person wants to maintain a sense of racial superiority, then he or she must attempt to locate motives and explain the actions of the white partner in the interracial couple. A white person who crosses the color line threatens the assumption that racial superiority is essential to whites. The interracially involved white person is thus often recategorized as inherently flawed—as "polluted." In this way, racist and essentialist thinking remains unchallenged.

Frequently white families disown a relative who marries a person of color, but several people have told me that their families accepted them again once their children were born. The need to disown demonstrates the desire to maintain the facade of a pure white family.[4] By the time children are born, however, extended family members have had time to shift their racial thinking. Some grant acceptance by making an exception to the "rule," others by claiming to be color blind. Neither form of racial thinking, however, challenges the color line or white supremacy. In fact, both can be painful for the multiracial family members, who may face unending racist compliments such as "I'll always think of you as white." ...

Privileges granted to people with white skin have been institutionalized and made largely invisible to the beneficiaries. With overwhelming power in society, why do individual whites insist on border patrolling? As economic insecurity heightens and demographics show that whites are losing numerical majority status, the desire to scapegoat people of color, especially the poor, also heightens. As whites lose their economic footing, they claim white skin as a liability. Far from recognizing whiteness as privilege, they become conscious of whiteness only when defining themselves as innocent victims of "unjust" laws, including affirmative action.[5] In their insecurity they cling to images that promote feelings of superiority. This, of course, requires a racial hierarchy and a firm essentialist color line. Border patrolling helps to maintain the myth of purity and thus a color line created to ensure that whites maintain privileges and power.

Black Border Patrolling

Some blacks in interracial relationships discover, for the first time, a lack of acceptance from black communities. Others experienced border patrolling before their marriage, perhaps because of hobbies and interests, class, politics, educational goals, skin tone, vernacular, or friendship networks. Patrolling takes on new proportions, however, when they go the "other way" and marry a white person. While all relationships with individuals not seen as black are looked down on, relationships with whites represent the gravest transgression. Interracially married black women and men often believe they are viewed as having lost their identity and culture—that they risk being seen as "no longer really black." Before their interracial marriage, most called black communities their home, the place from which they gained a sense of humanity, where they gained cultural and personal affirmation. During their interracial relationship many discovered black border patrolling....

An overwhelming percentage of black-white couples involve a black male and a white female at a time when there are "more single women in the black community than single men."[6] Many black men are hindered by a racist educational system and job market that make them less desirable for marriage. Many others are scooped into the prison industrial complex. High-profile athletes and entertainers who marry white women confirm for many that black men who are educated and earn a good living sell out, attempting to buy white status through their interracial relationship.[7] Beyond issues of money and status, many black women see black male-white female interracial relationships "as a rejection of black women's beauty, [and] as a failure to acknowledge and reward the support that black women give black men."[8]...

Black women and men may both feel a sense of rejection when they see an interracial couple, but for each that sense of rejection comes from a different place. In a society in which women's worth is judged largely by beauty—more specifically, Eurocentric standards of beauty—black women are presumed to be the farthest removed from such a standard. Men's worth is judged largely by their educational and occupational status, two primary areas in which black men are undermined in a racist system. Black men with few educational and job opportunities lack status in the marriage market. Thus, when black men see a black woman with a white man, they may be reminded of the numerous ways in which the white-supremacist system has denied them opportunities. The privilege and power granted to whites, particularly to white males, is paraded in front of them; and they see the black women in these relationships as complicit with the oppressor....

Today there are no longer legal sanctions against interracial marriage, but de facto sanctions remain. At times, family and friends exert pressure to end the interracial relationship; at other times, pressure may come from the border patrolling of strangers. Even if the relationship is clandestine, thoughts of how friends, family members, co-workers, employees, and the general public might respond can deter people from moving forward in a relationship....

Border patrolling plays a central role in life decisions and the reproduction of the color line. As decisions are made to enter and remain in an interracial relationship, the color line is challenged and racial identities shift. Many blacks spoke of the growth they experienced because of their interracial relationship and border patrolling. Parsia explains, "I used to be real concerned about how I would be perceived and that as an interracially married female I would be taken less seriously in terms of my dedication to African American causes. I'm not nearly as concerned anymore. I would hold my record up to most of those in single-race relationships, and I would say, 'Okay, let's go toe to toe, and you tell me who's making the biggest difference,' and so I don't worry about it anymore." Identities, once grounded in the presumed acceptance of other black Americans, have become more reflective. Acceptance can no longer be assumed. Definitions of what it means to be black are reworked. Likewise, because of border patrolling, many whites in interracial relationships began to acknowledge that race matters. Whiteness becomes visible in their claims to racial identity.

REFERENCES

1. Gloria Anzaldúa, *Borderlands/La Frontera: The New Mestiza* (San Francisco: Spinsters/Aunt Lute, 1987).
2. Grace Elizabeth Hale, *Making Whiteness: The Culture of Segregation in the South, 1890–1940* (New York: Vintage, 1998).
3. Etienne Balibar and Immanuel Wallerstein, *Race, Nation, Class: Ambiguous Identities* (New York: Verso, 1991), 71.
4. Naomi Zack, *Race and Mixed Race* (Philadelphia: Temple University Press, 1994).
5. Charles Gallagher, "White Reconstruction in the University," *Socialist Review* 24, 1 and 2 (1995): 165–188.
6. Michael Eric Dyson, "Essentialism and the Complexities of Racial Identity," in *Multiculturalism: A Critical Reader*, ed. David Theo Goldberg (Cambridge, Mass.: Blackwell, 1994), p. 222.
7. Paul C. Rosenblatt, Terri A. Karis, and Richard D. Powell, *Multiracial Couples: Black and White Voices* (Thousand Oaks, Calif.: Sage, 1995), 155.
8. Dyson, "Essentialism," p. 222.

DISCUSSION QUESTIONS

1. Dalmage uses the concept of *borderism*. What does she mean and how does she explain that borders are "patrolled?"
2. In what different ways do Black people and White people do their border patrolling, according to Dalmage, and how has this affected her experience and that of others who live along "borders?"

BORDERISM
BORDER PATROLLING
BORDER PATROLLER

18

Are Asian Americans Becoming "White"?

MIN ZHOU

Asian Americans are often believed to be the "model minority"—a stereotype that ignores the actual difficulties different groups of Asian Americans experience in U.S. society. At the same time, Zhou argues that constructions of Asian Americans as "honorary whites" reproduce social stereotypes of Asian Americans as somehow not fully American.

"I never asked to be white. I am not literally white. That is, I do not have white skin or white ancestors. I have yellow skin and yellow ancestors, hundreds of generations of them. But like so many other Asian Americans of the second generation, I find myself now the bearer of a strange new status: white, by acclimation. Thus it is that I have been described as an 'honorary white,' by other whites, and as a 'banana' by other Asians ... to the extent that I have moved away from the periphery and toward the center of American life, I have become white inside."

—ERIC LIU,
THE ACCIDENTAL ASIAN (P. 34)

Are Asian Americans becoming "white?" For many public officials the answer must be yes, because they classify Asian-origin Americans with European-origin Americans for equal opportunity programs. But this classification is premature and based on false premises. Although Asian Americans as a group have attained the career and financial success equated with being white, and although many have moved next to or have even married whites, they still remain culturally distinct and suspect in a white society.

At issue is how to define Asian American and white. The term "Asian American" was coined by the late historian and activist Yuji Ichioka during the ethnic consciousness movements of the late 1960s. To adopt this identity was to reject the western-imposed label of "Oriental." Today, "Asian American" is an

SOURCE: "Are Asian Americans Becoming White?" by Min Zhou. From *Contexts* 3(1): 29–37. Copyright © 2004 by the American Sociological Association. All rights reserved. Reprinted by permission.

umbrella category that includes both U.S. citizens and immigrants whose ancestors came from Asia, east of Iran. Although widely used in public discussions, most Asian-origin Americans are ambivalent about this label, reflecting the difficulty of being American and still keeping some ethnic identity: Is one, for example, Asian American or Japanese American?

Similarly, "white" is an arbitrary label having more to do with privilege than biology. In the United States, groups initially considered nonwhite, such as the Irish and Jews, have attained "white" membership by acquiring status and wealth. It is hardly surprising, then, that nonwhites would aspire to becoming "white" as a mark of and a tool for material success. However, becoming white can mean distancing oneself from "people of color" or disowning one's ethnicity. Pan-ethnic identities—Asian American, African American, Hispanic American—are one way the politically vocal in any group try to stem defections. But these group identities may restrain individual members' aspirations for personal advancement.

VARIETIES OF ASIAN AMERICANS

Privately, few Americans of Asian ancestry would spontaneously identify themselves as Asian, and fewer still as Asian American. They instead link their identities to specific countries of origin, such as China, Japan, Korea, the Philippines, India or Vietnam. In a study of Vietnamese youth in San Diego, for example, 53 percent identified themselves as Vietnamese, 32 percent as Vietnamese American, and only 14 percent as Asian American. But they did not take these labels lightly; nearly 60 percent of these youth considered their chosen identity as very important to them.

Some Americans of Asian ancestry have family histories in the United States longer than many Americans of Eastern or Southern European origin. However, Asian-origin Americans became numerous only after 1970, rising from 1.4 million to 11.9 million (4 percent of the total U.S. population), in 2000. Before 1970, the Asian-origin population was largely made up of Japanese, Chinese and Filipinos. Now, Americans of Chinese and Filipino ancestries are the largest subgroups (at 2.8 million and 2.4 million, respectively), followed by Indians, Koreans, Vietnamese and Japanese (at more than one million). Some 20 other national-origin groups, such as Cambodians, Pakistanis, Laotians, Thai, Indonesians and Bangladeshis, were officially counted in government statistics only after 1980; together they amounted to more than two million Americans in 2000.

The sevenfold growth of the Asian-origin population in the span of 30-odd years is primarily due to accelerated immigration following the Hart-Celler Act of 1965, which ended the national origins quota system, and the historic resettlement of Southeast Asian refugees after the Vietnam War. Currently, about 60 percent of the Asian-origin population is foreign-born (the first generation), another 28 percent are U.S.-born of foreign-born parents (the second generation), and just 12 percent were born to U.S.-born parents (the third generation and beyond).

Unlike earlier immigrants from Asia or Europe, who were mostly low-skilled laborers looking for work, today's immigrants from Asia have more varied backgrounds and come for many reasons, such as to join their families, to invest their money in the U.S. economy, to fill the demand for highly skilled labor, or to escape war, political or religious persecution and economic hardship. For example, Chinese, Taiwanese, Indian, and Filipino Americans tend to be overrepresented among scientists, engineers, physicians and other skilled professionals, but less educated, low-skilled workers are more common among Vietnamese, Cambodian, Laotian, and Hmong Americans, most of whom entered the United States as refugees. While middle-class immigrants are able to start their American lives with high-paying professional careers and comfortable suburban lives, low-skilled immigrants and refugees often have to endure low-paying menial jobs and live in inner-city ghettos.

Asian Americans tend to settle in large metropolitan areas and concentrate in the West. California is home to 35 percent of all Asian Americans. But recently, other states such as Texas, Minnesota and Wisconsin, which historically received few Asian immigrants, have become destinations for Asian American settlement. Traditional ethnic enclaves, such as Chinatown, Little Tokyo, Manilatown, Koreatown, Little Phnom Penh, and Thaitown, persist or have emerged in gateway cities, helping new arrivals to cope with cultural and linguistic difficulties. However, affluent and highly-skilled immigrants tend to bypass inner-city enclaves and settle in suburbs upon arrival, belying the stereotype of the "unacculturated" immigrant. Today, more than half of the Asian-origin population is spreading out in suburbs surrounding traditional gateway cities, as well as in new urban centers of Asian settlement across the country.

Differences in national origins, timing of immigration, affluence and settlement patterns profoundly inhibit the formation of a pan-ethnic identity. Recent arrivals are less likely than those born or raised in the United States to identify as Asian American. They are also so busy settling in that they have little time to think about being Asian or Asian American, or, for that matter, white. Their diverse origins include drastic differences in languages and dialects, religions, cuisines and customs. Many national groups also bring to America their histories of conflict (such as the Japanese colonization of Korea and Taiwan, Japanese attacks on China, and the Chinese invasion of Vietnam).

Immigrants who are predominantly middle-class professionals, such as the Taiwanese and Indians, or predominantly small business owners, such as the Koreans, share few of the same concerns and priorities as those who are predominantly uneducated, low-skilled refugees, such as Cambodians and Hmong. Finally, Asian-origin people living in San Francisco or Los Angeles among many other Asians and self-conscious Asian Americans develop a stronger ethnic identity than those living in predominantly Latin Miami or predominantly European Minneapolis. A politician might get away with calling Asians "Oriental" in Miami but get into big trouble in San Francisco. All of these differences create obstacles to fostering a cohesive pan-Asian solidarity. As Yen Le Espiritu shows, pan-Asianism is primarily a political ideology of U.S.-born, American-educated, middle-class Asians rather than of Asian immigrants, who are conscious of their national origins and overburdened with their daily struggles for survival.

UNDERNEATH THE MODEL MINORITY: "WHITE" OR "OTHER"

The celebrated "model minority" image of Asian Americans appeared in the mid-1960s, at the peak of the civil rights and the ethnic consciousness movements, but before the rising waves of immigration and refugee influx from Asia. Two articles in 1966—"Success Story, Japanese-American Style," by William Petersen in the *New York Times Magazine*, and "Success of One Minority Group in U.S.," by the *US News & World Report* staff—marked a significant departure from how Asian immigrants and their descendants had been traditionally depicted in the media. Both articles congratulated Japanese and Chinese Americans on their persistence in overcoming extreme hardships and discrimination to achieve success, unmatched even by U.S.-born whites, with "their own almost totally unaided effort" and "no help from anyone else." (The implicit contrast to other minorities was clear.) The press attributed their winning wealth and respect in American society to hard work, family solidarity, discipline, delayed gratification, non-confrontation and eschewing welfare.

This "model minority" image remains largely unchanged even in the face of new and diverse waves of immigration. The 2000 U.S. Census shows that Asian Americans continue to score remarkable economic and educational achievements. Their median household income in 1999 was more than $55,000—the highest of all racial groups, including whites—and their poverty rate was under 11 percent, the lowest of all racial groups. Moreover, 44 percent of all Asian Americans over 25 years of age had at least a bachelor's degree, 18 percentage points more than any other racial group. Strikingly, young Asian Americans, including both the children of foreign-born physicians, scientists, and professionals and those of uneducated and penniless refugees, repeatedly appear as high school valedictorians and academic decathlon winners. They also enroll in the freshman classes of prestigious universities in disproportionately large numbers. In 1998, Asian Americans, just 4 percent of the nation's population, made up more than 20 percent of the undergraduates at universities such as Berkeley, Stanford, MIT and Cal Tech. Although some ethnic groups, such as Cambodians, Lao, and Hmong, still trail behind other East and South Asians in most indicators of achievement, they too show significant signs of upward mobility. Many in the media have dubbed Asian Americans the "new Jews." Like the second-generation Jews of the past, today's children of Asian immigrants are climbing up the ladder by way of extraordinary educational achievement.

One consequence of the model-minority stereotypes is that it reinforces the myth that the United States is devoid of racism and accords equal opportunity to all, fostering the view that those who lag behind do so because of their own poor choices and inferior culture. Celebrating "model minorities" can help impede other racial minorities' demands for social justice by pitting minority groups against each other. It can also pit Asian Americans against whites. On the surface, Asian Americans seem to be on their way to becoming white, just like the offspring of earlier European immigrants. But the model-minority image implicitly casts Asian Americans as different from whites. By placing Asian Americans above whites, this image still sets them apart from other Americans, white or nonwhite, in the public mind.

There are two other less obvious effects. The model-minority stereotype holds Asian Americans to higher standards, distinguishing them from average Americans. "What's wrong with being a model minority?" a black student once asked, in a class I taught on race, "I'd rather be in the model minority than in the downtrodden minority that nobody respects." Whether people are in a model minority or a downtrodden minority, they are still judged by standards different from average Americans. Also, the model-minority stereotype places particular expectations on members of the group so labeled, channeling them to specific avenues of success, such as science and engineering. This, in turn, makes it harder for Asian Americans to pursue careers outside these designated fields. Falling into this trap, a Chinese immigrant father gets upset when his son tells him he has changed his major from engineering to English. Disregarding his son's talent for creative writing, such a father rationalizes his concern, "You have a 90 percent chance of getting a decent job with an engineering degree, but what chance would you have of earning income as a writer?" This thinking represents more than typical parental concern; it constitutes the self-fulfilling prophecy of a stereotype.

The celebration of Asian Americans rests on the perception that their success is unexpectedly high. The truth is that unusually many of them, particularly among the Chinese, Indians and Koreans, arrive as middle-class or upper middle-class immigrants. This makes it easier for them and their children to succeed and regain their middle-class status in their new homeland. The financial resources that these immigrants bring also subsidize ethnic businesses and services, such as private after-school programs. These, in turn, enable even the less fortunate members of the groups to move ahead more quickly than they would have otherwise.

NOT SO MUCH BEING "WHITE" AS BEING AMERICAN

Most Asian Americans seem to accept that "white" is mainstream, average and normal, and they look to whites as a frame of reference for attaining higher social position. Similarly, researchers often use non-Hispanic whites as the standard against which other groups are compared, even though there is great diversity among whites, too. Like most immigrants to the United States, Asian immigrants tend to believe in the American Dream and measure their achievements materially. As a Chinese immigrant said to me in an interview, "I hope to accomplish nothing but three things: to own a home, to be my own boss, and to send my children to the Ivy League." Those with sufficient education, job skills and money manage to move into white middle-class suburban neighborhoods immediately upon arrival, while others work intensively to accumulate enough savings to move their families up and out of inner-city ethnic enclaves. Consequently, many children of Asian ancestry have lived their entire childhood in white communities, made friends with mostly white peers, and grown up speaking only English. In fact, Asian Americans are the most acculturated non-European group in the United States. By the second generation, most have lost fluency in their parents' native languages. David Lopez finds that in Los Angeles, more than three-quarters of second-generation Asian Americans (as opposed to one-quarter of second-generation Mexicans) speak only English at home. Asian Americans also intermarry extensively with whites and with members of other minority groups. Jennifer Lee and Frank Bean find that more than one-quarter of married Asian Americans have a partner of a different racial background, and 87 percent of those marry whites; they also find that 12 percent of all Asian Americans claim a multiracial background, compared to 2 percent of whites and 4 percent of blacks.

Even though U.S.-born or U.S.-raised Asian Americans are relatively acculturated and often intermarry with whites, they may be more ambivalent about becoming white than their immigrant parents. Many only cynically agree that "white" is synonymous with "American." A Vietnamese high school student in New Orleans told me in an interview, "An American is white. You often hear people say, hey, so-and-so is dating an 'American.' You know she's dating a white boy. If he were black, then people would say he's black." But while they recognize whites as a frame of reference, some reject the idea of becoming white themselves: "It's not so much being white as being American," commented a Korean-American student in my class on the new second generation. This aversion to becoming white is particularly common among second-generation college students who have taken ethnic studies courses, and among Asian-American community activists. However, most of the second generation continues to strive for the privileged status associated with whiteness, just like their parents. For example, most U.S.-born or U.S.-raised Chinese-American youth end up studying engineering, medicine, or law in college, believing that these areas of study guarantee a middle-class life.

Second-generation Asian Americans are also more conscious of the disadvantages associated with being nonwhite than their parents, who as immigrants tend to be optimistic about overcoming the disadvantages of this status. As a Chinese-American woman points out from her own experience, "The truth is, no matter how American you think you are or try to be, if you have almond-shaped eyes, straight black hair, and a yellow complexion, you are a foreigner by default.... You can certainly be as good as or even better than whites, but you will never become accepted as white." This remark echoes a commonly-held frustration among second-generation, U.S.-born Asians who detest being treated as immigrants or foreigners. Their experience suggests that whitening has more to do with the beliefs of white America, than with the actual situation of Asian Americans. Speaking perfect English, adopting mainstream cultural values, and even intermarrying members of the dominant group may help reduce this "otherness" for particular individuals, but it has little effect on the group as a whole. New stereotypes can emerge and un-whiten Asian Americans, no matter how "successful" and "assimilated" they have become. For example, Congressman David Wu once was invited by the Asian-American employees of the U.S. Department of Energy to give a speech in celebration of Asian-American Heritage Month. Yet, he and his Asian-American staff were not allowed into the department building, even after presenting their congressional Identification, and were repeatedly asked about their citizenship and country of origin. They were told that this was standard procedure for the Department of Energy and that a congressional ID card was not a reliable document. The next day, a congressman of Italian descent was allowed to enter the same building with his congressional ID, no questions asked.

The stereotype of the "honorary white" or model minority goes hand-in-hand with that of the "forever foreigner." Today, globalization and U.S.–Asia relations, combined with continually high rates of immigration, affect how Asian Americans are perceived in American society. Many historical stereotypes, such as the "yellow peril" and "Fu Manchu" still exist in contemporary American life, as revealed in such highly publicized incidents as the murder of Vincent Chin, a Chinese American mistaken for Japanese and beaten to death by a disgruntled white auto worker in the 1980s; the trial of Wen Ho Lee, a nuclear scientist suspected of spying for the Chinese government in the mid-1990s; the 1996 presidential campaign finance scandal, which implicated Asian Americans in funneling foreign contributions to the Clinton campaign; and most recently, in 2001, the Abercrombie & Fitch t-shirts that depicted Asian cartoon characters in stereotypically negative ways, with slanted eyes, thick glasses and heavy Asian accents. Ironically, the ambivalent, conditional nature of their acceptance by whites prompts many Asian Americans to organize pan-ethnically to fight back—which consequently heightens their racial distinctiveness. So becoming white or not is beside the point. The bottom line is: Americans of Asian ancestry still have to constantly prove that they truly are loyal Americans.

REFERENCES

Liu, Eric. 1988. *The Accidental Asian.* New York: Random House.

DISCUSSION QUESTIONS

1. What barriers to a pan-ethnic identification among recent Asian immigrants does Zhou identify?
2. What does Zhou argue are the sociological explanations for the economic success of the so-called "model minority"?

19

Blinded by Whiteness:

The Development of White College Students' Racial Awareness

MARK A. CHESLER, MELISSA PEET, AND TODD SEVIG

The research on White university students on which this article is based explains some of the formation of racial identity for White people. This research also explores the challenges to constructions of Whiteness that White college students may experience. The research indicates the important of developing educational programs that enable White students to challenge dominant constructions of race.

Racial identity is the meaning attached to self as a member of a group or collectivity in racial situations, and individuals may express this identity differently in different circumstances (Cornell and Hartman 1998). Since identity is formed by class and gender as well as race, there are many ways of being white or any other race/ethnicity. Racial attitudes and changing attitudes are the statements of a person's preferred views or positions about others and about contemporary (or historic) policies and events. Attitudes are also shaped by one's social location and are expressed differently in different circumstances. Social and institutional structures and cultures provide the limits and opportunities for both the creation of racial identities and the formation and expression of racial attitudes.

Throughout, we present the voices of white students attending the University of Michigan, a university with a tradition of student, faculty, and administrative engagement with issues of racism and affirmative action. Recently, Michigan has become one of the nation's battlegrounds for competing narratives and institutional policies around racial matters. The data reported here were gathered from white students of varied backgrounds in individual and small-group interviews conducted between 1996 and 2000. Although they are not geographically, temporally, or in terms of cohort representative of other white students' racial consciousness, they are useful windows into the ways in which racial processes become visible and are expressed.

SOURCE: From *White Out: The Continuing Significance of Racism*, ed. by Ashley W. Doane and Eduardo Bonilla-Silva (New York: Routledge, 2003), pp. 215–227.
Copyright © 2003. Reprinted by permission of Routledge/Taylor & Francis Books, Inc.

BACKGROUND

The social and cultural context of the modern university is one of racial plurality but also of racial separatism and tension. Students come to these settings from racially separated and often segregated neighborhoods and communities (Bonilla-Silva and Forman 2000; Massey and Denton 1993). For many, the university is the first place in which they have sustained contact with a substantial number of students of another race. Although there are more numerous formal and informal opportunities for racial interaction and growth in the university than in most secondary educational environs, white students' lives in these environs are often not very different from their separated lives in previous home and school communities (Hurtado et al. 1994).

In these collegiate circumstances, white students are often confronted for the first time with the need to think about their own racial location. Having been socialized and educated at home, in their neighborhoods, through the media, and in previous schooling to expect people of color to be different, less competent, and potentially threatening, most young white people are ignorant, curious, and awkward in the presence of "others." Some may be aware of their racial group membership and identity, but others may be relatively unaware. Furthermore, during this developmental stage of late adolescence and early adulthood, students' identities as racial beings, as well as their racial attitudes, are subject to challenge and change. Hence it is important to understand the potential developmental trajectory of students' views as they move from their communities of origin to and through diverse collegiate experiences.

Recent explorations of whiteness suggest that changes in the economic, political, and cultural landscape have promoted greater self-consciousness about race. As a result, for many students the invisibility of whiteness, the notion that white is normal and natural, has become harder to sustain. Challenges to white ignorance and/or privilege have also increased some whites' sense of threat to their place in the social order and to their assumptions about their lives and society (Feagin and Vera 1995; Pincus 2000; Winant 1997). Discussions of historic privilege, structural inequality, and racial oppression have caused some white students (and college administrators and faculties as well) to question their enmeshment in pervasive (if unintended) patterns of institutional discrimination. In addition, institutions that now see the education of a diverse citizenry as integral to their missions of education and public service are struggling to make changes in the demographics of their faculty and student bodies, curricular designs, pedagogical tactics, student financial aid programs, and support services.

CONTEMPORARY THEORIES OF WHITE RACIAL ATTITUDES AND IDENTITIES

In the context of these shifts and struggles, scholars have described and explained the genesis and nature of whiteness and white racial attitudes and experiences as well as the developmental aspects of white racial identity and consciousness.

When understood in the context of larger patterns of institutional racism and changing cultural narratives about race, these identity and attitude frames are useful guides—heuristic devices—to understanding white racial consciousness and conceptions of whiteness itself. However, almost all interpretations and typologies of white attitudes and identities focus on their views of "the other" rather than on views of oneself or one's own racial group. That is, surveys of racial attitudes generally ask white people about their views of or prospective behavior toward people of color or race-related policies, seldom inquiring into whites' views of their own racial selves or of their earned/unearned status (i.e., privileges).

Similarly, most white identity development models focus on how whites view people of color rather than themselves; thus their racial identity is conceived as a reflection of their views of "the other." The stance that overlooks one's own race and focuses on others' can itself be seen as a manifestation of the "naturalness" and dominance of whiteness. Certainly one's views of the other and of the self are interactive, and people learn about their racial identity and attitudes in an interactive context, but one's views of others (or of the meaning of others' race) and one's view of themselves (or of the meaning of their own race) are not the same thing....

The notion of white racial "identity stages" suggests a developmental process that generally proceeds as follows (Helms 1990; Rowe et al. 1994): 1) from racial unawareness or conformity to traditional racial stereotypes, sometimes called an "unachieved" racial identity; 2) through questioning of these prior familial and societal messages, with attendant confusion, dissonance, and perhaps even "over-identification" with the other and attendant rebellion; 3) to retrogressive reintegration, where white culture is idealized, others are rejected, and a racially "dominative" ideology holds forth; 4) into a generally liberal (sometimes called pseudo-) acceptance or tolerance of people of color, often accompanied by adherence to notions of "color blindness" or denial and conflict around remaining prejudices; and, 5) it is hoped to an antiracist stance, wherein understanding of others' oppression and one's own privilege is (more or less) fully integrated into a personal worldview called an "autonomous" or "integrative" white racial identity....

Given increased collegiate attention to racial injustice and the desire of some people and advocacy groups to challenge institutional racism, it is not surprising that some young white collegians are becoming more conscious of their racial membership and its privileges. Such consciousness is likely to be painful, as it requires acknowledging both systemic advantage and personal privilege and enmeshment (historically and contemporarily) in structural or institutional discrimination and oppression. A few scholars have pointed to the emergence of a "liberationist" or "antiracist" form of white racial attitudes, wherein white people acknowledge and grapple with their accumulated racial privilege and their role (intentional or not) in sustaining white advantage and the domination over people of color. The racial identity literature refers to this belief/action system as an integrated, autonomous, introspective, or antiracist racial consciousness....

WHAT DO WHITE STUDENTS BRING WITH THEM TO THE UNIVERSITY?

In interviews, white students [we interviewed] discussed the neighborhood and schools in which they grew up and the effect these largely segregated experiences had on their conceptions of themselves, race, and racism. The major themes that characterize their precollege experience are lack of exposure, subtle and overt racism, racial tokenism, and lack of successful role models of people of color:

> I never really think about the fact that I am white. I just think that it is fortunate that we don't have to think about it, you know what I mean? It is one of the perks of being white.

> I consider myself white, but I don't think about it. The only time I think about it is when we have to do these dumb forms and think about what race we are.

According to Janet Helms (1990:3), racial identity is "a sense of group or collective identity based on one's perception that he or she shares a common racial heritage with a particular group." If the students above never thought about being white and didn't feel a sense of shared racial heritage, they could not possibly develop a self-conscious racial identity; they were at the unaware stage.

White students consistently indicated that their lack of prior contact with people of color, even in the midst of liberal rhetoric, failed to prepare them to engage meaningfully about race:

> I grew up in a very white community, and the church was really white. We talked about other cultures, but it was all about boys and girls are equal and worthy and so are people of different colors. It was all about "everything's OK."

> Where I grew up, everybody was white, and even though I knew (on some level) that not everyone was white, we never really had to deal with it, and so we didn't.

A few students reported coming from more diverse neighborhoods and schools, but they too indicated a relatively low level of sustained interaction or conscious educational attention to issues of diversity and intergroup relations. In these "more diverse" settings, racial segregation was still the normative experience for white students (as well as for students of color):

> [The city] is very segregated in terms of housing, and there's all different kinds of people who live here. But there isn't a tremendous amount of communication and social interaction between the groups ... unless you played sports or you were involved in something else, because it was tracked. Almost all of the kids on the college track were white and almost all of the kids on the other tracks were black ... and then there were also Asian kids and they were generally in the white track.

This lack of meaningful contact with people from other races was often coupled with various forms of both subtle and overt racism. If fact, many students' comments indicate that intergroup separation supported the home and media-based racism they were exposed to, creating and sustaining conditions wherein remnants of "old-fashioned racism" and an identity stage of unawareness and acceptance of stereotypes could be maintained:

> So I grew up with my dad particularly being really racist, he didn't really say much about any other group except Black people. "Nigger" was a common word in my family. I knew that that was not a good thing in terms of race. I knew that there was the black side of town, there was the black neighborhood, and then the rest of it was white, and that's what I grew up in.... But we never had any personal interactions with anybody [from the black neighborhood].

> My whole town was white except for a few families who migrated from Mexico to work. I had the clear sense that they weren't supposed to be there. They were like some unspoken exception that was supposed to be invisible.

In addition to the lack of contact in school and neighborhood and the various forms of racism that students were exposed to, several students indicated that when they did learn about people from other races, they were usually token efforts of inclusion:

> The only thing I learned in school was that [George] Washington Carver was a black man and he discovered peanuts or something like that. I think we might have peripherally dealt with Martin Luther King. But four years, two years of history, two years of government, we really didn't touch on African-American or any other issues at all ... that just didn't even exist as far as anybody was concerned. In elementary school we dealt with the Indians. You know, you put your hand on a piece of paper and you draw around it and you cut it out and you make a turkey, or you make little Indian hats and things like that with feathers.

Finally, even students who experienced token efforts of inclusion as unsatisfactory found little opportunity to formulate openly meaningful questions about race. Several students commented that when they did have racial questions and concerns during their high school years, they were simply told that there was "nothing to talk about":

> The message that I got from the white teachers at the school and other people was that the way not to be racist was to just pretend that you don't see any differences between people. And so everybody had feelings about race, but nobody talked, there was no place to talk about those things. And you only have to just treat everybody as an individual and everything will be fine.

> In my high school government class I asked a question about the Civil Rights movement and racism. The answer I got was basically that it was bad back then, but now everything was fine.

Growing up with everyday processes of segregation, lacking contact with racially (or socioeconomically) different peers, being exposed to various forms of racism and racial tokenism, and not being educated meaningfully about race and racism deeply affect white students' social identity—their sense of themselves as well as their relations with others. In their homes, schools, and communities these students acquired habitual attitudes, expectations, and ways of making meaning about their world. White students were socialized to not see themselves as having a race and did not understand their own (and their communities') exclusionary attitudes and behaviors....

WHITE STUDENTS' EXPERIENCES ON CAMPUS: NEW CHALLENGES TO WHITENESS

Students' precollege socialization forms a grid of attitudes and expectations about race and whiteness that is often reenacted and reified through their collegiate experience. As several white students reported, once in college they still did not think about themselves as being white—even in the presence of diversity; no one and no program invited or required them to. Hence, as the racial majority on campus and the dominant group within the larger society, the experience of knowing themselves as white was primarily reactive. That is, white students' numerical and cultural dominance protected them from having to know or understand others' experiences. Consequently, in order to "see" their race, they had to have a critical encounter or be consciously challenged to think and reflect about the particular experiences (perhaps privileges) that they had as a result of their racial position. Unless this challenge occurs at a conscious level, their own racial identity remains unknown and invisible during their college years.

Even when white students do have a critical encounter that raises their awareness of their race, they may not have the skills and consciousness (or instructional and experiential assistance) to deal with or act on it productively. Compare, for instance, the level of insight conveyed in these two excerpts:

> I don't understand why all the black students sit together in the dining hall. They complain about people being racist, but isn't that racist?

> Something I see is that the different races tend to stick with people like themselves. Once, in a class, I asked why all the black students sit together in the dormitory cafeteria. A black student then asked: "Well, why do you think all the white kids sit together?" I was speechless. I thought that was a dumb question until I realized that I see white people sitting together as normal and black people sitting together as a problem....

These comments reflect larger social assumptions about race relations on campus, wherein the prevailing myth has been that minority students are "self-segregating" and the exclusionary behaviors of the majority white group remain unseen (Tatum 1997). However, longitudinal research with over 200,000 students from 172 institutions found that it was white students who displayed the most exclusionary behaviors—particularly when it came to dating (Hurtado et al. 1994). Thus the view that minority students are self-segregating is clearly a skewed perspective that does not take into consideration the separatist and/or exclusionary behaviors of white students. It also fails to account for the ways in which institutional norms and cultures help students misinterpret patterns of interracial interaction.

Other white behaviors took the form of promoting or reacting to patterns of racial marginalization and separation in daily interactions in classrooms, social events, or casual encounters. The result, of course, continues to be minimal opportunity for sustained interaction:

> My black friend invited me to a party with her. And the first thing I could think of was how many white people are usually there. I remembered thinking, this is probably going to be uncomfortable, and I would rather just go out with my white friends. I'm feeling apprehensive about meeting their friends and therefore spending time with them.

> I used to feel very guilty thinking I don't have many diverse friends. I thought: "I have to go out and get a black friend."

Some white students reported finding these and other situations so discomforting that they began to express resentment against students of color. This type of resentment is supported by the discourse of whites as victims:

> I think white males have a hard time because we are constantly blamed for being power-holding oppressors, yet we are not given many concrete ways to change. Then we just feel guilty or rebel.

> I think that black people use their race to get jobs. I've seen it happen. My friend should have had this job as a resident advisor, but a black guy got it instead. There's no way the black guy was qualified.

The particular reference to "my friend" in the excerpt above is referred to by Eduardo Bonilla-Silva as one of the main "story lines of color blindness" (2001:159). Views such as these, expressing the emergence of a self-interested form of racial awareness, are consistent with Lawrence Bobo's [Editor's note: See article II in this volume] discussion of the group-position frame of racial attitudes.

Hence we encounter the view of the white person as the "new victim" of racism or as the target of "reverse discrimination" (Gallagher 1995; Pincus 2000). Victimhood, like all racial identities and views, is historically situated, and current public discourse about affirmative action and other race-based remedies stimulates and supports its development and expression. A lack of understanding of one's own prejudices, the realities of racial discrimination, and the advantages whites have leads to the view that minority advance is unmerited and a reflection of special privilege. The result often is aversive or self-interested racism that facilitates the interpretation of interracial encounters or circumstances as overprivileging minorities and victimizing whites. This also is referred to as the reintegrative or dominative stage of white racial identity.

The inability to understand racial membership is compounded by denial of any racial prejudice or racism. As a result of professed innocence about the meaning and implications of their own racial status and privileges, white students are often "blind" to the reality and status of students of color and regard themselves as "color-blind." If white students do not understand the personal or structural implications of being white and are unable to see how their racial behaviors affect others, they blindly negotiate racial encounters with the sense that all that matters is their good intentions. Their structural position of racial dominance, together with precollege socialization and color-blind ideology, makes it very difficult to distinguish between good intentions (or innocence) and a reflective consciousness that can enact just racial encounters:

> I am a pretty open person and someone who wouldn't even think about race, who would try to be color-blind.

> When I was asked in a class to describe my beliefs about race, it was easy. I said that I think that the whole idea of race has gone too far, that we need to stop thinking about race and start remembering that everyone is an individual.

Robert Terry (1981) identifies this pervasive color-blind ideology as an attempt to ignore or deny the relevance of race by emphasizing everyone's "humanness." Others have pointed out that the changing discourse of affirmative action—from a need to remediate past injustice to a concern about reverse discrimination—has affected how white people construct racial meaning. The new discourse of white victimhood not only acts to obscure the experiences of students of color but also further reinforces barriers to white students' ability to acknowledge their own racial identity as members of the dominant or privileged group.

Despite these reports of unawareness, negativity, blindness, and victimhood, there are also signs that some white students develop more sophisticated and progressive views of race. As they encounter themselves and others, some white students report moving out of the stage of "conformity" or "dissonance," going beyond "color blindness," and acknowledging their racism, prejudices, and stereotyped assumptions or expectations. This occurs partly as a function of structured educational experiences and informal contacts:

> It took me a long time to be able to get to a point where I can say that I have prejudices.

> Something I learned is that people have stereotypes. I learned that having stereotypes about other groups is part of the environment that we grow up in.

For a number of white students, these realizations led to a sense of shame or guilt: several scholars have also referred to these responses as the symbolic or emotional "costs" of white racism (Feagin and Vera 1995; Rose 1991);

> But I was so guilt-ridden, just horribly liberal guilt-ridden, paralyzed and unable to act. I was totally blowing every little minor interaction that I had with people of color way out of proportion and thinking that this determines whether or not I'm a good white person or a bad white person, and whether I'm racist or not. I saw how hard it was for me to stop doing that and start being more productive. And how hard it was for me to not be scared.

Such strong feelings, when combined in sensitive ways (as contrasted with self-pitying or defensive ways) with new educational input, helped some white students understand some of the privileges that were normally accorded them as a function of their white skin color (and associated socioeconomic and educational status):

> I learned that being white, they're so many privileges that I didn't even know of ... like loans from the bank, not being stopped by the police, and other things me and white kids can get away with.

> I had not noticed the extent to which white privilege has affected and continues to affect many aspects of my everyday life. I thought "I" had accomplished so much, but how much of where I am is due to my accumulated privilege—my family, economic status, school advantages? ...

Innovative educational programs must be designed and implemented to address these issues in students' racial identities and attitudes. However, even such innovations will not be effective or sustained without parallel changes in the operations of departments and the larger collegiate or university environment. Without changes in this broader organizational landscape, it is unlikely that individual white students' attitudes will change or that their racial identities will continue to "progress"—or that such change programs, if initiated, can be maintained. Moreover, students' consciousness and the academy itself are enmeshed in our society's continuing struggle with racial discrimination and racial privilege. There are real limits for any change toward more liberationist or antiracist white identities or racial attitudes within a highly racialized and racist society and higher educational system.

REFERENCES

Bonilla-Silva, Eduardo, and Tyrone A. Forman, 2000. "'I Am Not a Racist But ...': Mapping White College Students' Racial Ideology in the USA." *Discourse & Society* 11: 50–85.

Cornell, Stephen and D. Hartman, 1998. *Ethnicity and Race: Making Identities in a Changing World.* Thousand Oaks, CA: Pine Forge.

Feagin Joe R., and Hernán Vera. 1995. *White Racism: The Basics.* New York: Routledge.

Gallagher, Charles A. 1995. "White Reconstruction in the University." *Socialist Revolution* 94(1-2): 165–187.

Helms, Janet E., ed. 1990. *Black and White Racial Identity: Theory, Research and Practice.* New York: Greenwood.

Hurtado, S., Dey, E., & L. Trevino, 1994. "Exclusion or Self-Segregation: Interaction Across Racial/Ethnic Groups On Campus." Presented to meetings of the American Educational Research Association. New Orleans, LA.

Massey, Douglas S., and Nancy A. Denton, 1993. *American Apartheid: Segregation and the Making of the Underclass.* Cambridge, MA: Harvard University Press.

Pincus, Fred. 2000. "Reverse Discrimination vs. White Privilege: An Empirical Study of Alleged Victims of Affirmative Action." *Race and Society* 3:1–22.

Rose, L. 1991. "White Identify and Counseling White Allies about Racism. Pp. 24–47 in *The Impact of Racism on White Americans.* 2nd ed., ed. B. Bowser and R. Hunt. Thousands Oaks, CA: Sage.

Rowe, W., S. Bennet, and D. Atkinson. 1994. "White Racial Identity Models: A Critique and Alternative Proposal." *The Counseling Psychologist* 22:129–146.

Tatum, Beverly Daniel. 1997. *Why Are The Black Kids Sitting Together in the Cafeteria? And Other Conversations about Race.* New York: Basic.

Terry, Robert W. 1981. "The Negative Impact on White Values." Pp. 119–151 in *Impacts of Racism on White Americans,* ed. Benjamin P. Bowser and Raymond G. Hunt. Beverly Hills, CA: Sage.

Winant, Howard 1997. "Behind Blue Eyes: Whiteness and Contemporary U.S. Racial Politics." Pp. 40–53 in *Off White: Readings on Race, Power and Society.* Ed. Michelle Fine, Lois Weis, Linda C. Powell, and L. Mun Wong. New York: Routledge.

DISCUSSION QUESTIONS

1. What are the different phases that Chesler and his colleagues identify as affecting the development of White people's racial identity?
2. Chesler and his colleagues identify definite patterns in the attitudes of White students as they encounter new racial experiences in the college setting. Have you seen evidence of these same patterns in your observations of White students on your campus? How are they similar and/or different?

Part IV
Student Exercises

1. Design a brief study of interracial relationships for which you interview students on your campus about their attitudes toward interracial marriage. If possible, interview students from different racial-ethnic backgrounds. What degree of support do you find for dating someone of a different race? Is there a different level of support for marrying someone of a different race? Do these attitudes vary between men and women and/or between those from different racial-ethnic backgrounds? What factors do you think most influence people's attitudes toward interracial relationships?
2. Think back to the first time you remember recognizing your own racial identity. What were the circumstances? What did you learn? Now ask the same question of someone whose race is different from your own. How do the two experiences compare and contrast? How do the answers illustrate how racial identity is formed in different contexts and with different meanings depending on the group's experience?

PART V

INTRODUCTION

Race, Nation, and Citizenship

Up until now, we have been exploring the various definitions of race and looking at race largely in terms of beliefs and identities. But race is also part of larger social structures that form the organization of society. Here we turn to looking at how race is embedded in the social institutions of a nation. We will look at how race was a critical factor in the early building of the United States by granting or denying rights to different people. Rights are critical to belonging to a nation. As Chief Justice Earl Warren wrote in 1958, dissenting on a denationalization case, "Citizenship *is* man's basic right, for it is nothing less than the right to have rights" (Ngai 2003: 10).

Key to the realization of rights is the notion of citizenship, which enables people to participate fully in the society. The rules of citizenship—that is, who is granted this status and what the paths are to achieving it—are essential in shaping the relationship between groups in a society. In the United States, as in many nations, rules of citizenship developed along lines of racial and ethnic hierarchies, thus giving some groups access to social and political power and others, not. People who were not granted citizenship rights were pushed to the margins of the society, where they had little leverage in shaping the major social institutions that influenced their lives. Marginalized racial groups struggled for inclusion in the larger political body, a struggle that is ongoing in many respects. A legacy of exclusion has limited people's access to political, economic, and social resources and has shaped the majority group's thinking about racially different groups and their "place" in society.

Membership in different racial and ethnic groups was linked with rights and obligations even during the early days of colonial America, but as the nation became independent from England, it developed its own economic, social, and political institutions. Race was central to thinking about the nature of the country and the development of its citizenry. Important state documents, such as the U.S. Constitution and early laws about immigration, reveal the prevailing attitudes about race. "We the people" did not refer to an inclusive group; rather, the founding fathers were very deliberate in extending rights only to White men with property. As a group, American Indians—the indigenous populations—were seen as outside the developing society. Only people who paid certain taxes were granted full rights of citizenship. The Constitution also identified "other persons," that is, enslaved people, most of African descent. This initial national vision of people of color as unworthy of citizenship meant that they had to battle for inclusion in the nation. Moreover, their acceptance was not assured because of dominant ideas about their presumably limited abilities.

We can see the vision the founding fathers had for the future of this young nation in the procedures established for how new immigrants could become citizens. The 1790 Naturalization Act offered citizenship to "free white persons" who were of good moral character; the 1802 Act added a five-year residency requirement. The United States was viewed as a site of Anglo-Saxon settlement. The idea of a nation with a White majority shaped its very construction, including the exclusion and treatment of racial groups. Later, such barriers extended to eastern and southern European immigrants, who were viewed as racially inferior in the late nineteenth and early twentieth centuries. Yet, these newcomers, even though they were socially marginalized, were "White" politically and on a path to full participation in the nation, as Karen Brodkin relates in her family's experiences with social acceptance presented in Part II. We can contrast the experiences of eastern and southern European immigrants with those of American Indians, African Americans, Asian immigrants, and other people of color who faced extended hardships because of their lack of rights. These excluded groups would have to change many laws to become citizens. For example, immigrants from Asia were ineligible for citizenship because they were not "free white persons." The denial of rights made all racial minorities vulnerable to having their labor exploited and their land, if they had any, taken for the benefit of others.

For example, in the early twentieth century, although most were literate, Japanese immigrants could not become citizens and vote. They found their employment options limited, so some made work by serving their own community. Many used their agricultural skills to develop farms outside of cities, where they provided fruits and vegetables for the growing urban population. As their settlements grew, so did opposition from the Native and immigrant White population. Californians first passed laws in 1913 limiting Japanese people to being able to lease land for only three years. Later the Alien Law Act in 1920 prohibited them from leasing or owning land, so Japanese people either changed occupations or had White people purchase land for them (Takaki 1989). Their labor was critical for the growth of the nation, but denying them citizenship limited their political power and occupational choices.

Citizenship means that individuals are participants in the development of a nation's political and social institutions. They can use this power to shape institutions for their own benefit, while groups without power have to work within social institutions they did not design. In terms of the economy, those with power have historically appropriated the labor of others for their own benefit. As the nature of the economy shifts, a few more people might gain access to political power—such as the extension of the vote to White men without property in the mid-nineteenth century. However, the practice of denying racial-ethnic groups the rights of citizenship continued well into the twentieth century.

You can think of citizenship in two ways. First, there is an actual system of rights (the right to vote, the right to sit on a jury, the right to own property, the right to be counted for purposes of representation, and so forth) by which people are able to play a role in shaping institutions. Sometimes you can succeed within the system; other times you cannot, but you are still a participant. Citizenship is also symbolic—meaning the right of belonging to a nation or a community (Glenn 2002). For example, at the ceremony that opened the Lincoln Memorial in Washington, DC in 1922, Black Americans had to witness the event from behind a rope. The pattern of segregation, like separate schools for people of color and the denial of access to libraries and museums, communicated that they were not full citizens in the nation.

The articles in this part explore issues of citizenship, patterns of exclusion, and their implications for people's participation in the nation. Today, many people born in the United States take citizenship for granted, but many groups struggled for decades for essential rights. Evelyn Nakano Glenn ("Citizenship and Inequality") explores how people of color battled the limitations of the Constitution and opened up options for others. She finds that, even when laws are changed, racial practices continue to shape social policy. Glenn's article also shows how citizenship is integrally linked to the labor people provide. Overall, her selection provides us with insights into the contested nature of citizenship.

C. Matthew Snipp ("The First Americans: American Indians") explains how early national leaders considered American Indians to be candidates for extinction. Thus, Native Americans' path to incorporation into the nation involved removal from their original homelands and then later forced assimilation. We see how their inability to speak for themselves as citizens in the United States made them vulnerable to laws that were supposed to help, but really harmed American Indian people. Only in the twentieth century could tribes gain some measure of control and real sovereignty to improve their status, but their rights are still being contested in many states.

Who counts as a citizen and who does not is also linked to a sense of national identity and the history of exclusionary practices. Suzanne Oboler ("It Must Be a Fake! Racial Ideologies, Identities and the Question of Rights") sees the impact of our historical legacy in contemporary relations, because racial ideologies serve as barriers to building a nation that is a community of equals. She discusses the shift from an ideology of racism based in biology to a more social form, where new ethnic labels, in particular the Hispanic, monitor group progress but also limit full participation and rights.

Issues of citizenship continue today. People of various statuses live within U.S. borders. We see some of the additional complexities of citizenship by looking at the status of refugees. Refugees are people who flee their nation of origin to seek asylum in another nation, as in times of war, or political or religious persecution. The United States had a history of ambivalence about accepting war refugees early in the twentieth century, but new norms and expectations have developed since World War II (Daniels 2004). Immigration is a segment of foreign policy that not only reflects the desired racial composition of the state but is affected by the politics and practices of immigrants' birthplaces. The events of September 11, 2001, have changed the political landscape; thus immigrants from places that are seen as hostile to the United States face particular challenges. Most vulnerable are those who are not full citizens but are resident aliens and refugees. Tram Nguyen ("We Are All Suspects Now") discusses the plight of refugees from Somalia, who were once welcomed, but now face uncertain futures as foreigners from a nation viewed as a threat. This article also reveals the global context in which issues of citizenship and nation are now played out. The question remains: How can we respect individual human rights, build a society of equals, and also protect all of our citizens? These are questions we will likely grapple with for decades.

REFERENCES

Daniels, Roger 2004. *Guarding the Golden Door.* New York: Hill and Wang.

Glenn, Evelyn Nakano. 2002. *Unequal Freedom: How Race and Gender Shaped American Citizenship and Labor.* Cambridge: Harvard University Press.

Ngai, Mae M. 2003. *Impossible Subjects: Illegal Aliens and the Making of Modern America.* Princeton, NJ: Princeton University Press.

20

Citizenship and Inequality

EVELYN NAKANO GLENN

Glenn explores the concept of citizenship, as it has developed in reference to racial and ethnic groups in American society. She points out that citizenship includes both formal rights and the informal sense of belonging—forms of citizenship that have historically been denied to different groups in different ways. Her article shows how central the construction of citizenship is to being seen as having full "personhood."

HISTORICAL DEBATES ABOUT CITIZENSHIP IN THE UNITED STATES

... The concept of citizenship is, of course, historically and culturally specific. The modern, western notion of citizenship emerged out of the political and intellectual revolutions of the seventeenth and eighteenth centuries, which overthrew the old feudal orders. The earlier concept of society organized as a hierarchy of status, expressed by differential legal and customary rights, was replaced by the idea of a political order established through social contract. Social contract implied free and equal status among those who were party to it. Equality of citizenship did not, of course, rule out economic and other forms of inequality. Moreover, and importantly, equality among citizens rested on the inequality of others living within the boundaries of the community who were defined as non-citizens. The relationship between equality of citizens and inequality of non-citizens had both rhetorical and material dimensions. Rhetorically, the "citizen" was defined and, therefore, gained meaning through its contrast with the oppositional concept of the "non-citizen" as one who lacked the essential qualities needed to exercise citizenship. Materially, the autonomy and freedom of the citizen were made possible by labor (often involuntary) of non-autonomous wives, slaves, children, servants, and employees....

A specifically sociological conception of citizenship as membership is offered by Turner (1993:2), who defines it as "a set of practices (judicial, political, economic, and cultural) which define a person as a "competent" member of

SOURCE: From *Social Problems* 47(1): 1–20, 2000. Copyright © 2000 by The Society for the Study of Social Problems, Inc. All rights reserved. Reprinted by permission.

society...." Focusing on social practice takes us beyond a juridical or state conception of citizenship. It points to citizenship as a fluid and decentered complex that is continually transformed through political struggle.

Membership entails drawing distinctions and boundaries for who is included and who is not. Inclusion as a member, in turn, implies certain rights in, and reciprocal obligations toward, the community. Formal rights are not enough, however; they are only paper claims unless they can be enacted through actual practice. Three leading elements in the construction of citizenship, then, are membership, rights and duties, and conditions necessary for practice.

These three elements formed the major themes that have run through debates, contestation, and struggles over American citizenship since the beginning. First, membership: who is included or recognized as a full member of the imagined community (Anderson 1983), and on what basis? Second, what does membership mean in terms of content: that is, what reciprocal rights and duties do citizens have? Third, what are the conditions necessary for citizens to practice citizenship, to actually realize their rights and carry out their responsibilities as citizens?

MEMBERSHIP

Regarding membership, there are two major strains of American thought regarding the boundaries of the community. One tradition is that of civic citizenship, a definition based on shared political institutions and values in which membership is open to all those who reside in a territory. The second is an ethno-cultural definition based on common heritage and culture, in which membership is limited to those who share in the heritage through blood descent (Smith 1989; Kettner 1978).

Because of its professed belief in equality and natural rights (epitomized by the Declaration of Independence), the United States would seem to fit the civic model. However, since its beginnings, the U.S. has followed both civic and ethno-cultural models. The popular self-image of the nation, expressed as early as the 1780s, was of the United States as a refuge of freedom for those fleeing tyranny. This concept, later elaborated in historical narratives and sociological accounts of America as a nation of immigrants, blatantly omitted Native Americans, who were already here, Mexicans, who were incorporated through territorial expansion, and Blacks, who were forcibly transported. This exclusionary self-image was reflected at the formal level in the founding document of the American polity, the U.S. Constitution. The authors of the Constitution, in proclaiming a government by and for "we the people," clearly did not intend to include everyone residing within the boundaries of the U.S. as part of "the people." The Constitution explicitly named three groups: Indians, who were identified as members of separate nations, and "others," namely slaves; and finally, "the people": only the latter were full members of the U.S. community (Ringer 1983).

Interestingly, the Constitution was silent as to who was a citizen and what rights and privileges they enjoyed. It left to each state the authority to determine qualifications for citizenship and citizens' rights, e.g., suffrage requirements, qualifications for sitting on juries, etc. Individuals were, first, citizens of the states in which they resided and only secondarily, through their citizenship in the state, citizens of the United States. The concept of national citizenship was, therefore, quite weak.

However, the Constitution did direct Congress to establish a uniform law with respect to naturalization. Accordingly, Congress passed a Naturalization Act in 1790, which shaped citizenship policy for the next 170 years. It limited the right to become naturalized citizens to "free white persons." The act was amended in 1870 to add Blacks, but the term "free white persons" was retained. As Ian Haney Lopez (1996) has documented, immigrants deemed to be non-white (Hawaiians, Syrians, Asian Indians), but not Black or African, were barred from naturalization. The largest such category was immigrants from China, Japan, and other parts of Asia, who were deemed by the courts to be "aliens ineligible for citizenship." This exclusion remained in force until 1953....

It was Black Americans, both before and after the Civil War, who were the most consistent advocates of universal citizenship. Hence, it is fitting that the Civil Rights Act of 1865 and the Fourteenth Amendment, ratified in 1867 to ensure the rights of freed people, greatly expanded citizenship for everyone. Section 1 of the Fourteenth Amendment stated that "All persons born or naturalized in the United States and subject to the jurisdiction thereof, are citizens of the United States and of the State wherein they reside. No State shall make or enforce any law which shall abridge the privileges or immunities of citizens of the United States; nor shall any State deprive any person of life, liberty, or property, without due process of law; nor deny to any person within its jurisdiction, the equal protection of the law."

In these brief sentences, the Fourteenth Amendment established three important principles for the first time: the principle of national citizenship, the concept of the federal state as the protector and guarantor of national citizenship rights, and the principle of birthright citizenship. These principles expanded citizenship for others besides Blacks. To cite one personal example, my grandfather, who came to this country in 1894, was ineligible to become a naturalized citizen because he was not white, but his daughter, my mother, automatically became a citizen as soon as she was born. Birthright citizenship was tremendously important for second and third generation Japanese Americans and other Asian Americans who otherwise would have remained perpetual aliens, as is now the case with immigrant minorities in some European countries. Foner (1998) was on the mark when he said that the Black American struggle to expand the boundaries of freedom to include themselves, succeeded in changing the boundaries of freedom for everyone.

The battle was not won, once and for all, in 1867. Instead, the nation continued to vacillate between the principle of the federal government having a duty to protect citizens' rights and states' rights. By the end of the Reconstruction period in the 1870s, the slide back toward states' rights accelerated as all branches of the federal government withdrew from protecting Black rights and allowed southern states to impose white supremacist regimes (Foner 1988). In the landmark 1896,

Plessy v. Ferguson decision, the Supreme Court legitimized segregation based on the principle of "separate but equal." This and other court decisions gutted the concept of national citizenship and carved out vast areas of life, employment, housing, transportation, and public accommodations as essentially private activity that was not protected by the Constitution (Woodward 1974). It was not until the second civil rights revolution of the 1950s and 1960s, that the federal courts and Congress returned to the principles of national citizenship and a strong federal obligation to protect civil rights....

MEANINGS OF RIGHTS AND RESPONSIBILITIES

Just as with the question of membership, there has not been a single understanding of rights and duties. American ideas on rights and duties have been shaped by two different political languages (Smith 1989). One, termed liberalism by scholars, grew out of Locke and other enlightenment thinkers. In this strain of thought, embodied in the Declaration of Independence, citizens are individual rights bearers. Governments were established to secure individual rights so as to allow each person to pursue private as well as public happiness. The public good was not an ideal to be pursued by government, but was to be the outcome of individuals pursuing their own individual interest. The other language was that of *republicanism*, which saw the citizen as one who actively participated in public life. This line of thought reached back to ancient Greece and Rome where republicanism held that man reached his highest fulfillment by setting aside self-interest to pursue the common good. In contrast to liberalism, republicanism emphasizes practice and focuses on achieving institutions and practices that make collective self-government possible.

There has been continuing tension and alternation between these two strains of thought, particularly around the question of whether political participation is essential to, or peripheral to, citizenship. In the nineteenth century, when many groups were excluded from participation, the vote was a mark of standing; its lack was a stigma, a badge of inferiority, hence, the passion with which those denied the right fought for inclusion. Three great movements—for universal white manhood suffrage, for Black emancipation, and for women's suffrage—resulted in successive extensions of the vote to non-propertied white men, Black men, and women, between 1800 and 1920 (Roediger 1991; Foner 1990; DuBois 1978). According to Judith Shklar (1991), once the vote became broadly available, it ceased to be a mark of status. Judging by participation rates, voting is no longer an emblem of citizenship. One measure of the decline in the significance of the vote has been the precipitous drop in persons voting. In 1890, fully 80% of those who were eligible, voted. By 1924, four years after the extension of suffrage to women, participation had fallen to less than half of those registered, and it has been declining ever since....

In the U.S., participation is discouraged by not making registration easy. Requiring re-registration every time one moves is only one example of the obstacles placed upon a highly mobile population. While Americans today would object if the right to vote was taken away, the majority don't seem to feel it necessary to exercise the vote. The U.S. obviously has the capacity to make registration automatic or convenient, but has not made major efforts to do so. Needless to say, it is in the interest of big capital and its minions that most people don't participate politically.

CONDITIONS FOR PRACTICE

This leads to the third main theme, the actual practice of citizenship and the question of what material and social conditions are necessary for people to actually exercise their rights and participate in the polity.

The short answer to this question for most of American history has been that a citizen must be *independent*—that is, able to act autonomously.... Independence has remained essential, but its meaning has undergone drastic transformation since the founding of the nation.... In the nineteenth century, as industrialization proceeded apace and wage work became common even for skilled artisans and white collar men, the meaning of independence changed to make it more consistent with the actual situation of most white men. It came to mean, not ownership of productive property, but nothing more than ownership of one's own labor and the capacity to sell it for remuneration (Fraser and Gordon 1994). This formulation rendered almost all white men "independent," while rendering slaves and women, who did not have complete freedom to sell their labor, "dependent." It was on the basis of the independence of white working men that the movement for white Universal Manhood Suffrage was mobilized (Roediger 1991). By the 1830s, all states had repealed property requirements for suffrage for white men, while simultaneously barring women and Blacks (Litwak 1961).

By the late 19th century, capitalist industrialization had widened the economic gap between the top owners of productive resources and the rest, making more apparent the contradiction between economic inequality and political democracy. Rising levels of poverty, despite the expansion in overall wealth, raised the question of whether low-income, unemployment, and/or lack of access to health care and other services, diminished citizenship rights for a large portion of the populace. Growing economic inequality also raised the issue of whether some non-market mechanism was needed to mitigate the harshness of inequities created by the market.

One response during the years between World War I and World War II was rising sentiment for the idea of what T. H. Marshall called social citizenship. In Marshall's words (Marshall 1964:78), social citizenship involves "the right to a modicum of economic security and to share in the full social heritage and to live the life of a civilized being, according to the standards prevailing in the society." Economic and social unrest after World War I spurred European states to institute programs to ensure some level of economic security and take collective responsibility for "dependents"—the aged, children, the disabled, and others unable to work.... The significance of the "redistributive" mechanisms of the Welfare State, according to Marshall, was that they enabled working class people to exercise their civil and political rights.... Compared to Western Europe, the concept of social citizenship has been relatively weak in the U.S. Welfare state researchers have pointed out that, although 1930s New Deal programs such as Social Security and unemployment insurance greatly increased economic security, they continued a pattern of a two-tiered system of social citizenship from the 1890s (Nelson 1990; Skocpol 1992; Fraser and Gordon 1993; Fraser and Gordon 1994). The upper tier consisted of "entitlements" based on employment or military service, e.g., unemployment benefits, old-age insurance, and disability payments, which were relatively generous and did not require means-testing. The lower tier consisted of various forms of "welfare," such as Aid to Dependent Children (changed to Aid to Families of Dependent Children—AFDC—in 1962), which were relatively stingy and entailed means testing and surveillance by the state.

White men, as a class, have drawn disproportionately on first tier rights by virtue of their records of regular and well-paid employment. White women, more often, had to rely on welfare, which is considered charity, a response to dependence, rather than a just return for contributions. Latino and African American men were generally excluded from employment-based benefits because of their concentration in agriculture, day jobs, and other excluded occupations; Latina and African American women, in turn, have often been denied even second tier benefits (Oliver and Shapiro 1995; Mink 1994). Moreover, in contrast to the situation in most European countries, there has been little sense of collective responsibility for the care of dependents. Thus, raising children is not recognized as a contribution to the society and, therefore, as a citizenship responsibility that warrants entitlements such as parental allowances and retirement credit, which are common in Europe (Glenn 2000; Sainsbury 1996).

As with the other issues of citizenship, the 1980s and 1990s [saw] a neoconservative turn with a concerted effort to roll back even attenuated social citizenship rights. Government funding of social services has been vilified for draining money from hardworking citizens to support loafers and government bureaucrats who exert onerous control over people's lives. And, after years of attacks on Black and other single mothers "dependent" on welfare, Congress passed the Personal Responsibility and Work Opportunity Act in 1996, which dismantled AFDC and replaced it with grants to the states that limited eligibility for federal benefits for a lifetime maximum of five years. The aim was, as Representative Richard Armey put it, that the poor return to the natural safety nets—family, friends, churches, and charities. States instituted stringent work requirements and limited total lifetime benefits. By limiting total years that they can stay on welfare, proponents argued the new regulations would wean single mothers from unhealthy dependence on the state. In contrast to an overly indulgent government, the market is expected to exert a moral force, disciplining them and forcing them to be "independent" (Roberts 1997; Boris 1998)....

Thus, we must ask: if people have a responsibility to earn, then, don't they also have a corresponding right to earn?—i.e., to have a job and to earn enough to support themselves at a level that allows them, in Marshall's (1964:78) words, "to participate fully in the cultural life of the society and to live the life of a civilized being according to the standards prevailing in the society."

REFERENCES

Anderson, Benedict. 1983. *Imagined Communities: Reflections on the Origins and Spread of Nationalism*. London: Verso.

Boris, Eileen. 1998. "When work is slavery." *Social Justice* 25, 1:18–44.

DuBois, Ellen Carol. 1978. *Feminism and Suffrage*. Ithaca: Cornell University Press.

Foner, Eric. 1988. *Reconstruction: America's Unfinished Revolution*. New York: Harper and Row.

——— 1990. "From slavery to citizenship: Blacks and the right to vote." In *Voting and the Spirit of American Democracy*, ed. Donald W. Rogers. Urbana and Chicago: University of Illinois Press.

——— 1998. *The Story of American Freedom*. New York: Norton.

Fraser, Nancy, and Linda Gordon. 1993. "Contract versus charity: Why is there no social citizenship in the United States?" *Socialist Review* 212, 3:45–68.

——— 1994. "A genealogy of dependence: Tracing a keyword of the U.S. welfare state." *Signs* 19, 2:309–336.

Glenn, Evelyn. 2000. "Creating a caring society." *Contemporary Sociology* 29, 1: 84–94.

Haney Lopez, Ian F. 1996. *White By Law: The Legal Construction of Race*. New York: New York University Press.

Kettner, James. 1978. *The Development of American Citizenship, 1608–1870*. Chapel Hill: University of North Carolina Press.

Litwak, Leon F. 1961. *North of Slavery: The Negro in the Free States, 1790–1860*. Chicago: University of Chicago Press.

Marshall, T. H. 1964. *Class, Citizenship, and Social Development*. Garden City, NY: Doubleday and Company.

Mink, Gwendolyn. 1994. *The Wages of Motherhood: Inequality in the Welfare State, 1917–1942*. Ithaca: Cornell University Press.

Nelson, Barbara. 1990. "The origins of the two-channel welfare state: Workman's compensation and mother's aid." In *Women, the State and Welfare*, ed. Linda Gordon, 123–151. Madison: University of Wisconsin Press.

Oliver, Melvin L., and Thomas M. Shapiro. 1995. *Black Wealth/White Wealth: A New Perspective on Racial Inequality*. New York: Routledge.

Ringer, Benjamin B. 1983. *We the People and Others: Duality and America's Treatment of Its Racial Minorities*. New York: Tavistock.

Roberts, Dorothy. 1997. *Killing the Black Body: Race, Reproduction, and the Meaning of Liberty*. New York: Pantheon.

Roediger, David. 1991. *The Wages of Whiteness: Race and the Making of the American Working Class*. London: Verso.

Sainsbury, Diane. 1996. *Gender, Equality and Welfare States*. Cambridge, UK: Cambridge University Press.

Shklar, Judith. 1991. *American Citizenship: The Quest for Inclusion*. Cambridge, MA: Harvard University Press.

Skocpol, Theda. 1992. *Protecting Soldiers and Mothers: The Origins of Social Policy in the United States*. Cambridge, MA: Harvard University Press.

Smith, Rogers M. 1989. "'One United People': Second-class female citizenship and the American quest for community." *Yale Journal of Law and the Humanities* 1:229–293.

Turner, Brian S. 1993. "Contemporary Problems in the Theory of Citizenship." In *Citizenship and Social Theory*. ed. Brian S. Turner. 1–19. London: Sage Publications.

Woodward, C. Vann, ed. 1974. *The Strange Career of Jim Crow*, 3rd rev. New York: Oxford University Press.

DISCUSSION QUESTIONS

1. What elements does Glenn identify as constructing citizenship?
2. During what era did social citizenship rights expand and why?
3. Thinking about your own citizenship status, what do you see as the rights and obligations linked to your position in the nation?

21

The First Americans

American Indians

C. MATTHEW SNIPP

Snipp's analysis of different periods in the treatment of American Indians by the U.S. government shows how citizenship rights have been denied to Native Americans, even though the specific processes by which this has happened have changed over time.

By the end of the nineteenth century, many observers predicted that American Indians were destined for extinction. Within a few generations, disease, warfare, famine, and outright genocide had reduced their numbers from millions to less than 250,000 in 1890. Once a self-governing, self-sufficient people, American Indians were forced to give up their homes and their land, and to subordinate themselves to an alien culture. The forced resettlement to reservation lands or the Indian Territory (now Oklahoma) frequently meant a life of destitution, hunger, and complete dependency on the federal government for material needs.

Today, American Indians are more numerous than they have been for several centuries. While still one of the most destitute groups in American society, tribes have more autonomy and are now more self-sufficient than at any time since the last century. In cities, modern pan-Indian organizations have been successful in making the presence of American Indians known to the larger community, and have mobilized to meet the needs of their people (Cornell 1988; Nagel 1986; Weibel-Orlando 1991). In many rural areas, American Indians and especially tribal governments have become increasingly more important and increasingly more visible by virtue of their growing political and economic power. The balance of this [reading] is devoted to explaining their unique place in American society.

THE INCORPORATION OF AMERICAN INDIANS

The current political and economic status of American Indians is the result of the process by which they were incorporated into Euro-American society (Hall 1989). This amounts to a long history of efforts aimed at subordinating an

SOURCE: From Silvia Pedraza and Rubén G. Rumbaut, eds., *Origins and Destinies: Immigration, Race, and Ethnicity in America*. 1st Edition by Pedraza/Rumbaut. 1996.
Reprinted with permission of Wadsworth, a division of Thomson Learning: www.thomsonrights.com. Fax 800 730-2215.

otherwise self-governing and self-sufficient people that eventually culminated in widespread economic dependency. The role of the U.S. government in this process can be seen in the five major historical periods of federal Indian relations: removal, assimilation, the Indian New Deal, termination and relocation, and self-determination.

Removal

In the early nineteenth century, the population of the United States expanded rapidly at the same time that the federal government increased its political and military capabilities. The character of Indian-American relations changed after the War of 1812. The federal government increasingly pressured tribes settled east of the Appalachian Mountains to move west to the territory acquired in the Louisiana Purchase. Numerous treaties were negotiated by which the tribes relinquished most of their land and eventually were forced to move west.

Initially the federal government used bargaining and negotiation to accomplish removal, but many tribes resisted (Prucha 1984). However, the election of Andrew Jackson by a frontier constituency signaled the beginning of more forceful measures to accomplish removal. In 1830 Congress passed the Indian Removal Act, which mandated the eventual removal of the eastern tribes to points west of the Mississippi River, in an area which was to become the Indian Territory and is now the state of Oklahoma. Dozens of tribes were forcibly removed from the eastern half of the United States to the Indian Territory and newly created reservations in the west, a long process ridden with conflict and bloodshed.

As the nation expanded beyond the Mississippi River, tribes of the plains, southwest, and west coast were forcibly settled and quarantined on isolated reservations. This was accompanied by the so-called Indian Wars—a bloody chapter in the history of Indian-White relations (Prucha 1984; Utley 1963). This period in American history is especially remarkable because the U.S. government was responsible for what is unquestionably one of the largest forced migrations in history.

The actual process of removal spanned more than a half-century and affected nearly every tribe east of the Mississippi River. Removal often meant extreme hardships for American Indians, and in some cases this hardship reached legendary proportions. For example, the Cherokee removal has become known as the "Trail of Tears." In 1838, nearly 17,000 Cherokees were ordered to leave their homes and assemble in military stockades (Thornton 1987, p. 117). The march to the Indian Territory began in October and continued through the winter months. As many as 8,000 Cherokees died from cold weather and diseases such as influenza (Thornton 1987, p. 118).

According to William Hagan (1979), removal also caused the Creeks to suffer dearly as their society underwent a profound disintegration. The contractors who forcibly removed them from their homes refused to do anything for "the large number who had nothing but a cotton garment to protect them from the sleet storms and no shoes between them and the frozen ground of the last stages of their hegira. About half of the Creek nation did not survive the migration and the difficult early years in the West" (Hagan 1979, pp. 77–81). In the West, a band of Nez Perce men, women, and children, under the leadership of Chief Joseph, resisted resettlement in 1877. Heavily outnumbered, they were pursued by cavalry troops from the Wallowa valley in eastern Oregon and finally captured in Montana near the Canadian border. Although the Nez Perce were eventually captured and moved to the Indian Territory, and later to Idaho, their resistance to resettlement has been described by one historian as "one of the great military movements in history" (Prucha 1984, p. 541).

Assimilation

Near the end of the nineteenth century, the goal of isolating American Indians on reservations and the Indian Territory was finally achieved. The Indian population also was near extinction. Their numbers had declined steadily throughout the nineteenth century, leading most observers to predict their disappearance (Hoxie 1984). Reformers urged the federal government to adopt measures that would humanely ease American Indians into extinction. The federal government responded by creating boarding schools and the allotment acts—both were intended to "civilize" and assimilate American Indians into American society by Christianizing them, educating them, introducing them to private property, and making them into farmers. American Indian boarding schools sought to accomplish this task by indoctrinating Indian children with the belief that tribal culture was an inferior relic of the past and that Euro-American culture was vastly superior and preferable. Indian children were forbidden to wear their native attire, to eat their native foods, to speak their native language, or to practice their traditional religion Instead, they were issued Euro-American clothes, and expected to speak English and become Christians. Indian children who did not relinquish their culture were punished by school authorities. The curriculum of these schools taught vocational arts along with "civilization" courses.

The impact of allotment policies is still evident today. The 1887 General Allotment Act (the Dawes Severalty Act) and subsequent legislation mandated that tribal lands were to be allotted to individual American Indians, … and the surplus lands left over from allotment were to be sold on the open market. Indians who received allotted tribal lands also received citizenship, farm implements, and encouragement from Indian agents to adopt farming as a livelihood (Hoxie 1984, Prucha 1984).

For a variety of reasons, Indian lands were not completely liquidated by allotment, many Indians did not receive allotments, and relatively few changed their lifestyles to become farmers. Nonetheless, the allotment era was a disaster because a significant number of allottees eventually lost their land. Through tax foreclosures, real estate fraud, and their own need for cash, many American Indians lost what for most of them was their last remaining asset (Hoxie 1984).

Allotment took a heavy toll on Indian lands. It caused about 90 million acres of Indian land to be lost, approximately two-thirds of the land that had belonged to tribes in 1887 (O'Brien 1990). This created another problem that continues to vex many reservations: "checkerboarding." Reservations that were subjected to allotment are typically a crazy quilt composed of tribal lands, privately owned "fee" land, and trust land belonging to individual Indian families. Checkerboarding presents reservation officials with enormous administrative problems when trying to develop land use management plans, zoning ordinances, or economic development projects that require the construction of physical infrastructure such as roads or bridges.

The Indian New Deal

The Indian New Deal was short-lived but profoundly important. Implemented in the early 1930s along with the other New Deal programs of the Roosevelt administration, the Indian New Deal was important for at least three reasons. First, signaling the end of the disastrous allotment era as well as a new respect for American Indian tribal culture, the Indian New Deal repudiated allotment as a policy. Instead of continuing its futile efforts to detribalize American Indians, the federal government acknowledged that tribal culture was worthy of respect. Much of this change was due to John Collier, a long-time Indian rights advocate appointed by Franklin Roosevelt to serve as Commissioner of Indian Affairs (Prucha 1984).

Like other New Deal policies, the Indian New Deal also offered some relief from the Great Depression and brought essential infrastructure development to many reservations, such as projects to control soil erosion and to build hydroelectric dams, roads, and other public facilities. These projects created jobs in New Deal programs such as the Civilian Conservation Corps and the Works Progress Administration.

An especially important and enduring legacy of the Indian New Deal was the passage of the Indian Reorganization Act (IRA) of 1934. Until then, Indian self-government had been forbidden by law. This act allowed tribal governments, for the first time in decades, to reconstitute themselves for the purpose of overseeing their own affairs on the reservation. Critics charge that this law imposed an alien form of government, representative democracy, on traditional tribal authority. On some reservations, this has been an ongoing source of conflict (O'Brien 1990). Some reservations rejected the IRA for this reason, but now have tribal governments authorized under different legislation.

Termination and Relocation

After World War II, the federal government moved to terminate its longstanding relationship with Indian tribes by settling the tribes' outstanding legal claims, by terminating the special status of reservations, and by helping reservation Indians relocate to urban areas (Fixico 1986). The Indian Claims Commission was a special tribunal created in 1946 to hasten the settlement of legal claims that tribes had brought against the federal government. In fact, the Indian Claims Commission became bogged down with prolonged cases, and in 1978 the commission was dissolved by Congress. At that time, there were 133 claims still unresolved out of an original 617 that were first heard by the commission three decades earlier (Fixico 1986, p. 186). The unresolved claims that were still pending were transferred to the Federal Court of Claims.

Congress also moved to terminate the federal government's relationship with Indian tribes. House Concurrent Resolution (HCR) 108, passed in 1953, called for steps that eventually would abolish all reservations and abolish all special programs serving American Indians. It also established a priority list of reservations slated for immediate termination. However, this bill and subsequent attempts to abolish reservations were vigorously opposed by Indian advocacy groups such as the National Congress of American Indians. Only two reservations were actually terminated, the Klamath in Oregon and the Menominee in Wisconsin. The Menominee reservation regained its trust status in 1975 and the Klamath reservation was restored in 1986.

The Bureau of Indian Affairs (BIA) also encouraged reservation Indians to relocate and seek work in urban job markets. This was prompted partly by the desperate economic prospects on most reservations, and partly because of the federal government's desire to "get out of the Indian business." The BIA's relocation programs aided reservation Indians in moving to designated cities, such as Los Angeles and Chicago, where they also assisted them in finding housing and employment. Between 1952 and 1972, the BIA relocated more than 100,000 American Indians (Sorkin 1978). However, many Indians returned to their reservations (Fixico 1986). For some American Indians, the return to the reservation was only temporary; for example, during periods when seasonal employment such as construction work was hard to find.

Self-Determination

Many of the policies enacted during the termination and relocation era were steadfastly opposed by American Indian leaders and their supporters. As these programs became stalled, critics attacked them for being harmful, ineffective, or both. By the mid-1960s, these policies had very little serious support: Perhaps inspired by the gains of the Civil Rights movement, American Indian leaders and their supporters made "self-determination" the first priority on their political agendas. For these activists, self-determination meant that Indian people would have the autonomy to control their own affairs, free from the paternalism of the federal government.

The idea of self-determination was well received by members of Congress sympathetic to American Indians. It also was consistent with the "New Federalism" of the Nixon administration. Thus, the policies of termination and relocation were repudiated in a process that culminated in 1975 with the passage of the American Indian Self-Determination and Education Assistance Act, a profound shift in federal Indian policy. For the first time since this nation's founding, American Indians were authorized to oversee the affairs of their own communities, free of federal intervention. In practice, the Self-Determination Act established measures that would allow tribal governments to assume a larger role in reservation administration of programs for welfare assistance, housing, job training, education, natural resource conservation, and the maintenance of reservation roads and bridges (Snipp and Summers 1991). Some reservations also have their own police forces and game wardens, and can issue licenses and levy taxes. The Onondaga tribe in upstate New York have taken their sovereignty one step further by issuing passports that are internationally recognized. Yet there is a great deal of variability in terms of how much autonomy tribes have over reservation affairs. Some tribes, especially those on large and well-organized reservations have nearly complete control over their reservations, while smaller reservations with limited resources often depend heavily on BIA services. …

CONCLUSION

Though small in number, American Indians have an enduring place in American society. Growing numbers of American Indians occupy reservation and other trust lands, and equally important has been the revitalization of tribal governments. Tribal governments now have a larger role in reservation affairs than ever in the past. Another significant development has been the urbanization of American Indians. Since 1950, the proportion of American Indians in cities has grown rapidly. These American Indians have in common with reservation Indians many of the same problems and disadvantages, but they also face other challenges unique to city life.

The challenges facing tribal governments are daunting. American Indians are among the poorest groups in the nation. Reservation Indians have substantial needs for improved housing, adequate health care, educational opportunities, and employment, as well as developing and maintaining reservation infrastructure. In the face of declining federal assistance, tribal governments are assuming an ever-larger burden. On a handful of reservations, tribal governments have assumed completely the tasks once performed by the BIA.

As tribes have taken greater responsibility for their communities, they also have struggled with the problems of raising revenues and providing economic opportunities for their people. Reservation land bases provide many reservations with resources for development. However, these resources are not always abundant, much less unlimited, and they have not always been well managed. It will be yet another challenge for tribes to explore ways of efficiently managing their existing resources. Legal challenges also face tribes seeking to exploit unconventional resources such as gambling revenues. Their success depends on many complicated legal and political contingencies.

Urban American Indians have few of the resources found on reservations, and they face other difficult problems. Preserving their culture and identity is an especially pressing concern. However, urban Indians have successfully adapted to city environments in ways that preserve valued customs and activities—powwows, for example, are an important event in all cities where there is a large Indian community. In addition, pan-Indianism has helped urban Indians set aside tribal differences and forge alliances for the betterment of urban Indian communities.

These alliances are essential, because unlike reservation Indians, urban American Indians do not have their own form of self-government. Tribal governments do not have jurisdiction over urban Indians. For this reason, urban Indians must depend on other strategies for ensuring that the needs of their community are met, especially for those new to city life. Coping with the transition to urban life poses a multitude of difficult challenges for many American Indians. Some succumb to these problems, especially the hardships of unemployment, economic deprivation, and related maladies such as substance abuse, crime, and violence. But most successfully overcome these difficulties, often with help from other members of the urban Indian community.

Perhaps the greatest strength of American Indians has been their ability to find creative ways for dealing with adversity, whether in cities or on reservations. In the past, this quality enabled them to survive centuries of oppression and persecution. Today this is reflected in the practice of cultural traditions that Indian people are proud to embrace. The resilience of American Indians is an abiding quality that will no doubt ensure that they will remain part of the ethnic mosaic of American society throughout the twenty-first century and beyond.

REFERENCES

Cornell, Stephen. 1988. *The Return of the Native: American Indian Political Resurgence*. New York: Oxford University Press.

Fixico, Donald L. 1986. *Termination and Relocation: Federal Indian Policy, 1945–1960:* Albuquerque, NM: University of New Mexico Press.

Hagan, William T. 1979. *American Indians*. Chicago, IL: University of Chicago Press.

Hall, Thomas D. 1989. *Social Change in the Southwest, 1350–1880*. Lawrence, KS: University Press of Kansas.

Hoxie, Frederick E. 1984. *A Final Promise: The Campaign to Assimilate the Indians, 1880–1920*. Lincoln, NE: University of Nebraska Press.

Nagel, Joanne. 1986. "American Indian Repertoires of Contention." Paper presented at the annual meeting of the American Sociological Association, San Francisco, CA.

O'Brien, Sharon. 1990. *American Indian Tribal Governments*. Norman, OK: University of Oklahoma Press.

Prucha, Francis Paul. 1984. *The Great Father*. Lincoln, NE: University of Nebraska Press.

Snipp, C. Matthew and Gene F. Summers. 1991. "American Indian Development Policies," pp. 166–180 in *Rural Policies for the 1990s*, edited by Cornelia Flora and James A. Christenson. Boulder, CO: Westview Press.

Sorkin, Alan L. 1978. *The Urban American Indian*. Lexington, MA: Lexington Books.

Thornton, Russell. 1987. *American Indian Holocaust and Survival: A Population History since 1942*. Norman, OK: University of Oklahoma Press.

Utley, Robert M. 1963. *The Last Days of the Sioux Nation*. New Haven: Yale University Press.

Weibel-Orlando, Joan. 1991. *Indian Country, L.A.* Urbana, IL: University of Illinois Press.

DISCUSSION QUESTIONS

1. How do the four different periods of state policy that Snipp identifies reveal different ways that the state has managed American Indian affairs?
2. Explain what Snipp means when he concludes that American Indians have been resilient even in the face of massive state-based social control?

22

"It Must Be a Fake!"

Racial Ideologies, Identities, and the Question of Rights

SUZANNE OBOLER

Citizenship is, as Oboler shows, both a political and historical construction. How "Hispanics" are defined thus either confers or denies citizenship rights and shapes understandings of Hispanic identity.

About two and a half years ago Luis Gutiérrez, a Puerto Rican congressman from Chicago, was standing in line with his sixteen-year-old daughter and his niece, waiting to get into the Capitol to show them his office there. They had just been to "a tribute to all the veterans of the all–Puerto Rican 65th Army Infantry Regiment of the Korean War, including the 743 soldiers who were killed and the 2,797 who were wounded in that conflict." As a result, his daughter and niece were carrying small Puerto Rican flags. Gutiérrez told them to roll the flags up, thinking (mistakenly, as it turned out) that they were not allowed to bring flags into the Capitol. The girls did roll up the flags, but "they got caught in the rollers of the conveyer belt and unfurled." A Capitol police security aide, Stacia Hollingsworth, saw the unfurled flags and, according to the congressman, "yelled in [my] ear: 'Those flags cannot be displayed!'"

The *Chicago Tribune* journalists reporting the incident, David Jackson and Paul de la Garza, tell us that "Gutiérrez was embarrassed, but told his daughter to get rid of the flags, saying, "You know what the rules are."[1]

Overhearing him, Hollingsworth asked: "Who are you that you know what the rules are?" When he told her he was Luis Gutiérrez, a member of Congress, she replied, "I don't think so."

So Gutiérrez showed her his congressional ID card. Her immediate response was to say, "It must be a fake." And then she added, "Why don't you and your people just go back to the country you came from?"

According to Jackson and de la Garza, "Gutiérrez was stunned. 'It wasn't like on a side street in Chicago,' he said. 'This was in the middle of the gallery. In the Capitol. Where I work. Can you imagine how humiliating this was in front of my 16-year-old daughter?' At that point, a Capitol Police dignitary protection officer rushed over, recognized Gutiérrez and pulled the aide aside.

SOURCE: From *Hispanics/Latinos in the United States: Ethnicity, Race, and Rights*, ed. by Jorge Gracia and Pablo De Greiff (New York: Routledge, 2000), pp. 125–139. Copyright © 2000. Reprinted by permission of Routledge/Taylor & Francis Books, Inc.

He told Gutiérrez he saw what happened and suggested that Gutiérrez file a complaint."[2]

The exchange between Congressman Gutiérrez and the Capitol security aide raises at least four related issues that characterize the situation of Puerto Ricans and, more generally, of Latinos in the United States today. First—and although it is important to note that his racial characteristics are never explicitly stated in the article—Gutiérrez clearly did not look like a member of the U.S. Congress, which is largely made up of white males. Second, since he did not look like a congressman, he could not be trusted. Hence the aide assumed he had "faked" his congressional ID card. Third, Gutiérrez's visual features marked him as foreign to the image of people who belong in the United States. As such, he was told that he was neither recognized as a U.S. citizen nor welcome in this country. And finally, this relatively insignificant tale exemplifies the lack of awareness of the long historical presence and citizenship status both of Puerto Ricans (officially U.S. citizens since 1917) and, more generally, of the majority of the population today officially known as "Hispanics" in the United States.

Minimally, Congressman Gutiérrez's experience suggests the extent to which racism in the United States ensures that, unlike white Americans, Latinos constantly have to prove their citizenship and to insist on their rights—including their "right to have rights" as citizens of this society. In fact, the very symbolism of this exchange having taken place at the entrance to the building housing the U.S. Congress is illuminating, for it points to the ongoing emphasis on racial features and phenotype in defining membership in the nation's legislature, where the very meaning of national belonging is negotiated and the experience of representative democracy is recorded into the laws that reinforce the belief in a community of equals in the United States.

One conclusion we can draw from the Gutiérrez family's experience is that in spite of both the end of legal segregation brought about by the 1954 *Brown v. Board of Education* decision and of the civil rights movements of the 1960s and 1970s, racism continues to interfere with the possibility of creating a community of equals—and its modern synonym, citizenship—in the United States. Indeed, this encounter reminds us yet again that while citizenship may be commonly understood as a legal status, it is above all a political reality. As such, it cannot be fully understood without taking into account the specificity of the context within which it is understood and differentially experienced in people's daily lives....

The aim of this essay is to describe and clarify the national and global context within which we can discuss the ethnic identity, culture, and group rights of Hispanics/Latinos. I argue that it is a context in which, increasingly, racism and xenophobia shape both the meaning and social value attributed to individuals' ethnic identities and to their lived experience of national belonging in contemporary U.S. society. Insofar as citizenship is the political expression of national belonging, my aim is to clarify the contemporary role of racism in the decline of citizenship and in ensuring the impossibility of belonging to a national community of equals, both in U.S. society and in the broader international context....

CITIZENSHIP AS A POLITICAL AND HISTORICAL CONSTRUCTION

In considering the experience of Congressman Gutiérrez or, more generally, of Latinos in the United States, it is important to keep in mind that the concept of citizenship was never defined by the founding fathers in the Constitution. Instead, they spoke of "the people of the United States" and rarely mentioned the word citizen. Therefore, the political reality and social value of citizenship became contingent on a series of laws and/or court cases that at various times in the nation's history either reinforced or challenged each other. At the same time, both the laws and the courts ultimately aimed at specifying the role and implications of "race" in determining who could be a citizen, as well as in clarifying the responsibility of the state to the citizenry.

The Dred Scott case of 1857, for example, argued that the founding fathers did not mean to include blacks when they spoke of "the people of the United States." The subsequent Civil Rights Act of 1866 was specifically designed to reverse the Dred Scott decision, and was followed by the Fourteenth Amendment, ratified in 1868, which, for the first time in U.S. history, created a *national* citizenry and established the principle of equality under the law for all people born in the United States....

Less than two decades later, however, the *Plessy v. Ferguson* decision in 1896 effectively challenged that amendment. Ruling that "legislation is incapable of eradicating racial instincts," the Supreme Court established racial segregation as the law of the land for the next sixty years. The *Brown v. Board of Education* decision ended legal segregation in 1954, thus countering the Plessy ruling by pointing to the ways that the psychological damage created by segregation prevented black children's access to equal opportunity. But it took the subsequent civil rights movements of the 1960s to create the various civil rights acts specifically aimed at enforcing the *Brown* decision.

From this perspective, the policies enacted in the last thirty years of the twentieth century represent a new attempt to create a national community of equals—an attempt grounded in the explicit acknowledgement of the historical role of race in shaping the political reality and social value of citizenship....[3]

Like citizenship, race is also not a static concept. As a result of the discrediting of scientific racism underlying Nazism during World War II, its contemporary meanings and social value have gradually changed both in the United States and abroad. Certainly at the end of the twentieth century, the world as a whole has witnessed the disappearance of legal discrimination and, consequently, the seeming attenuation of racism in the political sphere of every society. Yet, paradoxically, the end of legal discrimination has signalled the unchallenged entrenchment of racism in social relations, particularly in the private sphere.

In the United States, the dualistic black/white biological racism that justified legal segregation until 1954 has been undermined over the past four decades by the emergence of a new ideology of "social racism," embedded in a new kind of social relations that are reminiscent of those found in Latin America.

The emergence of this ideology of social racism in the United States is particularly apparent in the growing adherence to the idea of racial mixture (or biracialism—in Latin America, it is called *mestizaje*), leading some to stress that social class rather than race is the key to understanding and solving the ongoing problem of poverty and deprivation of large sectors of the population, including racial minorities, in the United States. Reinforced by those who point to what one prominent mainstream magazine defined in 1990 as "the browning of America," this perceived belief in the "declining significance of race" has since been reinforced by the growing emphasis on the need to explicitly (re)define and stress American nationality as the basis for national unity.

The emphasis on national unity and the simultaneous insistence on the insignificance of biologically determined racial characteristics have long defined the meaning and social value of "race" and, hence, race relations in Latin America....

In the course of the twentieth century, Latin American racial ideologues increasingly modified and eventually rejected the notion of scientific racism that their U.S. colleagues consistently articulated at various inter-American conferences on eugenics.[4] Contrary to the biological determinism that historically has pervaded U.S. race relations, Latin American intellectuals and scientists alike understood "race" in social terms—specifically in terms of the belief in the existence of higher and lower cultures, which could clearly be assimilated into a national socioracial hierarchy organized and (in)visibly marked by skin color and phenotype. While the choice of terms, like their connotations and uses, continues to be debated throughout the continent, recent Latin American scholarship leaves no doubt that this hierarchy of cultural differences, which ensures both that "everyone knows their place" and the impossibility of forging national communities of equal citizens, has historically been grounded in what the Peruvian anthropologist Marisol de la Cadena has defined as "silent racism."

As I suggest in the following pages, this Latin American ideology of social racism, with its emphasis on the unifying force of nationality to the detriment of racial considerations, appears to be increasingly accepted in U.S. society, superimposed on—although not replacing—the biologically based black/white dualism that has been dominant for much of the nation's history....

THE LABEL "HISPANIC" AND THE QUESTION OF RIGHTS

... The effects of differentiating and, in effect, racializing the entire U.S. population through ... ethnic categories have been contradictory. Undoubtedly it has allowed us to track the progress toward political inclusion of racial minorities, as well as of women, since the end of legal segregation. But it has simultaneously reinforced the belief in the superiority of whiteness and "white privilege," making explicit the continuing existence of a socioracial hierarchy in a society that historically, and to this day, proclaims its adherence to the belief in equality for all. In fact, unlike past perceptions and beliefs that U.S. society was a "melting pot," there is today an implicit acknowledgment of an organized socioracial hierarchy, with whites at the top and blacks and Latinos alternating at the bottom....

In short, the official creation of these ethnic categories has ensured that, as in Latin America, everyone "knows his (or her) place" in U.S. society. And as in Latin America, the outcome is the impossibility of establishing an expanded community of equals in the United States. This assertion is reinforced by the following five interrelated points.

1. The vagueness of the census definition has led to many debates in this country concerning who is a Hispanic and on what grounds. This debate includes questions such as whether citizens from Latin America's sovereign nations currently living in the United States are "as Latino" as those born in the United States. Should this distinction be made? Given the vagueness of the wording and its consequences for public policy, social and race relations, and individuals' daily lives, it is essential that we acknowledge that in the United States, the term *Hispanic*—as originally conceived by the state in the 1970s and currently understood—is first and foremost a bureaucratic invention, used for census data collection. Like its grassroots alternative designation *Latino*, the term Hispanic does not refer to, and is in no way tied to, an actual historical, territorial, or cultural background or identity of any of the national-origin groups or ethnic populations it encompasses in the United States. Instead, it comprises the populations of all the Spanish-American nations and of Spain....
2. The term Hispanic, like other ethnic labels, is here to stay. And from this point of view, the Hispanic (or Latino) experience and identity in the United States cannot be understood outside of the context of the relations that colonized citizens (such as the Puerto Ricans) and conquered peoples (such as sectors of the Chicano population) have historically had with the U.S. government. This context conflated race and nationality and, in 1977, allowed for the official designation of the ethnic label Hispanic which homogenized all people of Latin American descent. Nor can it be understood outside of the context of the historical and very specific differences that mark U.S. relations with each of the various Latin American nations. These relations invariably differentiate the sociopolitical experiences of each national population in this country. But it is also important to note that this is an unprecedented historical moment in the history of the hemisphere and of its populations, for it is the first time that there has been a significant meeting of the various national populations of Latin America in one country, which, perhaps ironically, happens to be the United States.
3. The emphasis on ethnicity, and more particularly on ethnic labeling, is directly related to the distribution and withdrawal of resources and opportunities. Yet the establishment of these official categories has not significantly improved either the social and economic conditions or society's attitudes and perceptions toward people of Latin American descent. Indeed, according to

a recent news release by the National Council of La Raza, "Hispanics now have the highest poverty rate of any major ethnic or racial group in the U.S."—albeit still closely followed by African Americans.

This points to the contradictory role that labels are playing today. On one hand, these labels do allow us to track and compare poverty and illiteracy rates among racial groups, to measure the nation's progress toward what Johnson called "equality as a fact and as a result." On the other, the labels are not improving the social or economic conditions in which people live. Instead, the label "Hispanic" marks all Latinos as culturally and socially inferior, as having "bad values" that are perceived to be related to their "foreign"—un-American—origins.... Hence, as the case of Congressman Gutiérrez suggests, on the basis of their "un-American" cultural and linguistic difference, as well as of their racial markings, the label is in fact serving to locate all "Hispanics" as a group in a hierarchy in which ... inequalities are naturalized on the basis of racial, gendered, and cultural characteristics.

4. The label "Hispanic," like the categories "Asian American" and "African American," exemplifies the impact of globalization in "minoritizing" all populations from Asia, Africa, and Latin America in the United States. The minoritization of the Third World and the simultaneous emphasis on ethnic-group belonging rather than on citizenship (that is, Hispanic first, American second) has resulted in a variety of complex responses by both Latinos and non-Latinos to the growth of the Latino population in the United States. These are visible in the heated and often acrimonious debates on, and subsequent passage of, anti-immigrant, anti-affirmative-action, and anti-bilingual-education propositions in California and elsewhere, as well as in the proposals for similar bills in Congress. One of the consequences of this is that it now seems natural that the burden and responsibility of protecting both the human and the citizenship rights of individuals lie solely with the particular "ethnic group" to which they ostensibly belong, rather than with the national society as a whole, or with the state, for that matter....
5. Finally, it is important to note that the growth of the populations of Latin American descent in the United States and its racialization as a homogeneous "Hispanic ethnic group" are taking place in a larger global context, which I believe frames the entire debate on the ethnic identity, culture, and rights of Latinos. Clearly, the international context of this post–Berlin Wall decade has immersed all democracies in a process of expanding the scope of citizenship. Yet there has been relatively little, if any, sign of a significant and structured general debate within or among the older democracies about how to define the very notion of a collectivity—of a national citizenry—in the new global context. Instead, the historically inherited structures of citizenship rights—like the very political reality of citizenship itself—are being brought into question, with little effort made toward creating new international agreements and institutions (or at least reinforcing those that exist) in order to fully guarantee the human and political rights of the world's population....

CONCLUSION

Ethnic labels such as "Hispanic" allow us to identify a racial hierarchy that, now rationalized in essentializing "cultural" terms, accounts for the ongoing (in)visibility of people of Latin American descent. In so doing, it is reinforcing inequalities not only within the United States, but also—as a result of the consequent minoritization of the entire Third World—between the Western developed nations and the developing world. From this perspective, while the state-imposed categories increasingly undermine the possibility of constructing a community of equals, they simultaneously highlight the process by which the United States is moving in the direction of a rigid, class-based society—a society in which, as in Latin America, the lack of social mobility, like the concomitant widening gap between the rich and the poor, can be explicitly rationalized along ethnic and racial lines.

Ultimately, the persistence of racism and of the ongoing racialization practices in the United States and abroad has put us in a quandary. On the one hand, we are confronted with the question of the very viability of focusing the analysis of the concept of rights—whether we are referring to group or individual rights—exclusively within the old parameters of national boundaries. On the other, given the absence of legitimized international institutions that protect the human rights of all individuals, regardless of citizenship, we need to find new ways of reinforcing the institutions of citizenship, even while we simultaneously create new ways of safeguarding human rights in an increasingly transnational world.

REFERENCES

1. David Jackson and Paul de la Garza, "Rep. Gutiérrez Uncommon Target of a Too Common Slur," *Chicago Tribune*, April 18, 1996, 1.
2. I first read a summary of this story in Kevin R. Johnson's thought-provoking essay "Citizens as Foreigners," in Richard Delgado and Jean Stefancic, eds., *The Latino/a Condition: A Critical Reader* (New York: New York University Press, 1998), 198–201.
3. Meta Mendel-Reyes, *Reclaiming Democracy: The Sixties in Politics and Memory* (New York: Routledge, 1995); William H. Chafe, "The End of One Struggle, the Beginning of Another," in Charles W. Eagles, ed., *The Civil Rights Movement in America* (Jackson: University Press of Mississippi, 1986), pp.127–48.
4. Nancy Leys Stepan, *"The Hour of Eugenics": Race, Gender and Nation in Latin America* (New York: Cornell University Press, 1996), 171–96.

DISCUSSION QUESTIONS

1. What is the difference between biological racism and social racism?
2. What are the advantages and disadvantages of the national government's employing ethnic labels?

23

We Are All Suspects Now

TRAM NGUYEN

Nguyen studies a group of African immigrants (from Somalia) now living in Minneapolis. Using a case study of one such family, the article shows the global context for immigration—in this case, a context marked by war, refugee flight, and military intervention. This family's experience shows the struggles that many immigrants have in becoming established as citizens of the United States.

It was February 2003 in Minneapolis, and Abdullah Osman zipped his jacket as he crossed Cedar Street. His wife Sukra's brown eyes lit up as he entered their apartment, and his three-year-old daughter, Maria, in a pink sweatsuit with her braids bouncing, wrapped her limbs around his leg. Sukra asked if it was warm outside. He replied with a firm no, returning her smile with his eyes. "The sun must be lying then," she said, disappointment on her face.

Sukra, a petite woman in a patterned headscarf, grew up at the equator and was still adjusting to the Minnesota cold. After nearly a decade of separation, she and Abdullah were reunited less than five years ago in Minneapolis.

Unlike the stark exterior of their housing project, the inside of the Osmans' one-bedroom apartment burst with color and felt like an oasis. A Persian rug covered the linoleum living room floor. Throughout the tidy room lay Maria's toys—a purple and pink bike with training wheels, a large stuffed bear, other things that roll and squeak. Next to a computer desk stood an entertainment center, where a television was turned on to morning cartoons. A poster of the Kenyan city of Mombasa—a city that once provided its own refuge for Sukra and her family—hung above the doorway. The apartment's most distinguishing feature was an intricately patterned sectional foam sofa that circled the perimeter of the living room. "We have to have space for a lot of relatives," explained Sukra, her long headscarf draped over her upper body.

In 1999, Sukra, twenty-four, was able to join Abdullah in Minneapolis. She spoke and wrote Somali and Kiswahili, and became fluent in English after six months in the United States. She worked as an education specialist at a public elementary school, where she prepared lesson plans, translated materials for Somali students, and served as a liaison to Somali parents. Soft-spoken with a gentle smile, Sukra had earned respect within the school. "The Somali kids will

SOURCE: From *We are All Suspects Now* by Tram Nguyen. Copyright © 2005 Tram Nguyen and the Applied Research Center. Reprinted by permission of Beacon Press, Boston.

listen to me even more than their teachers," she said proudly, "especially when it comes to discipline."

Abdullah worked in construction in 1996, until a wrist injury forced him to seek less-physical work. For several years he was a bus driver for Minneapolis Public Schools, and supplemented his income by driving a taxi during evenings, week-ends, and school holidays. "When people first came, they mostly worked at the meat factory," he recalled. "Now there are more jobs. And the state helps people here if they are in need. We love Minnesota."

The Cedar-Riverside neighborhood where they lived is the heart of the largest Somali community in the United States. It's a neighborhood where cafés and organic food stores, several independent theaters, and music venues that cater to the local college crowd sit alongside money-transfer agencies, halal grocers, and a Somali mall with fabrics, furniture, and other East African products. On a Sunday morning, the streets and cafés were filled with men drinking spicy tea and conversing in Somali. Many read the local Somali newspaper, which contained news of current events in Somalia and local job postings. Tall, handsome, and sharply dressed in a sweater, dark blue jeans, and black leather shoes, Abdullah, thirty-three, walked between the snow-drifts and ice patches that covered the sidewalks. He greeted almost every passerby with a warm smile, typically accompanied by a hug or a handshake.

Toward the end of 1991 the State Department began resettling refugees from Somalia's civil war, and chose Minneapolis–St. Paul as one of several resettlement destinations. Community organizations and social service agencies now estimate the city's Somali population at 35,000, more than 70 percent of whom entered the country through the U.S. refugee resettlement program. Most others have been granted, or are seeking, asylum. With its robust local economy and liberal political culture, Minneapolis was a place where Somali refugees felt accepted and found opportunity. "If you want an education, you can get it," Abdullah marveled. His optimism was natural and infectious. "You can get money, food. You can raise your children, and you can find people to help them if you're having trouble. My child, Maria, is lucky. A lot of people don't have what I have. My brothers' children in Somalia aren't so lucky."

For Somalis, Minneapolis had been a safe haven, where they could reestablish family connections, and an economic base from which to support relatives abroad. But after September 11, the haven was becoming a more complicated place....

When Abdullah arrived in Minneapolis in 1996, there were 200 to 300 Somalis in the city. Now, more than twice that number live in the Osmans' building alone, and the community has spread throughout the suburbs and around the state.

"I moved into that building right away, and I've lived in the same apartment ever since," Abdullah said, signaling across the street to the Cedar-Riverside Plaza. Five concrete towers rise nearly forty stories above the Mississippi River and the campus of the University of Minnesota. From a distance they resemble the drab, institutional public housing projects built in inner cities during the 1950s and 1960s. Although the plaza's sixteen hundred units are occupied mostly by Somalis, local cab drivers refer to the complex as the United Nations towers because of the diversity of nationalities it contains. For thousands of new immigrants from all over the world, the plaza has represented the beginning of their pursuit of an American dream.

Abdullah's long-sleeved shirt hid most of the scars that covered his body, which he received during a decade of war, violence, and flight across borders and oceans. Until 1990 Abdullah lived in Mogadishu, a large industrialized Red Sea city that was Somalia's capital and a business and cultural hub for East Africa. His family was from the majority clan. "It was a lot like race and minorities in the U.S," according to Abdullah. While he and his brother, sister, and father enjoyed a middle-class lifestyle, other minority clans were more likely to be poor. Sukra was from a minority clan, and despite her feelings for Abdullah, a relationship between them was forbidden in their communities. As a teenager Abdullah helped Sukra's mother care for their family, which allowed him to spend time with her pretty daughter....

"Most people want to go back," said Ali Galaydh [a professor of international development], "especially the older people. My mom is from a nomadic clan, and before they would go to a new location, they would always send a scout. My mom was sitting on a rocking chair and looking out the window at the cold, and she said to me, who scouted this place? She has papers, but she wants to go back to our village in Somalia." While Galaydh had not abandoned hope, he was sober in his assessment of Somalia's prospects for the near future. "Some people pray for world peace," he said, "I pray to Almighty Allah that the weather in Minnesota would be more clement."

His sad eyes belied the academic detachment in his voice as he traced Somalia's route to civil war. "Because of the superpower rivalry, the U.S.S.R. wanted a foothold in the Horn of Africa. Somalia was a strategic location—the U.S. already had a presence in Ethiopia, and it was on the Red Sea. It was seen as the gateway to Africa."

The Soviet Union invested heavily in the armed forces of Somalia, and the military soon became the dominant force in the country. In 1969 the president was assassinated. The new leader, Siad Barre, dissolved the national assembly, banned political parties, and established a Supreme Revolutionary Council with the power to rule by decree.

"With Barre, Somalia fell into one-man rule," explained Galaydh. "There was disenchantment with a government that killed religious leaders, with the loss of democratic culture, and with corruption." Throughout the 1980s, discontent with Somalia's government intensified. Several armed movements formed, mostly operating from Ethiopia through hit-and-run tactics. The United States, meanwhile, continued to support Barre's regime, pouring hundreds of millions into arms in return for the use of military bases from which it could intervene in the Middle East.

"I then realized what was going on," Galaydh continued. "Siad Barre would find the resistance fighters, and he would not only punish them but also their families. He would kill their next of kin and destroy their property." Galaydh and several other ministers defected in 1982. When resistance fighters captured urban areas in the north in 1988 and 1989, Siad Barre responded with extreme force—including aerial bombings of civilian neighborhoods. "That's when people really started to leave. They fled to the Gulf States, Ethiopia, Djibouti, the U.K., the U.S., and Canada. That was before the total collapse."

In January of 1991 a civil uprising forced Barre to flee the capital. Anarchy ensued. In Mogadishu, armed militias vying for power launched artillery with little regard for civilian casualties.

While Galaydh was safe in the United States, Abdullah and Sukra were still in Mogadishu. "There were tanks rolling through the streets," Abdullah remembered. "We were just trying to survive. All of us were shot—my brother, sister, and my father." Abdullah pointed to a scar on his leg where a bullet passed through as they were fleeing. "There were no doctors, no hospitals. You could get bandages, but you had to treat the wounds yourselves. You would just wrap yourself up and try to continue. We ran so we wouldn't get killed." Abdullah's family escaped to a boat on the Red Sea and an uncertain future as refugees. "There were two to three hundred people packed onto a boat intended for a hundred and fifty. We only drank water for five days," said Abdullah.

After a seven-day journey at sea, their boat was the first to arrive at the undeveloped site of a United Nations refugee camp. It was a barren and remote location in the Kenyan desert, far from any population center. "We spent two weeks sleeping under a tree in hard rain before the UN officials arrived," Abdullah said. "When the UN came they counted us, then gave us tents and water tanks and two blankets a person." UN rations included corn, wheat, flour, and cooking oil. Occasionally they received kidney beans and a little sugar. While there were no cities nearby, neighboring farmers raised goats. "You would trade a pound of flour for a cup of goat's milk, and that's how you fed your children," Abdullah said.

Sukra's family remained in Somalia. "I can remember my dad saying, 'they just want to overthrow the government, then things will get better.' But all of your belongings could be stolen at any time. If you have girls, they get raped." Each new ruler was worse than the last, and soon Sukra's family had to flee Mogadishu. "They were bombing everything," she said. "We had to step over dead people, and sometimes we had to step on them, to get out." She raised her foot and pointed her toe down, as if she could still feel the flesh under her shoes. "There were empty houses everywhere, with only dead bodies inside."

Different armed factions occupied the roads that led out of the city. The factions were largely based on clanship, and often showed little mercy for people from other clans. "All of the clans have different dialects, and you had to try to guess which clan the men were from and try to speak in their dialect," Sukra remembered. "Maybe someone recognizes you, or may be they don't believe you. If so, you're probably going to be dead." Still not yet a teenager, Sukra watched her brother and her sister die as they tried to escape Mogadishu.

Sukra and her family traveled overland to Chisimayu, another coastal city hundreds of miles south. The war followed them. In 1992, after three months imprisoned in their homes by militia forces, they escaped to the Red Sea and boarded a boat for the Kenyan city of Mombasa. "They packed eighty people on a forty-person boat. We had no food, no personal belongings," Sukra said.

"The boat would stop every day, and then charge us another thousand Somali rupees to continue on." They arrived at a UN camp on the border between Kenya and Somalia. That camp became home for the next seven years.

By 2002, there were more than 300,000 Somali refugees living in UN camps in twelve countries around the world. Refugee camps in Kenya are by far the most extensive, hosting over 140,000 Somali refugees. Many camps issued each person three kilograms of maize every fifteen days, only eight hundred calories per day per person. Maize has to be cooked, which requires firewood. People had to leave the camps to get firewood, and they were often raped or robbed. Wheat and oil, which they were supposed to receive, were scarce. There were well-stocked markets in most camps, so fortunate people relied on money sent from abroad.

Most refugee camps in Kenya were connected by the *taar*, or telegram. But Abdullah and Sukra had to rely on mail or an occasional phone call, for which Kenyan Telecom would charge a significant fee. In the early 1990s, soon after arriving in Kenya, Abdullah's older brother had been granted refugee status and passage to the United States. In 1996 he was able to sponsor Abdullah, his younger brother, his older sister, and their father to come to Minnesota through the refugee resettlement program. When Abdullah arrived, he found a temporary construction job and moved into his apartment in the Cedar-Riverside Plaza.

Abdullah soon found permanent work cleaning rental cars for Avis at the Minneapolis airport, and began sending money not only to his family, but to Sukra and her mother as well. "Abdullah told me, 'I want you to learn English because you will need it when you come to America,'" Sukra recounted. "So I took English classes in the camp from some North Americans, for a small fee."

In 1999, Sukra was able to get support from family members abroad to finance her passage to the United States. She was nineteen, and it was the first time she had ever been out of sight of her family. She had no documents. "I was too afraid to even look up during the trip. I didn't want to be found out," Sukra said. "But you know you're going there to help your family, so you hold onto that." Sukra arrived in San Diego, California, where she applied for asylum. After five months Sukra's asylum petition was accepted, and she moved to Minnesota to meet Abdullah. They were married six months later.

"This is home to us. This is where we came, this is where our community is," Sukra said. She and Abdullah continued to send money back to Kenya to other relatives in refugee camps, including Abdullah's mother. In January of 2001, Sukra gave birth to Maria. "I was happy. I had my family, a job, money to pay rent, to buy clothes. We have families in Kenya and Somalia that need food, so we need to work," said Abdullah. "I felt like a man, like I had life, opportunity."

On Friday, June 16, 2001, Abdullah woke up at 10:00 A.M. to start his taxi shift. He found his first fare on Cedar Street, where he would drive a Somali man to a car repair shop three miles away. The man asked Abdullah to wait outside while he checked to see if the repairs were complete. While he waited, two men came to his window.

"One of them asked me to lend them twenty dollars," Abdullah recounted. "He said they would give me their address and I could come and pick up the money later. I told them my shift had just started so I didn't have any money." Abdullah had never met either of the men, but when they started knocking on his window, he decided it was time to leave.

"I thought they were just crazy," Abdullah said. "I didn't think they would attack me."

When Abdullah tried to get out of the car to get his passenger, one man slammed his door shut on him, then opened it and pulled him to the ground. The man hit him repeatedly. "He was much bigger and stronger then me. He had me pinned on my back and was punching me in the head," Abdullah said. "He hit me in the face, in my eyes and in my mouth." The man took the fifty dollars that were in Abdullah's shirt pocket to pay for the daily taxi rental. Then the assailant's keychain, attached to a razorblade, fell out of his pocket.

"We both saw it and reached for it, and I got it first. He was holding my wrist, but I shook it free and slashed at his arm," Abdullah said. When the razor cut the man's shoulder, he released Abdullah and ran down the street.

Three other men, who had been watching the attack, began to approach Abdullah. Frightened and in shock, Abdullah hurried to his cab and drove away. He realized that he should go back to the scene and contact the police. By the time he returned to the repair shop, the police were already there. "I was bleeding from my eyes, and from my arm," Abdullah remembered, displaying a six-inch scar between his elbow and wrist from when the man threw him to the ground. "But the people [at the scene] said I was the one who tried to kill somebody. The police threw me against the car and handcuffed me. They never asked me my side of the story."

Abdullah was held in jail for three days. He was not charged with a crime and was released. But two months later, while Abdullah was at work, the sheriff showed up at the Osmans' apartment door with a warrant for his arrest.

"I asked to see the warrant because I didn't believe it," recalled Sukra. "They said they didn't send notification because they thought he would flee. I said, from what? He didn't do anything wrong, so why would he flee? And where would he go? They searched the house, and even picked up the sofas to see if he was hiding."

After a call from the sheriff, Abdullah turned himself in that day. He remained in jail for eleven days until his family and friends were able to raise the $10,000 bail. His court date was set for April 2002. Abdullah hired a criminal-defense attorney based in downtown Minneapolis, who quickly learned that Abdullah's assailant had been found guilty of seven prior felonies, mostly for assault. "My attorney told me to sleep easy," Abdullah recalled. "I had never had any problems with the law. He told me that because I was working at the time, and it was self-defense, and that I was a family man, that the case would be easily dismissed." Abdullah went back to work, and life returned to normal for a brief month. Then came September 11....

On October 1, 2001, which was the start of the 2002 fiscal year, President Bush refused to sign the annual Presidential Determination that allows an allotted number of refugees to enter the United States. The virtual moratorium lasted until November 21, when Bush issued an order allowing 70,000 people to be resettled in the coming year, 10,000 less than the number set in 2001. Of this number, only 27,000 refugees were actually admitted. Five thousand Somali refugees had been admitted from October 2000 to October 2001. In the following year, less than 200 Somali refugees entered the United States. Asylum applications continue to face difficulty.

While identifying countries that might harbor terrorists, Secretary of State Colin Powell told the Senate Foreign Relations Committee at a 2002 hearing, "A country that immediately comes to mind is Somalia because it is quite a lawless place with out much of a government. Terrorism might find fertile ground there and we do not want that to happen." This focus on Somalia paralleled the restrictions on refugee acceptance and asylum. According to Craig Hope, director of the Episcopal Church's refugee resettlement office, "No one is going to say it's because of September 11, but that's the reason. They're being screened and watched carefully."

Soon after, Attorney General John Ashcroft's Justice Department launched a series of aggressive campaigns against immigrants. In Minnesota, the Somali community began to hear of arrests. Omar Jamal, a vocal Somali activist and director of the Somali Justice Advocacy Center in St. Paul, was accused of concealing his previous Canadian residency when he applied for asylum in 1997. Convicted in federal court a few years later, Jamal faced deportation. Another crucial blow to the community was the arrest of Mohamed Abshir Musse, a seventy-six-year-old former Somali general. His visa expired in September 2002 despite his application for renewal, and when he reported to the immigration office to comply with the new special registration program, officials ordered his deportation. Called by one former U.S. ambassador "the greatest living Somali," he has been credited for saving the lives of innumerable American soldiers during violent conflicts in the 1990s....

Like other immigrants, Abdullah and Sukra Osman felt the chilling effect of September 11 on their adopted city.

"I was very afraid. As a person who has been through war, you know there is going to be isolation, and people targeted," said Sukra. "And it happened here. A community leader was killed, a girl got beaten up, someone was thrown off a bridge. We couldn't go out at night. I was afraid to go on the freeway to my job, because I thought someone would see me and harass me. And it's not like Somalia, where you can blend in and escape. Here you are so different from everyone else, so you can't hide. You stand out."

There are issues of prejudice that have arisen since September 11, because of the generalization of Muslims." Somalis were even afraid to send money to relatives overseas. "People were worried," explained Noor. "If I send money, will I be targeted? Will I be linked to al-Qa'ida? Will I be targeted as a terrorist? People were afraid that the questioning process would be extended to them—that they might lose their refugee status and the benefits that come with it."...

September 11 affected the three most important things for Somalis in the United States—feeling safe in a place they can call home, being able to reunite with family members, and supporting relatives who live in poverty abroad. For Abdullah and many others, their life and liberty are also at stake.

DISCUSSION QUESTIONS

1. Nyugen's essay shows that racial profiling—indeed, the social construction of race—also occurs within the context of global politics. How do you see this, given what is reported here about the immigrant experiences of Somalis?
2. In what ways was Abdullah subjected to *racial profiling?* Under what conditions do groups become vulnerable to racial profiling?

Part V
Student Exercises

1. Voting is an important sign of citizenship, yet a very low percentage of voters actually vote. Voting among young people has been particularly low. Interview ten students from other classes that you are taking, and ask whether or not they vote and why. Then ask what their reasons are for participating or not in this act of citizenship. Having done so, explain why you think voter turnout among the youth is low. Is race a significant factor in youth voting?
2. Can you imagine ever making the decision to permanently relocate to another nation? What might make you do so? What rights would you expect in your new home? What would you do were you not fully accepted? Having imagined this, think about how this exercise might compare (or not) to the experiences of groups examined in this part.

PART VI

INTRODUCTION

Immigration, Race, and Ethnicity

The United States, it is said, is a nation of immigrants. Virtually everyone living here has an immigrant experience somewhere in their family's history—with the exception of indigenous Native Americans and African Americans whose ancestors were forced into slavery. At times, immigrants are celebrated in America. Other times, they are chastised and defined as "outsiders" or "aliens." Even our national symbols have sometimes shone, other times, lost their luster. Ellis Island, the point of entry for most immigrants coming into the East Coast, operated from 1892 until 1954. It was in disrepair for decades, but then refurbished and opened as a museum in 1990. Like Angel Island, the major entry point on the West Coast, these sites are now marketed as tourist destinations. The Statue of Liberty, also a tourist destination, is a national symbol, yet also viewed from a distance by New York residents as they ride the ferries that connect the islands that make up most of New York City. The story of immigration to the various shores is a complicated one because not all arrivals have been treated equally. The legacy of immigration shapes how we think about our borders and how we define ourselves as a nation.

Patterns of immigration are global trends—people are pushed and pulled between nations for economic, political, and religious reasons. Once independent of England, the United States found immigrants arriving from many different shores. Germans and other Europeans who entered the United States after the War of 1812 found the Native White population a little distant, but most of the immigrants were able to settle in rural and opening sections of the country like Ohio and Illinois, where they could replicate the lives they left in Europe. Although these newcomers were foreign to native-born White Americans, they were mostly farmers and craftspeople whose way of life did not threaten much of the established population. Over time, these residents were considered Americans.

The Irish, the majority of whom settled in cities in the 1820s to 1850s, found there were few jobs and faced harsh discrimination. Irish immigrant men worked in the harbors and traveled to work on canals and other industrial sites, while women did domestic work. With little money and few work opportunities, these immigrants clustered in urban centers like New York and Boston but eventually found a niche in society, gaining citizenship and playing a key role in urban and state politics. By the time new immigrant groups came from eastern and southern Europe during the development of our industrial economy (1880s to 1920s), the Irish were a group with some degree of political power in some northeastern cities.

As industrialization accelerated in the early twentieth century, new immigrants from eastern and southern Europe entered the nation but were defined as racially inferior. Alarmed at the prospect of downgrading the "stock" of White Americans, people took actions that reflected their beliefs about racial hierarchies. Immigrants from southern and eastern Europe participated in expanding industries (e.g., steel, oil, glass, and railroads) that transformed the nation. Dramatic changes were welcomed by big businesses and growing corporations, but new immigrants were seen as a threat to many native-born White Americans, who used their political power to shape who could become future citizens.

The U.S. Congress has the right to regulate immigration as part of its mandate to govern foreign affairs (Ngai 2003). These regulations reflect economic and political power, as well as pressure from voting citizens, who are often conflicted about the growing presence of "foreign" people. In 1882, Congress passed the Chinese Exclusion Act that limited Chinese immigrants to merchants, thus excluding the working class. This legislation marks "the moment when the golden doorway of admissions to the United States began to narrow and initiated a thirty-nine-year period of successive exclusions of certain kinds of immigrants, 1882–1921, followed by twenty-two years, 1921–1943, when statutes and administrative actions set narrowing numerical limits for those immigrants who had not otherwise been excluded" (Daniels 2004:3). Not only were native-born White people uneasy with Asian immigrants but immigrants from eastern and southern Europe were also suspect.

Immigrant labor was essential to the industrial owners whose factories made the United States a manufacturing power. Yet these newcomers were perceived as racially different from the native-born White Americans. Their participation in industrial work changed the nation from a land of independent farmers and craftsmen to a nation state where economic power was concentrated in the hands of a few robber barons. Many native-born White Americans wanted to limit immigration as a means of holding on to their established ways. Industrial employers, on the other hand, wanted to keep the borders open so that they would always be able to hire people at low wages, replacing them with newer immigrants when their current workers wanted higher wages. National leaders, including presidents, sided with the big industries because such businesses were important to expanding the nation's economic and political power. Over time, regulation became the rule with immigration inspectors, as part of the Immigration and Naturalization Service (INS), admitting some and rejecting others, depending on the nation's labor needs and relations with other countries.

The Johnson-Reed Act in 1924 limited immigration to 150,000 people per year and mandated a quota system based on the nation of origin. This Act also barred the admission of people who were deemed ineligible for citizenship, thus ending immigration for most Asians. The 1924 law, with many modifications, remained the framework of U.S. immigration policy until 1965. By establishing national origin quotas, powerful groups insured that people from Northern and Western Europe could continue to enter the United States. Thus, because they held citizenship rights, Irish politicians got a higher quota for those from Ireland. Meanwhile, the quotas for Italians, Polish, and those from eastern European nations were lower than the numbers entering prior to 1920. Still, all European immigrant groups had more advantages than African Americans, who were already in the United States, and were highly restricted in their social and economic options.

In 1965, the Hart-Celler Act opened the door to immigrants, offering the same quota to all eligible nations. Many Italian and Eastern European families were reunited, and other groups had new access to this democratic nation. By the end of the twentieth century, the majority of immigrants were coming from Latin American, Asia, and the Caribbean. They entered a political economy that was changing—this time from a manufacturing-based economy to one focused on service work—both highly skilled service work (such as doctors, financial planners, and college professors) and less-skilled, low-wage service work (such as janitors, domestic workers, and fast-food workers). Workers are necessary for these tasks and issues of citizenship and race have emerged as key to economic and civic opportunities. Many of the new immigrants are *transnationals*, meaning that parts of their families live in their nation of origin and that they maintain close ties with their homelands.

After decades of limited immigration, new policies and a new economy have made the United States a major destination for immigrants from around the globe. Our laws have changed old practices, but introduced new dilemmas. In this part, we explore various issues with regard to immigration and the experiences of immigrants.

Mae M. Ngai ("Impossible Subjects: Illegal Aliens and the Making of Modern America") looks at how new legislation has created opportunities to be part of global changes as people move to increase their economic options. She explains that recent laws have created new categories of "illegal aliens" as well as militarizing our borders, particularly those with Mexico. Now, the nation is debating what to do with millions of undocumented workers. Ngai's historical perspective puts this debate in context. She also discusses the implications of recent changes, such as the abolition of the Immigration and Naturalization Service (INS) and its absorption into the Department of Homeland Security.

Nancy Foner ("From Ellis Island to JFK: Education in New York's Two Great Waves of Immigration") compares the pursuit of education by earlier immigrants (1880–1920) with those coming to New York City since the 1965 law. The majority of new immigrants, from Latin American, Asia, and the Caribbean, are often compared to the earlier waves of Europeans entering during industrialization. Foner clarifies how the economic progress of earlier, uneducated immigrants translated into educational advantages for their children. Immigrants entering the country in a postindustrial economy are now arriving when higher education is essential for economic security. Many parents make sacrifices so that their children gain those credentials, a trend that promotes their children's incorporation into the mainstream.

New immigrants share much with older ones, but new communication technologies also enable people to remain closely connected to their home nations. Peggy Levitt ("Salsa and Ketchup: Transnational Migrants Straddle Two Worlds") shows how families negotiate their place in communities when their ethnic identity makes them part of two worlds—the community in which they live and their community of origin. Her study of immigrants from the Dominican Republic and Gujarat (a state on the west coast of India) demonstrates how they create community, even while maintaining strong ties to their initial homelands. Levitt describes the adaptations that families have to make in the context of racial and ethnic stratification.

While some immigrants travel alone, many bring their children or give birth here, thus making the children citizens. One in every five children in our nation has either one or two immigrant parents. Margie K. Shields and Richard E. Behrman ("Children of Immigrant Families: Analysis and Recommendations") detail the diversity among these children. Some children have educated parents who enter legally to work in various professional and technical positions. However, others have less educated parents and face different challenges. Depending upon their specific backgrounds and immigration status, the transition to the United States can be very difficult—and its impact can affect the success or failure of future generations.

We see some of the adaptations to the racial context of the United States in Johanna Lessinger's study of Indian immigrants ("Class, Race and Success: Two Generations of Indian Americans Confront the American Dream"). Strong attachments to their homeland lead many such immigrants to reject or ignore the racial dynamics they find in the United States. However, their children, more often challenged with questions of their racial identity as they grow up, are more likely to see themselves as part of a South Asian diaspora. This perspective changes how the younger generation confronts the racial discrimination in their lives.

At this writing, the United States Congress has not passed new legislation about immigration, but it will likely do so in the future. Together, the articles in this part introduce you to how issues of the past inform the debates about immigration today.

REFERENCES

Daniels, Roger 2004. *Guarding the Golden Door.* New York: Hill and Wang.

Ngai, Mae M. 2003. *Impossible Subjects: Illegal Aliens and the Making of Modern America.* Princeton, NJ: Princeton University Press.

24

Impossible Subjects: Illegal Aliens and the Making of Modern America

MAE M. NGAI

Ngai reviews the changes in immigration patterns in the United States in recent years and also documents the changes in policy that have become more restrictive and punishing, even while serving the needs of the contemporary labor market. She also assesses alternatives that would have a less punitive effect on immigrants and their families.

If the Johnson-Reed Act ushered in the most restrictionist era in American immigration law, the Hart-Celler Act, which ended that period, altered and refined but in no way overturned the regime of restriction. Certainly, patterns of immigration changed dramatically in the period after 1965, as the abolition of quotas based on national origin opened the way for increased immigration from the third world. Yet Hart-Celler's continued commitment to numerical restriction, especially its imposition of quotas on Western Hemisphere countries, ensured that illegal immigration would continue and, in fact, increase. During the late twentieth century, illegal immigration became perceived as the central and singularly intractable problem of immigration policy and became a lightning rod in domestic national politics generally. Moreover, legal reform in the area of aliens' rights of due process was decidedly uneven over the past thirty years. Together, these trends have contributed to yet another ethno-racial remapping of the nation.

Notwithstanding the Hart-Celler Act's intentions to keep migration at modest levels, legal immigration into the United States has climbed steadily since 1965. Immigration rose sharply after 1990, when Congress raised the numerical ceiling on immigration by 35 percent in response to the 1980s boom in the U.S. economy and concomitant demands for labor in low-wage sectors and in some high ones as well. By the mid-1990s immigration approached one million a year. Refugee admissions also increased in the mid-and late 1970s from Southeast Asia in the wake of the Vietnam War and in the early 1990s from Russia and former Soviet-bloc countries. Just as important, if not more so, since 1980 Europeans have accounted for only 10 percent of annual legal admissions; Mexico and Caribbean nations account for half the new immigrants and Asia for 40 percent.

SOURCE: From *Impossible Subjects: Illegal Aliens and the Making of Modern America.* (Princeton: Princeton University Press, 2004).

Paralleling the increase in legal admission, illegal immigration also increased dramatically after 1965.... Unauthorized migration increased radically in the late 1960s and mid-1970s as a direct result of the imposition of quotas on Western Hemisphere countries, especially Mexico. New illegal-migrant streams also emerged, comprising undocumented migrants displaced from war-torn countries in Central America and from the instabilities of economic transition in China. During the 1990s removals increased 40 percent from the previous decade; by the turn of the twenty-first century the INS was removing some 1.8 million illegal aliens a year. Yet the INS estimates that the illegal population still accretes by 275,000 persons a year.

During the 1980s and 1990s Americans manifested a schizophrenic attitude towards illegal immigration. On the one hand, undocumented workers provided labor for low-wage agricultural, manufacturing, and service industries, as well as in the informal economy of domestic work, housing construction, and other services that support the lifestyle of not just the prosperous but many middle-class Americans. On the other hand, critics worried that the nation had lost control of its borders and was being overrun by undesirable illegal aliens; estimates of the illegal population in the early 1980s ranged from 2 to 8 million. After much contentious debate, Congress passed the Immigration Reform and Control Act of 1986. IRCA provided amnesty for some 2.7 million undocumented immigrants and sought, unsuccessfully, to curb future illegal entries by imposing sanctions against employers who knowingly hire illegal workers and by greater enforcement efforts.

A stunning militarization of the U.S.-Mexico border was accomplished during the 1990s. Congress authorized a doubling of the Border Patrol's force, the erection of fences and walls, and the deployment of all manner of high-tech surveillance on land and by air. Enforcement at the southwest border now costs taxpayers $2 billion a year. The militarization of the border has not stopped illegal entry, but it has made it more difficult and dangerous. The INS's Operation Gatekeeper, initiated in 1994, pushed illegal entries from the San Diego–Tijuana border area to remote desert sections of California and Arizona, which has increased the likelihood that migrants will die of exposure before reaching safety. Others succumb as a result of desperate attempts to enter in the sealed holds of ships, trucks, and boxcars. Many pay thousands of dollars to smugglers or guides to make these dangerous journeys, and those who survive the trip then often remain indentured for years in sweatshops before they can pay their debts. Illegal aliens are at once familiar and invisible to middle-class Americans: their labor is desired but the difficulties of their lives for the most part go unnoticed. The INS now estimates there are at least 5 million undocumented immigrants residing in the United States—of which over half are from Mexico—and debate rages on about what to do about it.

All told, net annual immigration (legal and illegal immigration less emigration) over the last decade averaged about 700,000. At less than one-half of 1 percent of the total U.S. population that is not a terribly large number, but the rate of increase and the shift to third-world immigration have had considerable consequence. The foreign-born now account for nearly 10 percent of the population, which is much greater than in 1970, when they were less than 5 percent, but still less than the historical high of 14 percent before World War I. And, while concentrated in six states, immigrants live and work in every region of the country and are thus a visible presence throughout the United States.

The new demographics have both enhanced the politics of diversity and multiculturalism and provoked nativist sentiment and campaigns, like California's Proposition 187 and the English-only movement, suggesting that Asians and Latinos continue to be constructed as foreign racial others, even as their positions in the American racial landscape have changed. In the last quarter of the twentieth century, Latinos and Latinas and Asian Americans were the fastest growing ethno-racial groups in the United States. The 2000 Census counted 32.8 million Latinos and Latinas in the United States; these diverse communities—Mexican, Cuban, Puerto Rican, Dominican, Salvadoran—collectively account for 12 percent of the total U.S. population. Because Mexicans and Central Americans make up the vast majority of undocumented immigrants today, they experience persistently high levels of poverty and disfranchisement and embody the stereotypical illegal alien. The growth of a dynamic transborder economy and culture in the Southwest has engendered new, hybrid cultures and identities, complicating ideas about national difference and belonging. At the same time, some Latinos and Latinas have achieved a measure of structural assimilation into the mainstream of American society. The business and professional classes (particularly among Cuban Americans and Mexican Americans) have garnered considerable political influence, particularly in Texas and Florida; and Latinos and Latinas such as Ricky Martin, Jennifer Lopez, and Cameron Diaz have broken into the top echelons of the entertainment industry.

The end of Asiatic exclusion, the post-1965 influx of Asian immigrant professionals and, more recently, the arrival of wealthy elites from Hong Kong, South Korea, and Singapore have repositioned Asian Americans as "model minorities." While Asian Americans have in many ways overcome the status of alien citizenship that defined them during exclusion era—through access to naturalized citizenship and occupational and residential mobility—the model minority stereotype elides the existence of large numbers of working-class immigrants, undocumented workers, and refugees. Moreover, the model minority idea reproduces Asian Americans' foreignness. Critics contend it is a new, pernicious form of "yellow peril" that rests on essentialized notions of Asian culture and breeds new forms of race discrimination (occupational segregation and glass ceilings, reverse quotas against Asians in college admissions).

The growth in the size of Asian and Latino and Latina populations has made it manifestly clear that the question of race in the United States is not just about the status of African Americans or black-white relations. Yet, in an important sense, contemporary multiracial politics remain informed by the historically dominant black-white paradigm, and are also implicated in reproducing it. For example, for Afro-Caribbean and African immigrants, foreignness becomes subsumed by racial "blackness," as these immigrants join, variegate, and complicate established African American communities.

Social scientists, journalists, and politicians have used both Asian and Latino and Latina immigrant success narratives to discipline African Americans. Using cultural stereotypes to occlude economic structures, the strategy deploys immigrant exemplars of hard work, thrift, and self-reliance against an alleged "culture of poverty" among native-born minorities (African Americans as well as Puerto Ricans) to explain the persistence of unemployment and poverty among the latter. In a twist to this argument, some liberals blame immigrants for undercutting native-born workers' wages and for displacing African Americans from jobs in the lower strata of the workforce. While economists debate the costs and benefits of immigrant labor to the economy, the more troubling aspect of the new liberal restrictionism is that it posits minority groups in a zero-sum calculus and regards immigrants as outsiders whose claims on society are less deserving than those of citizens.

...What do these trends portend for the central figure of this study, the illegal alien? As in the mid-twentieth century, illegal immigration today results from the confluence of two conditions: macroeconomic structures that push and pull migration from developing countries to low-wage sectors of the United States, and positive domestic law, which sets qualitative terms and quantitative limits on immigration. The first dynamic is also what makes enforcement of the second difficult and is why enforcement alone has never been a solution for illegal immigration.

It may be that illegal immigration will persist as long as the world remains divided into sovereign nation-states and as long as there remains an unequal distribution of wealth among them. Yet, this is not to say that no solutions exist short of open borders, on the one hand, or a totalitarian police state, on the other. We might consider, instead, strategies aimed at altering the push-and-pull dynamics of migration from the developing world to the United States. Trade and investment policies that strengthen the economies of sending nations would lessen pressures on emigration. Raising the numerical ceiling on legal migration, reestablishing a statute of limitations on deportation, enforcing wage and hour standards, and facilitating collective bargaining for workers in agriculture and low-wage industries would counter the reproduction of undocumented workers as an exploited underclass.

Amnesty for undocumented immigrants has diverse political support, including from the presidents of the United States and Mexico, business and organized labor, and human rights activists. Amnesty is a humane gesture that recognizes the claims to legitimate membership that come with settlement. Still, it is a limited reform that addresses only past illegal migration. Some advocates of amnesty (now called "regularization") propose to curb future unlawful entries by channeling would-be illegal migrants into legal, temporary guest-worker programs. But guest-worker programs pose the moral problem of creating a caste of second-class persons that we exploit economically but deny full membership to in the polity. Historical experience also suggests that guest workers do not all quietly return home when their services are no longer needed. Many remain, and in so doing become illegal aliens.

Some legal scholars propose eliminating or vastly lowering restrictions on migration from Mexico and Canada, a plan that has antecedents in the pre-1965 pol- of Pan-Americanism. They argue for extending the North American Free Agreement, which has already lowered national barriers in the hemisphere to ease the flow of capital and products, to migration, citing the European Union as a more comprehensive model of integrated regionalism. In fact, some critics believe NAFTA's one-sided emphasis on free trade will lead to greater pressures on illegal migration. Some observers fear that the elimination of tariffs on American agricultural products entering Mexico in 2003 threatens to "crush the ability of millions of Mexican farmers to survive and drive them north."

Indeed, while the hardened nationalism that characterized the world order for so much of the last century has dramatically relaxed in many realms of economic and cultural exchange, it remains in force in other fields, notably politics. Nation-states remain committed to restrictive immigration policies, so that states' regulation of the transnational movement of bodies differs greatly from that of capital, goods, and information. The principle of universal human rights—that there is a moral code higher than positive domestic law—has grown in prominence over the last half-century and has been key to recognizing the claims of refugees and asylum-seekers, particularly in a number of European states. Yet human rights as an international or multinational legal regime carries little authority over sovereign nation-states. As Seyla Benhabib wrote, "There are still no global courts of justice with the jurisdiction to punish sovereign states for the way that they treat refugees, foreigners, and aliens." In fact, notwithstanding the European Convention on Human Rights, the European Union appears to be shifting emphasis to joint border operations against illegal migrants and unwanted asylum seekers.

Thus, even as migration patterns change according to new global conditions, they remain shaped by asymmetrical relations of economic and political power between nation-states. The endurance of "crustacean" immigration policies may not, in fact, be eccentric to twenty-first-century globalization but a constituent element of it, aimed at maintaining the privileged position of the most powerful countries. Transnational migrants today move about in a very different world than that of the mid-twentieth century, navigating faster and more dangerous circuits in pursuit of work and safety, but their journeys reprise past crossings in important ways. They travel with the ghosts of migrants past as they, too, traverse boundaries, negotiate with states over the terms of their inclusion, and alter the future histories of nations and the history of our world.

DISCUSSION QUESTIONS

1. In what ways has U.S. immigration policy moved from one of accommodation to punishment? Given Ngai's argument, why would you say this has occurred?
2. What are the macro-sociological conditions (that is, large-scale changes) that frame the current debates about immigration?

25

From Ellis Island to JFK

Education in New York's Two Great Waves of Immigration

NANCY FONER

Foner compares the experiences of two groups of immigrants—those who entered the United States in the first wave of mass immigration (1880–1920) and those who are contemporary. She shows how the specific economic and social conditions that groups experience at different points in time shape their opportunities and achievements, most notably, in the case she examines, in education.

New York City is in the midst of a profound transformation as a result of the massive immigration of the last four decades. More than two and a half million immigrants have arrived since 1965, mainly from Latin America, Asia, and the Caribbean, and they are now streaming in at a rate of over 100,000 a year. Immigrants already constitute over a third of the city's population. In the midst of these dramatic changes, commentators and analysts, popular and academic, in the press and in the journals, are comparing the new immigration with the old.

This is not surprising. Few events loom larger in the history of New York City than the wave of immigration that peaked in the first decade of the 20th century. Between 1880 and 1920, close to a million-and-a-half immigrants arrived and settled in the city, so that by 1910, fully 41 percent of all New Yorkers were foreign-born. The immigrants, mostly Eastern European Jews and southern Italians, left an indelible imprint on the city—indeed, a large and influential part of New York's current citizens are their descendants.

An elaborate mythology has grown around immigration of a century ago, and perceptions of that earlier migration deeply color how the newest wave is seen. For many present-day New Yorkers, their Jewish and Italian immigrant forebears have become folk heroes of a sort—and represent a baseline against which current arrivals are compared and, unfortunately, often fail to measure up.

Nowhere is the nostalgia for the past more apparent than when it comes to education. Sentimental notions about Eastern European Jewish immigrants' love affair with education and their zeal for the life of the mind have become part of our picture of the "world of our fathers." Jews are remembered as the "people of

SOURCE: From *Brandeis Review* 21(2): 32–37, 2001. Reprinted by permission of the author.

the book" who embraced learning on their climb up the social ladder. These memories set up expectations about what immigrants can and should achieve in the schools. If my grandparents and great-grandparents could succeed in New York City's schools a hundred years ago, without special programs to help them adjust, why—many people say—can't today's immigrants and their children do well when they get so much more assistance?

A comparison of New York's two great waves of immigration shows that inspirational tales about Eastern European Jews' rise through education and their success in New York's schools in the so-called golden immigrant age do not stand up against the hard realities of the time. Today, despite a dramatically different context and significant problems in the schools, many immigrant children are doing remarkably well.

In the years before World War I, most Eastern European Jews did not make the leap from poverty into the middle class through education. Those who made substantial progress up the occupational ladder in this period generally did so through businesses in the garment, fur, shoe, and retail trades and in real estate. It was only in later decades that large numbers of Eastern European Jewish children used secondary and higher education as a means of advancement.

A hundred years ago, most Jewish immigrant children left school with, at best, an eighth grade education; few went to high school, and even fewer graduated. In the first decade of the 20th century, well below five percent of the Russian Jewish children in New York City graduated from high school; less than one percent of Russian Jewish young people of college age ever reached the first year of college. By 1908, the City College of New York (CCNY) had already become a largely Jewish school, but Jewish undergraduates there and at other New York colleges were a select few. Only a tiny number graduated. In 1913, City College's entire graduating class had only 209 students, less than 25 of them Eastern European Jews. The 25 percent of Hunter College graduates who were Eastern European Jews in 1916 amounted to only 58 women. This was at a time when the Jewish population of New York City was almost a million! It was not until the 1930s that there were big graduating classes at City College that contained large numbers of Jews of Russian and Polish origin.

One reason so few Russian Jewish students went to high school or college is that there weren't many high schools or colleges at the time. In 1911, after a decade of expansion, the city had 19 high schools, but the high school student body was still only about a quarter of the size of the four preceding elementary school grades. CCNY and Hunter, the only two public colleges, had about 1,400 students in 1908. A high school degree wasn't necessary for the jobs employing most New Yorkers, and an eighth grade graduate could even get a white-collar job. A business career didn't require four years of college nor did teaching or the law. In any case, extended schooling was a luxury beyond the means of most new immigrant families, who needed their children's contributions to the family income. Even those who managed it rarely saw all their children go to high school.

The elementary school also did a poor job of educating immigrant children so that many were not prepared or motivated to continue on. The schools were severely overcrowded in the wake of the huge immigrant influx and the inability of school construction to keep up with demand. By 1914, enrollments had grown to almost 800,000, more than triple the figure for 1881. Educators of the time joked that teachers should have prior experience in a sardine factory before being hired to work in the New York schools.

There's a nostalgia for the "sink or swim" approach to learning English, but unfortunately, many in the past "sank" rather than "swam." Most non-English speaking children were placed in the lowest grade regardless of their age. (The special "steamer" classes introduced in 1904—in which students were totally immersed in English for a few months—catered to a mere fraction of students needing them, only 1,700 students throughout the city in 1908.) Teachers made promotion decisions, and many children who could not do the work were left back. In 1908, over a third of the Russian Jewish elementary school pupils in New York City were two or more years over age for their grade.

If Russian Jewish children in this early period were not the education exemplars often remembered, they did do much better than Italians, the other "greenhorns" at the time. This favorable comparison helps explain why Jewish academic achievements have stood out and received so much attention. Russian Jewish students' progress, however, was fairly similar to that of native white New Yorkers at the elementary school level—and they did less well than native whites in making it to high school and college.

It's difficult to compare immigrants' educational achievements today with those in the past because the context in which education is a path to mobility is so radically different: formal education, and a more extended education, is now more important in getting a job owing to educational upgrading and transformations in the world of work. In the last great immigration wave, high schools were just becoming mass institutions. Today, getting a high school diploma is the norm, achieved by more than 70 percent of New York City's adult population and essential for many low-skilled positions. College is no longer an institution for a tiny elite—in 1990, a quarter of New Yorkers 25 and older had a college degree. Today, college graduates compete for jobs that immigrants with a high school diploma could have obtained a century ago, and a college degree—or more—is required for the growing number of professional, technical, and managerial positions.

How are immigrant children doing today? Admittedly, there are many dropouts and failures—something more serious now when more education is needed to get a decent job. Many, perhaps most, immigrants attend New York City schools where student skill levels are low, dropout rates high, and attendance rates poor. Once again, the surge of immigrants has led to soaring public school enrollments—now over the million mark—and serious overcrowding. New language programs are inadequate to meet the enormous need. Whereas a hundred years ago, New York schools mainly had to cope with Yiddish- and Italian-speaking children, today they confront a bewildering array of languages; a Board of Education count indicates that more than 100 languages are spoken by students from over 200 countries. When high schools were institutions for a minority of the better and more motivated students, violence, crime, and student indifference—and hostility—were not issues the way they are today.

Despite these problems, a substantial number of students are making it, and some are doing exceptionally well. There are new educational opportunities: expanded, and more widely available, higher education; a host of new programs for teaching students English; and even special immigrant schools designed specifically for newcomer children. Perhaps most important, large numbers of immigrant children have highly educated parents, and some immigrant children themselves have previous experience in fine schools in their home country. This has translated into academic success for many newcomers.

Although data on immigrant students in New York City are woefully inadequate—the only Board of Education data on immigrants refer to "recent immigrant students" who have entered as U.S. school system for the first time in the past three years—they show immigrants comparing favorably with other students in several ways. Students who were recent immigrants to the public school system in middle school, graduate from high school on time by a slightly greater percentage than their native-born peers. They also have lower dropout rates. Although recent immigrants' median test scores in math, reading, and English are somewhat lower than those of other students, they improved their scores between 1989-1990 and 1990-91 more than the rest of the student body. At the City University of New York, immigrants make up a high, and growing, proportion of the student body: in 1998, 48 percent of CUNY's freshman class were foreign-born.

National studies based on large representative samples show immigrants often outperforming their native-born peers. In a study of eighth graders, children of immigrants—those born abroad and in the United States—earned higher grades and math scores than children of native-born parents even after the effects of race, ethnicity, and parental socioeconomic status were held constant. Another national study, which interviewed more than 21,000 10th and 12th graders in 1980 and followed them over a six-year period, compared immigrants and the native born in the aggregate as well as immigrants and the native born in four different ethnoracial groups (Asian, white, black, and Hispanic). Whichever way they were compared, immigrants were more likely to follow an academic track in high school than their native-born counterparts; once graduated, immigrants were also more likely to enroll in postsecondary education, to attend college, and to stay continuously through four years of college.

Like Jews of an earlier era, today's educational exemplars are Asians; white European immigrants are also doing comparatively well. Most striking is Asian (native as well as foreign born) overrepresentation in New York City's elite public high schools that select students on the basis of notoriously difficult entrance exams. In 1995, an astounding half of the students at the most selective high school of all, Stuyvesant, were Asian; at the Bronx High School of Science, 40 percent were Asian, and at Brooklyn Technical High School, 33 percent. This is at a time when Asians were 10 percent of the city's high school population. So many Korean students now attend Horace Mann School—one of New York's most competitive private secondary schools—that there is a Korean parents' group there.

Just how important is Asians' culture in accounting for their educational achievements? As among Jews in the past, it plays a role. But I would argue that social-class factors, then as well as now, are more important. And today, race must also be considered.

A major reason Eastern European Jews did so much better academically than Italians in the old days was their occupational head start. Jews were more urban and arrived with higher levels of vocational skills, which gave them a leg up in entering New York's economy. Because the Jewish immigrant population was, from the start, better off economically than the Italian, Jewish parents could afford to keep their children in school more regularly and for longer. The poorer, less skilled Italians were more in need of their children's labor to help in the family. That Jewish children were more likely to have literate parents was also a help; the children themselves often arrived with a reading and writing knowledge of one language, making it easier to learn to read and write English than it was for southern Italian immigrant children, who generally arrived with no such skills.

Today, educational background plays a much larger role in explaining why the children of Asian immigrants are doing so well. Relatively high proportions of Asian students have highly educated parents. Although they often experience downward occupational mobility in New York, highly educated parents have higher educational expectations for their children and provide family environments more conducive to educational attainment. If their children started school in the home country, they typically attended excellent—and rigorous—institutions. Well-educated parents, moreover, are usually more sophisticated about the way the American educational system works and have an easier time, and more confidence in, navigating its complexities—and steering their children into good schools—than those with less education. In large part because so many come from professional and middle-class backgrounds, Asian New Yorkers are also doing fairly well economically. A century ago, economic resources were important because they allowed immigrants to keep their children in school; now they make it possible (or easier) to send children to private schools or to move to areas in the city (and suburbs) with better schools.

As for culture, the high value Jews placed on education—and the fact that southern Italians' heritage made them less oriented to and more skeptical about the value of book learning—helps account for the different educational achievements of the two groups. Today, most immigrant parents, in all groups, arrive with positive attitudes to education and high educational expectations for their children. Asian immigrants have particularly high aspirations for their children, though it's hard to say how much these aspirations are due to the cultural values and resources they bring to America as opposed to social-class advantages. That several national studies show Asian students outperforming all other racial/ethnic groups even after taking into account family income, household composition, and parental education, strongly suggests that culture is a factor.

Among Chinese immigrant parents, for example, hard work and discipline, not innate intelligence, are the keys to educational success. If their children study longs hours, parents believe, they can get As, and they put intense pressure on their children to excel. Confucian teaching, it is said, emphasizes discipline, family unity, and obedience to authority, all of which contribute to academic success. Children who do poorly in school bring shame to Chinese families; those who do well bring honor. According to sociologist Min Zhou, Chinese immigrant parents denounce consumption of name-brand clothes and other "too American" luxuries, but do not hesitate to pay for books, after-school programs, Chinese lessons, private tutors, music lessons, and other educationally oriented activities. Chinese and Korean immigrants also have imported after-school institutions that prepare their children for high-school admissions and college-entry exams. In Chinatown, "school after school" has, according to Zhou, become an accepted norm. According to one survey, a fifth of Korean junior and senior high school students in New York City were taking lessons after school, either in a private institution or with a private tutor.

Finally, there is the role of race. At the turn of the 20th century, race was irrelevant in explaining why Jews did better academically than Italians. Both groups were at the bottom of the city's ethnic pecking order, considered to be inferior white races. Today, the way Asian, as opposed to black and Hispanic, immigrants fit into the racial hierarchy makes a difference in the opportunities they can provide their children. Because they are not black, Asian (and white) immigrants have greater freedom in where they can live and, in turn, send their children to schools. Asians have been able to move into heavily white neighborhoods with good schools fairly easily. Moreover, their children are less likely than black or Hispanic immigrants to feel an allegiance with native minorities and be drawn into an oppositional peer culture that emphasizes racial solidarity and opposition to school rules and authorities and sees doing well academically as "acting white."

What, then, in a broad sense, can be learned from this comparison? The remembered past is clearly not the same thing as what actually transpired, and it is wrong to judge today's immigrants by a set of myths rather than actual realities. We place an added burden on the newest arrivals if we expect them to live up to a set of folk heroes and heroines from a mythical golden age of immigration. As New York, and indeed the nation as a whole, continues to be transformed by the current wave of immigration, this is something important to keep in mind.

DISCUSSION QUESTIONS

1. List two of the major differences that Foner cites comparing immigrants entering New York City at the end of the nineteenth century and those entering now.
2. What factors account for the difference in educational success for Jewish and Italian immigrants to New York early in the twentieth century?

26

Salsa and Ketchup:

Transnational Migrants Straddle Two Worlds

PEGGY LEVITT

This study of two immigrant communities in Boston shows how immigrants form communities that sustain their ethnic identity. Even though the backgrounds of immigrants may be diverse, the formation of community ties also fosters a continuing link to their homes of origin, creating transnational identities that give them a dual identity.

The suburb, with its expensive homes with neatly trimmed lawns and sport-utility-vehicles, seems like any other well-to-do American community. But the mailboxes reveal a difference: almost all are labeled "Patel" or "Bhagat." Over the past two decades, these families moved from the small towns and villages of Gujarat State on the west coast of India, first to rental apartments in northeastern Massachusetts and then to their own homes in subdivisions outside Boston. Casual observers watching these suburban dwellers work, attend school, and build religious congregations might conclude that yet another wave of immigrants is successfully pursuing the American dream. A closer look, however, reveals that they are pursuing Gujarati dreams as well. They send money back to India to open businesses or improve family homes and farms. They work closely with religious leaders to establish Hindu communities in the United States, and also to strengthen religious life in their homeland. Indian politicians at the state and national level court these emigrants' contributions to India's political and economic development.

The Gujarati experience illustrates a growing trend among immigrants to the United States and Europe. In the 21st century, many people will belong to two societies at the same time. Researchers call those who maintain strong, regular ties to their homelands and who organize aspects of their lives across national borders "transnational migrants." They assimilate into the country that receives them, while sustaining strong ties to their homeland. Assimilation and transnational relations are not mutually exclusive: they happen simultaneously and influence each other. More and more, people earn their living, raise their family, participate in religious communities, and express their political views across national borders.

SOURCE: From *Contexts* 3(2): 20–26. Copyright © 2004 by the American Sociological Association. All rights reserved. Reprinted by permission.

Social scientists have long been interested in how newcomers become American. Most used to argue that to move up the ladder, immigrants would have to abandon their unique customs, language and values. Even when it became acceptable to retain some ethnic customs, most researchers still assumed that connections to homelands would eventually wither. To be Italian American or Irish American would ultimately have much more to do with the immigrant experience in America than with what was happening back in Italy or Ireland. Social scientists increasingly recognize that the host-country experiences of some migrants remain strongly influenced by continuing ties to their country of origin and its fate.

These transnational lives raise fundamental issues about 21st century society. What are the rights and responsibilities of people who belong to two nations? Both home- and host-country governments must decide whether and how they will represent and protect migrants and what they can demand from them in return. They may have to revise their understandings of "class" or "race" because these terms mean such different things in each country. For example, expectations about how women should balance work and family vary considerably in Latin America and in the United States. Both home- and host-country social programs may have to be reformulated, taking into account new challenges and new opportunities that arise when migrants keep one foot in each of two worlds.

TWO CASES: DOMINICANS AND GUJARATIS IN BOSTON

My research among the Dominican Republic and Gujarati immigrants who have moved to Massachusetts over the past three decades illustrates the changes that result in their origin and host communities. Migration to Boston from the Dominican village of Miraflores began in the late 1960s. By the early 1990s, nearly two-thirds of the 550 households in Miraflores had relatives in the Boston area, most around the neighborhood of Jamaica Plain, a few minutes from downtown. Migration has transformed Miraflores into a transnational village. Community members, wherever they are, maintain such strong ties to each other that the life of this community occurs almost simultaneously in two places. When someone is ill, cheating on their spouse, or finally granted a visa, the news spreads as fast on the streets of Jamaica Plain, Boston as it does in Miraflores, Dominican Republic.

Residents of Miraflores began to migrate because it became too hard to make a living at farming. As more and more people left the fields of the Dominican Republic for the factories of Boston, Miraflores suffered economically. But as more and more families began to receive money from relatives in the United States (often called "remittances"), their standard of living improved. Most households can now afford the food, clothing, and medicine for which previous generations struggled. Their homes are filled with the TVs, VCRs, and other appliances their migrant relatives bring them. Many have been able to renovate their houses, install indoor plumbing, even afford air conditioning. With money donated in Boston and labor donated in Miraflores, the community built an aqueduct and baseball stadium, and renovated the local school and health clinic. In short, most families live better since migration began, but they depend on money earned in the United States to do so.

Many of the Mirafloreños in Boston live near and work with one another, often at factories and office-cleaning companies where Spanish is the predominant language. They live in a small neighborhood, nestled within the broader Dominican and Latino communities. They participate in the PTA and in the neighborhood organizations of Boston, but feel a greater commitment toward community development in Miraflores. They are starting to pay attention to elections in the United States, but it is still Dominican politics that inspires their greatest passion. When they take stock of their life's accomplishments, it is the Dominican yardstick that matters most.

The transnational character of Mirafloreños' lives is reinforced by connections between the Dominican Republic and the United States. The Catholic Church in Boston and the Church on the island cooperate because each feels responsible for migrant care. All three principal Dominican political parties campaign in the United States because migrants make large contributions and also influence how relatives back home vote. No one can run for president in the Dominican Republic, most Mirafloreños agree, if he or she does not campaign in New York. Conversely, mayoral and gubernatorial candidates in the northeastern United States now make obligatory pilgrimages to Santo Domingo. Since remittances are one of the most important sources of foreign currency, the Dominican government instituted policies to encourage migrants' long-term participation without residence. For example, under the administration of President Leonel Fernández (1996-2000), the government set aside a certain number of apartments for Dominican emigrants in every new construction project it supported. When they come back to visit, those of Dominican origin, regardless of their passport, go through the customs line for Dominican nationals at the airport and are not required to pay a tourist entry fee.

RELIGIOUS TIES

The people from Miraflores illustrate one way migrants balance transnational ties and assimilation, with most of their effort focused on their homeland. The Udah Bhagats, a sub-caste from Gujarat State, make a different set of choices. They are more fully integrated into certain parts of American life, and their homeland ties tend to be religious and cultural rather than political. Like Gujaratis in general, the Udah Bhagats have a long history of transnational migration. Some left their homes over a century ago to work as traders throughout East Africa. Many of those who were forced out of Africa in the 1960s by local nationalist movements moved on to the United Kingdom and the United States instead of moving back to India. Nearly 600 families now live in the greater Boston region.

The Udah Bhagats are more socially and economically diverse than the Mirafloreños. Some migrants came from small villages where it is still possible to make a good living by farming. Other families, who had moved to Gujarati towns a generation ago, owned or were employed by small businesses there. Still others, from the city of Baroda, worked in engineering and finance before migrating. About half of the Udah Bhagats now in Massachusetts work in factories or warehouses, while the other half work as engineers, computer programmers or at the small grocery stores they have purchased. Udah Bhagats in Boston also send remittances home, but for special occasions or when a particular need arises, and the recipients do not depend on them. Some still own a share in the family farm or have invested in Gujarati businesses, like one man who is a partner in a computer school. Electronics, clothing, and appliances from the United States line the shelves of homes in India, but the residents have not adopted western lifestyles as much as the Miraflorenos. The Gujarati state government has launched several initiatives to stimulate investment by "Non-Resident Gujaratis," but these are not central to state economic development policy.

In the United States, both professional and blue-collar Gujaratis work alongside native-born Americans; it is their family and religious life that is still tied to India. Some Bhagat families have purchased houses next door to each other. In an American version of the Gujarati extended family household, women still spend long hours preparing food and sending it across the street to friends and relatives. Families gather in one home to do puja, or prayers, in the evenings. Other families lives in mixed neighborhoods, but they too spend much of their free time with other Gujaratis. Almost everyone still speaks Gujarati at home. While they are deeply grateful for the economic opportunities that America offers, they firmly reject certain American values and want to hold fast to Indian culture.

As a result, Udah Bhagats spend evenings and weekends at weddings and holiday celebrations, prayer meetings, study sessions, doing charitable work, or trying to recruit new members. Bhagat families conduct these activities within religious organizations that now operate across borders. Rituals, as well as charitable obligations, have been redefined so they can be fulfilled in the United States but directly supervised by leaders back in India. For example, the Devotional Associates of Yogeshwar or the Swadhyaya movement requires followers back in Gujarat to dedicate time each month to collective farming and fishing activities; their earnings are then donated to the poor. An example of such charitable work in Boston is families meeting on weekends to assemble circuit boards on sub-contract for a computer company. For the Udah Bhagats, religious life not only reaffirms their homeland ties but also erects clear barriers against aspects of American life they want to avoid. Not all Indians are pleased that Hindu migrants are so religious in America. While some view the faithful as important guardians of the religious flame, others, claim that emigrants abroad are the principal underwriters of the recent wave of Hindu nationalism plaguing India, including the Hindu-Muslim riots that took place in Ahmedabad in 2002.

THE RISE OF TRANSNATIONAL MIGRATION

Not all migrants are transnational migrants, and not all who take part in transnational practices do so all the time. Studies by Alejandro Portes and his colleagues reveal that fewer than 10 percent of the Dominican, Salvadoran, and Colombian migrants they surveyed regularly participated in transnational economic and political activities. But most migrants do have occasional transnational contacts. At some stages in their lives, they are more focused on their country of origin, and at other times more committed to their host nation. Similarly, they climb two different social ladders. Their social status may improve in one country and decline in the other.

Transnational migration is not new. In the early 1900s, some European immigrants also returned to live in their home countries or stayed in America while being active in economic and political affairs at home. But improvements in telecommunications and travel make it cheaper and easier to remain in touch than ever before. Some migrants stay connected to their homelands daily through e-mail or phone calls. They keep their fingers on the pulse of everyday life and weigh in on family affairs in a much more direct way than their earlier counterparts. Instead of threatening the disobedient grandchild with the age-old refrain, "wait until your father comes home," the grandmother says, "wait until we call your mother in Boston."

The U.S. economy welcomes highly-educated, professional workers from abroad, but in contrast to the early 20th century, is less hospitable to low-skilled industrial workers or those not proficient in English. Because of poverty in their country of origin and insecurity in the United States, living across borders has become a financial necessity for many less skilled migrant workers. At the same time, many highly skilled, professional migrants choose to live transnational lives; they have the money and know-how to take advantage of economic and political opportunities in both settings. These days, America tolerates and even celebrates ethnic diversity—indeed, for some people, remaining "ethnic" is part of being a true American—which also makes long-term participation in the homeland and putting down roots in the United States easier.

Nations of origin are also increasingly supportive of long-distance citizenship, especially countries that depend on the remittances and political clout of migrants. Immigrants are no longer forced to choose between their old and new countries as they had to in the past. Economic self-sufficiency remains elusive for small, non-industrialized countries and renders them dependent on foreign currency, much of it generated by migrants. Some national governments actually factor emigrant remittances into their macro-economic policies and use them to prove credit-worthiness. Others, such as the Philippines, actively promote their citizens as good workers to countries around the world. Transnational migrants become a key export and their country of origin's main connection to the world economy. By footing the bill for school and road construction back home, transnational migrants meet goals that weak home governments cannot. The increasingly interdependent global economy requires developing nations to tie themselves more closely to trade partners. Emigrant communities are also potential ambassadors who can foster closer political and economic relations.

THE AMERICAN DREAM GOES TRANSNATIONAL

Although few immigrants are regularly active in two nations, their efforts, combined with those of immigrants who participate occasionally, add up. They can transform the economy, culture and everyday life of whole regions in their countries of origin. They transform notions about gender relations, democracy, and what governments should and should not do. For instance, many young women in Miraflores, Dominican Republic no longer want to marry men who have not migrated because they want husbands who will share the housework and take care of the children as the men who have been to the United States do. Other community members argue that Dominican politicians should be held accountable just like Bill Clinton was when he was censured for his questionable real estate dealings and extramarital affairs.

Transnational migration is therefore not just about the people who move. Those who stay behind are also changed. The American-born children of migrants are also shaped by ideas, people, goods, and practices from outside—in their case, from the country of origin—that they may identify with during particular periods in their lives. Although the second generation will not be involved with their ancestral homes in the same ways and with the same intensity as their parents, even those who express little interest in their roots know how to activate these connections if and when they decide to do so. Some children of Gujaratis go back to India to find marriage partners and many second-generation Pakistanis begin to study Islam when they have children. Children of Mirafloreños born in the United States participate actively in fund-raising efforts for Miraflores. Even Dominican political parties have established chapters of second-generation supporters in the United States.

Transnational migrants like the Mirafloreños and the Udah Bhagats in Boston challenge both the host and the origin nations' understanding of citizenship, democracy, and economic development. When individuals belong to two countries, even informally, are they protected by two sets of rights and subject to two sets of responsibilities? Which states are ultimately responsible for which aspects of their lives? The Paraguayan government recently tried to intercede on behalf of a dual national sentenced to death in the United States, arguing that capital punishment is illegal in Paraguay. The Mexican government recently issued a special consular ID card to all Mexican emigrants, including those living without formal authorization in the United States. More than 100 cities, 900 police departments, 100 financial institutions, and 13 states accept the cards as proof of identity for obtaining a drivers' license or opening a bank account. These examples illustrate the ways in which countries of origin assume partial responsibility for emigrants and act on their behalf.

Transnational migration also raises questions about how the United States and other host nations should address immigrant poverty. For example, should transnationals qualify for housing assistance in the United States at the same time that they are building houses back home? What about those who cannot fully support themselves here because they continue to support families in their homelands? Transnational migration also challenges policies of the nations of origin. For example, should social welfare and community development programs discriminate between those who are supported by remittances from the United States and those who have no such outside support? Ideally, social programs in the two nations should address issues of common concern in coordination with one another.

There are also larger concerns about the tension between transnational ties and local loyalties. Some outside observers worry when they see both home country and U.S. flags at a political rally. They fear that immigrants' involvement in homeland politics means that they are less loyal to the United States. Assimilation and transnational connections, however, do not have to conflict. The challenge is to find ways to use the resources and skills that migrants acquire in one context to address issues in the other. For example, Portes and his colleagues find that transnational entrepreneurs are more likely to be U.S. citizens, suggesting that becoming full members of their new land helped them run successful businesses in their countries of origin. Similarly, some Latino activists use the same organizations to promote participation in American politics that they use to mobilize people around homeland issues. Some of the associations created to promote Dominican businesses in New York also played a major role in securing the approval of dual citizenship on the island.

These are difficult issues and some of our old solutions no longer work. Community development efforts directed only at Boston will be inadequate if they do not take into account that Miraflores encompasses Boston and the island, and that significant energy and resources are still directed toward Miraflores. Education and health outcomes will suffer if policymakers do not consider the many users who circulate in and out of two medical and school systems. As belonging to two places becomes increasingly common, we need approaches to social issues that not only recognize, but also take advantage of, these transnational connections.

REFERENCES

Guarnizo, Luis, Alejandro Portes, and William Haller. "Assimilation and Transnationalism: Determinants of Transnational Political Action among Contemporary Migrants." *American Journal of Sociology* 108 (2003): 1211–48.

Portes, Alejandro, William Haller, and Luis Guarnizo. "Transnational Entrepreneurs: The Emergence and Determinants of an Alternative Form of Immigrant Economic Adaptation." *American Sociological Review* 67 (2002): 278–298.

DISCUSSION QUESTIONS

1. How are twenty-first century immigrants different from previous generations?
2. Having read this article, do you think immigrants have to give up a former national identity to become assimilated into a new society?
3. Compare and contrast the appearances of Dominican and Gujarat immigrants.

27

Children of Immigrant Families: Analysis and Recommendations

MARGIE K. SHIELDS AND RICHARD E. BEHRMAN

While the debate about immigration rages on, some never give a human face to the subject of immigration. Shields and Behrman review the status of children of immigrant parents. Some of these are second-generation immigrants—that is, born in the United States. Others have come with family members. Either way, their futures are tangled up in the policies that affect immigrant families.

As the 21st century progresses, our nation will become increasingly dependent on the current generation of children, a generation that is dramatically more diverse than previous generations. Racial/ethnic minorities, in aggregate, are destined to become the numerical majority in the United States within the next few decades. This dramatic shift in demographics is being driven by immigration and fertility trends with the number of children in immigrant families growing rapidly in nearly every state across the country. According to the 2000 Census, 1 of every 5 children in the United States was a child of immigrants—that is, either a child who is an immigrant or who has at least one immigrant parent.

Regardless of how one might feel about our nation's immigration policies, there is no turning back the clock on the children of immigrants already living here, most of whom are U.S. citizens. Who these children grow up to be will have a significant impact on our nation's social and economic future. Will we have a cohesive society—or one rife with intergenerational and intercultural conflict? Will we have a prosperous economy—or one struggling with a labor force dominated by low-wage earners? Will we have a strong safety net for the elderly, poor, and disabled—or will the taxes to support historic entitlement programs become prohibitive?

...Compared with children of U.S.-born parents, children of immigrants are more likely to be born healthier and to live with both parents. They also are more likely to be living in poverty and to be without health insurance. Although indicators of child well-being vary widely based on the family's country of origin, the overall trends are dominated by the large number of immigrants from Mexico, Asia, Central America, and the Caribbean. Parents with limited English skills emigrating from these regions tend to be poorly educated and

SOURCE: From *The Future of Children* 14 (Summer 2004): 4–12.

have limited job prospects. Some are legal immigrants, some are refugees, and some are undocumented. Thus, while the children in these families often share the same hardships experienced by other children from low-income families, what is needed to help them overcome these hardships requires a greater understanding of each group's unique circumstances.

Investing in the healthy development of all our nation's children, including children of immigrants, is to invest in a brighter future—not just for these children themselves, but for our entire nation. All society benefits by providing this segment of our population with the education and supports they need today to become America's productive, engaged citizens of tomorrow.

STRENGTHS OF IMMIGRANT FAMILIES

Immigrant families generally come to America with many strengths, including healthy, intact families, strong work ethic and aspirations, and for many, a cohesive community of fellow immigrants from the same country of origin. These strengths can help to insulate children of immigrants from various negative influences in American society, but they are not always sufficient to keep children on pathways to success over time.

Healthy, Intact Families

According to several measures, children born to immigrant mothers are healthier than those born to U.S.-born mothers, on average. For example, infant mortality rates are lower among immigrant mothers, and their babies are less likely to be born with low birth weights. Also, children of immigrants are reported to experience fewer health problems across a wide range of conditions—from injuries and physical impairments, to infectious diseases and asthma.

Moreover, children in immigrant families are more likely than children in U.S.-born families to live with two parents in the home, with a father who works and a mother who does not work.... The percentage of children of immigrant families living in a single-parent household is only about 16%, compared with 26% for children of U.S.-born families.

Children of immigrants are also more likely to live with a large extended family that can help provide child care and other household support. Nearly 40% live with other relatives and non-relatives in their homes, compared with about 22% for children of U.S.-born families.... Although overcrowding can place a strain on resources, parental time, and even the ability to find a quiet place to do homework, large households also can provide many social and economic benefits.

Strong Work Ethic and Aspirations

Immigrant families generally come to America eager to improve their standard of living. Parents are willing to work hard, and they expect their children to do the same.... The parents in immigrant families are almost as likely to be working as those in U.S.-born families (97% versus 99%).

Children of immigrants typically are imbued with a strong sense of family obligation and ethnic pride, and with the importance of education. As a result, the children of immigrants tend to have high educational aspirations and are less likely than children of U.S.-born families to engage in risky behaviors such as substance abuse, early sexual intercourse, and delinquent or violent activity. Studies show that they also tend to spend more time doing homework and that they do better in school, at least through middle school. For example, although their reading test scores are somewhat lower, 8th-grade children of immigrants have slightly higher grades and math test scores than their counterparts of the same ethnicity in U.S.-born families.

According to the National Center on Education Statistics, the dropout rate is higher for children of immigrant parents than for children of U.S.-born parents, but the rate is calculated based on the number of 16- to 24-year-olds who are not enrolled in high school and have not graduated. As a result, the rate includes a large number of older foreign-born children—especially Hispanics—who never attended U.S. schools. The dropout rate for non-Hispanic children of immigrants is considerably lower than the U.S. average (6% versus 11%).

Community Cohesion

When immigrant families arrive in America, they often settle in communities with others from their same country of origin. Fellow immigrants in these communities can facilitate a family's adjustment, helping them learn to navigate new systems and institutions (such as schools) and to find jobs.... Such communities also can be supportive of the child's emotional and academic adjustment by reinforcing cultural values and parental authority, and by buffering them from the negative influences of mainstream society. The role of a cohesive, culturally-consonant community can make a critical difference in helping youth maintain positive aspirations despite the challenges they face as newcomers to this country.

CHALLENGES FACED BY CHILDREN OF IMMIGRANTS

Although possessing many strengths, immigrant families also confront many challenges. The children in these families often must navigate the difficult process of acculturation from a position of social disadvantage with limited language skills and minimal family and institutional support.

Less-Educated Parents

Children in immigrant families are far more likely than children in U.S.-born families to have parents who have not graduated from high school. Among all children with U.S.-born parents, 12% have mothers, and 12% have fathers, who are not high school graduates. In contrast, among children with foreign-born parents,

23% have mothers, and 40% have fathers, who are not high school graduates.... The lower level of parental educational attainment in immigrant families has major implications for child well-being and development. Poorly-educated parents are less able to help their children with homework, and less able to negotiate educational and other institutions to foster their children's success. Across a wide range of socioeconomic indicators, children whose parents have more education tend to fare better than those whose parents have less education.

Low-Wage Work with No Benefits

Over the past 30 years, the industrial base of the United States has shifted from manufacturing to services and, more recently, to technology and communication.... This shift has resulted in a widening of the wage gap between those with high levels of education and skills, and those without. For the most part, immigrant parents find themselves on the bottom side of this wage gap. They are over-represented among workers who are paid the least, and are most in need of training to improve their skills and earnings. Immigrants represent about 11% of the U.S. population, but they account for 20% of the low-wage labor force, often with limited access to benefits. They are more likely than U.S.-born workers to have only part-time and/or partial-year work (25% versus 21%), and they are less likely to have private, employer-provided health insurance for their children (55% versus 72%).

Language Barriers

Among all children in this country, 18% speak a language other than English at home. Among children in immigrant families, 72% speak a language other than English at home. While the ability to speak two languages has potential benefits, if no one in the household speaks English well, the family is likely to encounter difficulties finding higher wage employment, talking with children's teachers, and accessing health and other social services. Census data indicate that among children in immigrant families, 26% live in linguistically-isolated households where no one age 14 or older has a strong command of the English language.

Discrimination and Racism

Many children of immigrants and their families must contend with discrimination and racism.... Social position, racism, and segregation can set children of color and children of immigrants apart from mainstream populations, and how schools serving primarily children of color are likely to have fewer resources, lower teacher expectations, and patronizing attitudes toward students of non-mainstream cultures. They maintain that for these students, schools can come to be perceived as instruments of racial oppression, and efforts to advance through education as hopeless. Thus, while children from immigrant backgrounds enter school with very positive attitudes toward education, by adolescence they can become disillusioned, and their attitudes toward teachers and scholastic achievement can turn negative.

Poverty and Multiple Risk Factors

Poverty rates for children in immigrant families are substantially higher than for children in U.S.-born families. According to the official poverty measure, 21% of those with immigrant parents live in poverty, compared with 14% of those with U.S.-born parents. If families with incomes up to twice the poverty level are included, the differences are even more dramatic: 49% of those with immigrant parents live in poverty, compared with 34% of those with U.S.-born parents.

Poverty often means lack of access to quality health care and education resources, which can lead to children's poor health and school failure. In fact, studies indicate that the health and academic achievement of children of immigrants deteriorates as exposure to mainstream American culture increases, perhaps due to the negative effects associated with poverty, such as poor diet, destructive behaviors, and racial/ethnic stratification.... Poverty accentuates racial disparities in children's health, and poor health and poverty spiral together in a vicious circle that injures all children.

Negative developmental outcomes for children have been linked to a variety of risk factors, such as having a poorly-educated mother, and/or living in a household that is poor, linguistically-isolated, or headed by a single parent. Moreover, research suggests that while children are generally resilient to a single risk factor, the effects of multiple risk factors—regardless which they are—can work synergistically to undermine a child's healthy development. Data compiled by Donald Hernandez show that children in immigrant families are more than twice as likely as those in U.S.-born families to experience two or more risk factors.

Lack of Supports

The provision of supports for low-income families to enable parents to better care for their children is a longstanding tradition in this country, and studies show that work programs and supports that increase parental employment and income have positive impacts on indicators of child well-being. In terms of access to such programs, until fairly recently, legal immigrants were generally eligible under the same terms as citizens. But, ... the 1996 federal welfare reform law imposed a wide range of restrictions on immigrant eligibility for federal public assistance programs and the impact has been dramatic.

Between 1996 and 2001, the share of non-citizens receiving assistance from programs such as Temporary Assistance for Needy Families, Food Stamps, and Medicaid dropped significantly. Moreover, though targeted to excluding non-citizens, participation rates have fallen among citizen children as well—especially those living in families with non-citizen parents—even though they remain eligible for benefits. Reasons include parents' confusion or lack of knowledge about eligibility, language barriers, and fear of adverse immigration consequences. Although four of every five children of immigrants are U.S. citizens, many are in "mixed status" families—that is, living in households where some members are not citizens—and their parents fear involvement with government agencies. This is especially true for parents who are in the country illegally. In April 2004, it was estimated that of the 33 million foreign-born persons living in the United States, about 9.3 million are undocumented.

In contrast, immigrants afforded refugee status are provided access to a variety of supports enabling them to improve their economic stability and status more quickly. Refugees' relatively more secure economic circumstances likely contribute to the research findings that suggest that compared to other children with similar family characteristics, children of refugees do better in school at least until middle school.

VARIATION ACROSS DIFFERENT COUNTRIES OF ORIGIN

Under the surface of these overall trends, there is substantial variation in immigrant families' assets and challenges across different countries of origin. In general, those families emigrating from West and Central Europe, and from other English-speaking countries such as Canada, Australia, New Zealand, and India, for example, tend to have more advantages and face fewer challenges, compared with those emigrating from Mexico, Central America, the non-English-speaking Caribbean, and Indochina. To have a significant impact on improving the health and well-being of children in immigrant families, it is important to focus on the unique circumstances of the groups who are struggling the most to succeed in this country. There are some similarities among these groups, but also some significant differences. Individual and family characteristics, reasons for immigration, and the social context families find upon their arrival, all play important roles in understanding these differences....

CONCLUSIONS AND RECOMMENDATIONS

Studies show that on average, because of their lower incomes, larger households, and lack of English-language skills, immigrant families contribute less to public revenues and cost more in terms of use of services. Implementing programs that promote the healthy development of children in immigrant families and that provide them with opportunities for achievement more equal to those enjoyed by children in U.S.-born families clearly places an added financial burden on society. However, failure to implement such programs will also place a financial burden on society—a burden that is likely to grow over time as these children enter adulthood, and their lifetime earnings and tax contributions are less than what they might have been had they received more supports early in life. To assure a cohesive society, a prosperous economy, and a strong safety net for the elderly, poor, and disabled into the next century, more attention must be paid to meeting the developmental needs of the large number of children in immigrant families now living in this country, especially those who are at greatest risk of failure.

In some ways, the needs of children of immigrants are the same as for other vulnerable low-income children, and efforts to support the positive development of all disadvantaged youth would undoubtedly help to address a wide range of their challenges as well....

At the same time, current strategies aimed at addressing poverty in general are not always appropriate for this population as their situation is unique in several ways. For example, children in immigrant families tend to live in two-parent families with at least one working parent, so programs that are aimed at promoting marriage and greater work effort are less likely to be effective in boosting the incomes of these families. Instead, immigrant families are more likely to need help dealing with low education levels and lack of access to supports and programs due to their citizenship status. Most importantly, for many, efforts to enhance their English language skills are critical.

DISCUSSION QUESTIONS

1. How do children of immigrants compare on various measures of social well-being to children born in the United States?
2. What does the status of immigrant children suggest for new social policies?

28

Class, Race, and Success: Two Generations of Indian Americans Confront the American Dream

JOHANNA LESSINGER

Lessinger examines differences in the ethnic and transnational identities of two generations of Indian Americans. Her research also explores the myth of the model minority and how the experiences of these two groups, often very successful, are still shaped by race, ethnicity, and social class.

Asian Indian immigrants in New York City, like fellow Indians in the rest of the United States, have become noted for their unusually rapid and successful incorporation into the U.S. economy. As part of the large post-1965 migration from Asia, recent arrivals from India are often cited as an ethnic success story, as a group that has entered the professions or founded businesses and gone on to experience American-style suburban affluence within the first immigrant generation.... Unlike many other groups of newcomers, Indians are perceived as having "made it" quickly and with relative ease, without waiting for the second or third generation's painful climb into the professional middle class that has characterized the incorporation of many other immigrant groups, past and present.

This image of rapid economic assimilation of Asian Indians within two decades of arrival is made possible by the backgrounds of most of the post-1965, first-generation Indian immigrants. By and large born into India's urban, professional middle classes, these migrants are well-educated, particularly in technical fields, fluent in English, sophisticated about the U.S. job market, and increasingly transnational. Many have prepared themselves for out-migration years in advance, by seeking out particular kinds of higher education, professional training, or investment opportunities that will maximize their access to student or immigrant visas and their chances for well-paid, permanent work in the United States. By this time, prospective arrivals can tap into well-established employment and educational networks both in India and in the United States.

SOURCE: From *Migration, Transnationalization and Race in a Changing New York*, edited by Hector R. Cordero-Guzman, Robert C. Smith, and Ramon Grostoquel. Used by permission of Temple University Press, © 2001 by Temple University. All Rights Reserved.

Within a decade of arrival, a great many of these immigrants will be successfully placed in professional or entrepreneurial positions and will be strategizing to ensure that their children will be even better placed for professional careers, hastening, for instance, to get a two-year-old enrolled in a Montessori school "so we can be sure she goes to Harvard." Many of these children do indeed attend good universities and go into highly paid and prestigious professions. The pressure from their parents to take up science, medicine, law, or finance, to prosper professionally, to make money, and to succeed to their parents' class position is very great.

Recent immigrants, brown- to black-skinned, from one of the world's poorest countries, yet generally well-to-do and professionally established once here, Indian immigrants present an interesting problem in terms of the contradictions between the ethnic self-identity they have developed here and the ways in which they are incorporating into the larger society. In that incorporation, racial location, class status, and transnational Indian identity are still contentious and problematic to Indian immigrants in New York City and the tri-state area of New York, New Jersey, and Connecticut.

This [article] looks at some aspects of the process, particularly the issue of racial location and racial discrimination as experienced rather differently by two generations of Indian Americans. The first generation of Indian immigrants tends to see itself as white because of its pre- and postmigration class privilege and has been in some degree insulated by money, professional jobs, and suburban residence from some of the most direct expressions of U.S. racism and nativism. These experiences of class privilege and of racial location are somewhat different for their American-raised children, whose lives in school and college, as well as their entry into the work force, may have given them close encounters with U.S. racial realities. I argue that the second generation of Indian Americans is slowly developing a more sophisticated sense of self, less afraid to acknowledge a non-white identity in the context of New York City's ethnic politics. At the same time, I suggest that some of the first generation's transnationalism is fueled by either a rejection of or profound unease with U.S. racism and nativism....

ETHNICITY, CLASS, AND RACE

In the discourse of U.S. migration, Indians' economic success, their status as one of the country's several Asian "model minorities," has become a validation—both of the pay-off for hard work, optimism, and family values and of the conservative American myth that paints the United States as a land of boundless economic opportunity for all. The economic success of Asian Indians, like that of Korean and Chinese Americans, has also come to be read as a subtle reproach to those immigrant groups and native-born minorities, particularly African Americans and Hispanics, who have not been able to achieve the American dream as rapidly and effectively, who are seen as trapped in eternal poverty, social dysfunction, and inner-city misery.

Indian immigrants wholeheartedly embrace this view of themselves as an exceptionally well-educated, affluent, and well-integrated American ethnic group, moving at ease in the upper reaches of U.S. society. It is an identity that Indians celebrate repeatedly, in their media and through their community events. The ethnic leadership is drawn largely from elite individuals who approximate this ideal: doctors, engineers, university professors, bankers, large business owners. The same ideal of success is echoed in both press accounts and daily discourse in India, where the American migrant is simultaneously admired, envied, and despised—but always imagined as rich, powerful, and centrally located in his or her adopted society. The image of success is clearly central to Indian immigrants' sense of themselves, a cornerstone of both internal group structure and self-representation to the larger world of non-Indians, and an important aspect of their transnational standing in India.

At the same time, the first generation of Indian immigrants is still deeply imbued with strong feelings of attachment to India. Most still think of themselves as culturally Indian, devote time and effort to maintaining an Indian identity, and fret that their U.S.–born children are failing to achieve the same sense of attachment to an imagined homeland. Even those first-generation immigrants most determined to shed Indianness and to remake themselves as purely American often slip into a nostalgic yearning to "give something back" to India, if only to eradicate its embarrassing poverty and backwardness. The first generation is not, therefore, a population that is hastening to assimilate culturally. Many actively deny or resist cultural assimilation, feeling that they can enjoy the economic advantages of the United States and participate in its economic structure without major cultural concessions. Others feel that their partial Westernization, by virtue of elite Indian class background, is assimilation enough. In contrast, the second generation is more genuinely at ease with many aspects of American culture, from food to music to interpersonal relations. These younger people are more likely to stress the American in what many call their "hyphenated identity."

The widespread sense of passionate connection that Indian immigrants in the United States express toward India fuels the multitude of transnational structures that keep Indian immigrants here deeply involved with friends, relatives, business partners, political parties, religious institutions, and cultural movements "at home." Such transnational links are also exerting increasing pressure on Indian culture, accelerating a process of Westernization that has become increasingly American-inflected since the 1970s.

The images of success that Indian immigrants use to locate themselves socially within the United States contain implications about both class status and race—implications that are often confused and at odds with observed reality. In terms of class, first-generation immigrants take for granted the superiority of Indians. Many would like to think that the entire ethnic group is as well-off as the Silicon Valley millionaires who recently raised several million dollars for the Indian science and engineering colleges that educated them (see Dugger 2000). The Indian doctors who are building themselves Mogul-style trophy houses in the New Jersey suburbs (Gordon 2000) are admired and held up for emulation.

Yet the Indian immigrant population also contains large numbers of people holding far more modest jobs, particularly in New York City. For instance, 21 percent of the New York City Indian immigrant population held some form of blue-collar job in 1990, according to the Bureau of the Census. There are, in addition, large numbers of service workers and owners of small, almost marginal, businesses. The Indian proprietor of a grocery store or newspaper stand may own his own business and be able to support his family, but he clings to the margins of lower-middle-class respectability only by working seven days a week. The city's Indian taxi drivers, apartment house doormen, factory workers, deli countermen, and waiters are frankly working class. Moreover, their employment conditions and pay frequently parallel those of other new immigrants, both past and present, who have struggled to find and keep low-paid, insecure, and sometimes dangerous jobs. Looked at from this angle, Indian immigrant exceptionalism and privilege is far less visible. Nevertheless, the immigrant leadership continues to insist that the entire ethnic group must downplay such internal class cleavages, presenting the ethnic group "as a whole," with a "healthy image" (Abraham 1984, 1). There is, in fact, widespread insistence that the group, which the ethnic leadership refers to as "the community," is in fact a functioning moral community, despite widening geographical dispersal and clear class cleavages.

One of the effects of this stance is to hamper recognition of social problems experienced by Indian immigrants. Those Indian doctors, social workers, and progressive activists (see, for instance, Motwani and Barot-Motwani 1989, iv) who would like to mobilize community resources, as well as the resources of the larger society, to combat problems such as alcoholism, unemployment, wife battering, and AIDS among Indians, find their efforts snubbed, disparaged, and ignored. Organizations that try to build coalitions—whether with other Asian groups, with feminists, or with city social service agencies, find themselves sharply criticized as disloyal or as no longer Indian (Lessinger 1995, 129–54). These disputes, too, often take on a generational aspect, with younger people far less defensive than their elders about the possibility that Indians, like all other groups, suffer social problems. Indeed, younger people are far more likely to cite pride in being Indian American as a reason for the ethnic group to "take care of its own."

The other aspect of Indian immigrant self-identification, still more strongly marked by silence and denial, is that of race. The society of India experiences color consciousness and is deeply divided by class antagonism and caste prejudice, but it experiences nothing approaching the blunt dualism of hegemonic U.S. racial categories, which assign everyone to either "white" or "black" status, a categorization based on alleged skin color and geographical origins. This country's post-1965 influx of new immigrants has complicated but not displaced what Michael Omi and Howard Winant have called the United States' "racially bipolar vision" (1994, 154), so that the more newly arrived Hispanics and Asians form an awkward residual or intermediate category. For Asians, in particular, their place within this hierarchy is contradictory, since their "nonwhite" skins often inhabit "white" class or occupational locations.

In some sense the first generation of Indian immigrants in the United States is genuinely bewildered about where they might fit into this model, since so many Indians still identify themselves primarily on the basis of national origin, and their skin color varies widely. At another level, Indians, like other new immigrants, have been quick to absorb America's general prejudices against blacks, Hispanics, and the poor in general (Mazumdar 1989). The frankly racist views of ultra-conservative Indian American political commentator Dinesh D'Souza are simply an extreme variant of opinions circulating within sectors of the Indian immigrant population. At the same time, many first-generation Indian immigrants tacitly assume that they themselves are white, on the basis of class criteria. Since most have accepted the American folk model that tends to conflate race with class status and with social disabilities such as poverty, welfare dependency, and drug use, Indians assume that they—hardworking, family oriented, affluent, and upwardly mobile—must be white. Their imagined group class status makes it so. The same kind of conservative denial that leads the Indian immigrant community as a whole to blot out distasteful realities such as drug abuse and homosexuality also renders it helpless to respond in the face of racism and nativism.

These "honorary whites," or what Nazli Kibria calls "ambivalent nonwhites" (1998, 71), are imbued with bourgeois ideals of individual achievement and self-reliance. They are stunned to find that the larger U.S. society considers them nonwhite and distinctly Other, often disregards their vaunted economic achievements, and makes them potential victims of racial/ethnic prejudice. Only a handful of Indian immigrant intellectuals and progressives have begun to confront the ambiguities involved in the group's claim to whiteness (see, for example, Mazumdar 1989; Prasad 2000; Radhakrishnan 1994; Sethi 1993; Singh 1996).

Additionally, it is worth considering briefly how the transnationalism of Indian immigrants is shaped by their racial and class location within U.S. society. The more affluent Indian immigrants are those best positioned to maintain transnational ties, to move comfortably in and out of both societies. Those with money and paid vacation time can travel to India yearly, whether to maintain personal ties or to invest money, taking their children with them. (Such trips are partly designed to enhance the Indian self-identification of the young.) Indian immigrant professionals and large-scale entrepreneurs, with large savings and technical expertise, are able to take advantage of Indian government investment schemes offering favored treatment to overseas Indians legally defined as nonresident Indians or NRIs (see Lessinger 1992). In contrast, lower-middle-class or working-class Indians are less likely to visit India regularly, since both airfare and the concomitant exchange of gifts such travel entails are very expensive. If they do travel home, usually for purely familial reasons, they are certainly unable to start businesses there, fund their former colleges, or contribute to Indian political parties.

At the same time, the affluent are also far more likely to spend time and resources constructing a public Indian ethnic identity in the United States through membership in a network of Indian ethnic organizations, participation in ethnic public events, including planning and subsidizing them, and consumption in the form of tickets to expensive dinners and performances, travel to fancy restaurants in distant suburbs, or attendance at three-day conferences in resort hotels. And for all of the events, one must purchase the clothes to match. Inevitably, many of these events turn into celebrations of the economic achievement of the ethnic elite.

It is possible to argue that the kind of racial and cultural dislocation that many first-generation Indians feel in the United States, whether they fully understand it or not, may well fuel both their adherence to an Indian ethnic identity and their embrace of transnationalism. In particular, this sense of racial and cultural unease may explain the wide appeal within the Indian diaspora of the ultranationalist Hindu rightist groups that have established proselytizing and fund-raising units in the Indian immigrant communities of the United States, Canada, and Britain (see Kumar 1993, 10–12; Prashad 2000, 133–56; Rajagopal 1993, 13). The current anti-Muslim and anti-Christian rhetoric of the rightist groups in India parallels the discourses of racism and class prejudice found in North America and Britain and may validate social conservatism among immigrants. At the same time, the right's premise of a powerful and eternal (Hindu) Indian-ness is comforting to many whose sense of cultural security and identity has been battered by the process of immigration and partial acculturation....

FACING HARSH REALITIES

Indian immigrants in the United States do not, in general, suffer as much from overt racial discrimination or racial violence as do other nonwhite minority groups in this country (see U.S. Commission on Civil Rights 1992). In comparison with the large, urban, working-class South Asian population in Britain, battered by periodic explosions of racial prejudice (see Werbner and Anwar 1991; Robinson 1990), Indian Americans have a relatively peaceful existence, partially shielded by education, white-collar jobs, and well-to-do suburban neighborhoods. Nevertheless, despite a carefully constructed ethnic identity, Indians in the United States are not perceived as white.

For Indian immigrant children, the question arises early. Teachers in the New York City and suburban schools describe attempts to protect and comfort Indian American children who are subjected to merciless teasing from schoolmates because the Indians refuse to identify themselves as black (or Hispanic or Chinese—the recognized "other" categories in the area's dualistic racial model). Classmates accuse young Indians of being nonwhites who repudiated their color and tried to pass for white. "Who am I?" wept one Indian child to a teacher after a schoolyard fracas. "Am I black? Am I white? Am I Hispanic?"

Indian immigrant children who attend college, particularly at large universities, encounter for the first time a wider and more acceptable choice of identities: South Asians, people of color, Asian Americans. Through student activism, young Indian Americans may encounter their first forms of pan-ethnic identity construction by cooperating with Pakistani, Korean, or Chinese students. They may get to see ethnic or political coalitions in formation, as when a wide spectrum of student groups protests a campus outbreak of ethnic slurs. At family-oriented Indian immigrant conferences, young adults have confronted the parental generation over their prejudice toward Muslims and Pakistanis, or over their refusal to identify with other South Asians.

Older Indian immigrants have a far narrower range of experiences with racism, usually at work or in their neighborhoods, and they rarely have the chance to take action against it. They encounter a whole range of racist or anti-immigrant slights and slurs, some of which they do not at first recognize. People with strong Indian accents are told to "speak English—you're in America now," a comment that outrages those who pride themselves on their British-style education. Women frequently find their traditional clothing a source of unwelcome comment that ranges from well-meant but ignorant questions to muttered admonitions to "Go home and get dressed properly—why are you wearing your nightclothes on the street?" People who bring meals to work (for reasons of economy, ritual purity, and taste) are told their lunches are "weird" or "stinky." Well-intentioned suburban teachers, engaged in earnest efforts to validate their students' Indian identity against the racism of classmates, nevertheless fret that their Indian students "smell funny" and thus endanger the integrationist effort. African Americans know how wearing such petty insults can be, even if they do not threaten life and livelihood.

In suburban neighborhoods in which Indians are newcomers, prejudice may go beyond individual remarks to involve vandalized cars or mailboxes, insulting graffiti, shouted racial slurs from neighbors or passing carloads of youth. Indians' (and other new immigrants') efforts to transform their environment physically, and to build bigger houses, new businesses, temples, or community centers, often touch off litigation, accompanied by expressions of anti-Indian or anti-immigrant prejudice. Similar responses oppose the academic prowess of Asian children in local schools and parental demands for improved science teaching. The consensus among the white original inhabitants is that the newcomers should go back where they came from and stop trying to remake the community in their own (foreign) image.

What this material suggests, therefore, is that Indian Americans' experiences of race and racism are related to occupation, area of residence, and generation. The young, less able to insulate themselves from white Americans through self-employment, immersion in an all-Indian social milieu, or intense transnational involvement, have generally been obliged to see and experience their nonwhite status since childhood. Many succeed in confronting it openly. Because their American upbringing makes them more secure in this society, it is easier for younger people to acknowledge and live with the contradiction between nonwhite skin color and bourgeois class status....

REFERENCES

Abraham, Thomas. 1984. "Preface." P. 1 in *North American Directory and Reference Guide of Asian Indian Businesses and Independent Professional Practitioners*, edited by Thomas Abraham. New York: India Enterprises of the West.

Dugger, Celia. 2000. "Return Passage to India: Emigres Pay Back," *New York Times*, February 29, A-1.

Gordon, Alastair. 2000. "Raj Style Takes the Silk Road to the Suburbs." *New York Times*, January 27, D-1.

Kibria, Nazli. 1998. "The Racial Gap: South Asian American Racial Identity and the Asian American Movement." Pp. 69–78 in *A Part Yet Apart: South Asians in Asian America*, edited by Lavina Dhingra Shankar and Rajini Srikanth. Philadelphia: Temple University Press.

Kumar, Krishna. 1993. "Behind the VHP of America." *Frontline*, September 10, 10–12.

Lessinger, Johanna. 1992. "Investing or Going Home? A Transnational Strategy Among Indian Immigrants in the United States." Pp. 53–99 in *Towards a Transnational Perspective on Migration, Race, Class, Ethnicity, and Nationalism Reconsidered*, edited by Nina Schiller, Linda Basch, Cristina Blanc-Szanton. New York: New York Academy of Sciences.

———. 1995. *From the Ganges to the Hudson: Indian Immigrants in New York City*. Boston: Allyn and Bacon.

Mazumdar, Sucheta. 1989. "Race and Racism: South Asians in the United States." Pp. 25–38 in *Frontiers of Asian American Studies*, edited by Gail Nomura, Russell Endo, Stephen Sumida, and Russell Leong. Pullman: Washington State University Press.

Motwani, Jagat K., and Jyoti Barot-Motwani. 1989. "Introduction." Pp. i–iv in *Global Migration of Indian: Saga of Adventure, Enterprise, Identity, and Integration*, edited by Jagat Motwani and Jyoti Barot-Motwani. Commemorative volume. New York: First Global Convention of People of Indian Origin.

Omi, Michael, and Howard Winant. 1994. *Racial Formation in the United States from the 1960s to the 1990s*. New York: Routledge.

Prashad, Vijay. 2000. *The Karma of Brown Folk*. Minneapolis: University of Minnesota Press.

Radhakrishnan, R. 1994. "Is the Ethnic 'Authentic' in the Diaspora?" Pp. 219–33 in *The State of Asian America*, edited by Karin Aguilar–San Juan. Boston: South End Press.

Rajagopal, Arvind. 1993. "An Unholy Nexus: Expatriate Anxiety and Hindu Extremism." *Frontline*, September 10, 12–14.

Sethi, Rita Chaudhry. 1993. "Smells Like Racism: A Plan for Mobilizing Against Anti-Asian Bias." Pp. 235–50 in *The State of Asian America*, edited by Karin Aguilar–San Juan. Boston: South End Press.

Singh, Amritjit. 1996. "African Americans and the New Immigrants." Pp. 93–110 in *Between the Lines: South Asians and Postcoloniality*, edited by Deepika Bahri and Mary Vasudeva. Philadelphia: Temple University Press.

United States Commission on Civil Rights 1992. *Civil Rights Issues Facing Asian Americans in the 1990s*. Washington, D.C.: Government Printing Office.

DISCUSSION QUESTIONS

1. What social factors fuel the ethnic and transnational identities of Asian Indians in the United States?
2. How does Lessinger assess the myth of the model minority in the experience of Asian Indians?

Part VI
Student Exercises

1. Is there an immigrant experience in the history of your family? Do you know what it is? If so, what were the specific challenges faced by that generation? What were the economic and social conditions in the United States at the time of arrival? If your family history does not include such an experience, why not? How does your family's history relate to that of other groups?
2. Immigration is transforming many communities in the United States. Who are the immigrants in your community? Where do they live? What work do they do? Did their children attend school with you? After your own personal assessment, review the newspaper and the visual media coverage to understand popular images of these immigrants. Are immigrants presented as a positive influence in the community or a negative one? Do you think the media depictions of the immigrants are related to their nations of origin and to their race?

PART VII

INTRODUCTION

Race, Class, and Inequality

Racial stratification is the hierarchical pattern in society wherein different racial and ethnic groups have varying opportunities and life chances. Racial stratification emerges historically, but it is sustained by policies, practices, and beliefs that re-create and transform it in the present. The consequence is that, even if people think that race should not shape one's ability to succeed, the social structure of which racial stratification is a part continues to differentiate racial groups' experiences. Racial stratification also intersects with other forms of inequality in society—class, certainly, but also gender, age, sexual orientation, and other stratifying features of people's lives. As we will see in Part VIII, racial stratification is also embedded in the nation's social institutions.

Racial stratification involves people's access to resources, but it also includes the degree to which people exercise **power** over their own lives and that of others. Power enables the most privileged groups to shape institutions for their own benefit. We have seen this in terms of how dominant groups used their political influence to shape concepts of citizenship and formulate immigration policies. Power is central to how people work, get an education, and enable their families to advance, or not, in society.

In this part we will focus on racial stratification—how it emerged (some of which we have seen already) and how it is changing. Racial stratification is also part of a broader context of growing, not shrinking, inequality in the United States, despite beliefs to the contrary. Racial stratification is also now present in an increasingly global context, a phenomenon we have also seen in some of the earlier articles in this book. Here we will see that legacies of inequality—that is, the inequality created in the past, continue to differentiate opportunities for people in the present. At the same time, contemporary institutional practices shape the opportunities and resources of various groups.

In the early history of the United States, landowners and industrialists were dependent upon the labor of enslaved and indentured people, who lacked land

and had to work for others without compensation. With the balance of power to their advantage, these powerful leaders faced minimal resistance from workers, including enslaved people, because they had political and economic power that they used to promote ideas about their natural abilities to rule. The newspapers and documents of the time presented these arrangements as appropriate and justified by the assumed ability of propertied White men to rule over others, just as fathers ruled in families.

Without access to political and economic power, people of color faced major hardships. Not only did they lack citizenship rights but beliefs about them outside of their own communities were negative. As we could see in Part V, indigenous people and African Americans faced political actions that did not enable them to strive for the same livelihoods as native-born White Americans. Even White immigrants had more options than most people of color. The economic place for people in the nation was reflected in views about their appropriateness for citizenship. People of color were thought uneducable and were largely confined to agricultural and manual work like planting and harvesting crops, building canals and railroads, and laboring in mines, with women primarily also doing domestic labor. For example, during the first 200 or so years of the nation's growth, African Americans tended the nation's major export crops—tobacco, wheat, and cotton. Even though, under slavery, African Americans performed much of the skilled labor in the South, they were represented as fit only for menial work. Even after slavery was abolished, African Americans were not seen as a modern people. Many former slaves living in the South labored as sharecroppers, working land they did not own for a portion of the crop. They were frequently in debt, with money coming into households only from the labor of mothers and daughters, who typically did domestic work in the homes of White families (Marks 1989).

Asian immigrants and the small Asian American population also faced limited employment options, mostly in agriculture, railroad construction, and service work (Takaki 1989). Mexican Americans, citizens since the United States had won the Mexican American War (in 1848) and annexed most of what is now the Southwest, found that their experiences were shaped by wealth. Those with land and resources had a measure of power, but as more Anglos dominated the region, many lost their rights and resources. Long-term residents and new immigrants found employment in agriculture and the hard labor of building new industries like mining and railroads (Acuña 1981). All of these groups, deeply involved in building the nation, faced obstacles to earning a living and supporting their families because of unfair economic and political arrangements.

As the nation industrialized, African Americans and people of color sometimes gained work as strikebreakers when employers wanted to prevent White workers from unionizing. Ironically, most unions excluded them, as a way of protecting White workers. Over time, more and more African Americans, Mexican Americans, and some Chinese Americans found their way into industrial jobs, especially as European immigration waned after the 1924 immigration law. Beliefs about the limited abilities of people of color were modified because their labor was needed, but there were ceilings on their advancement. Men of color doing industrial work earned lower wages than White men in the same industries. The practice of a dual wage system was common in many industries. White workers could earn enough to support their wives and enable their children to remain in school, while families of color often required the labor of many family members to survive. Ideas about the latter's assumed inferiority were kept alive by residential segregation and segregated schools, segregated hospital wards, and segregated public accommodations, including seating on buses and trains. Yet, people resisted with the means available to them to change their situations, and eventually the Civil Rights Movement prompted changes in the law, even if not in practice.

All told, our nation's historical treatment of people of color has resulted in a system of racial stratification whereby racial and ethnic groups are arrayed along different lines of opportunities, resources, and overall social and economic well-being. These inequalities have also been created and sustained through government policies that overtly, or inadvertently, have promoted the advantage of some and the disadvantage of others. For example, following World War II, the G.I. Bill enabled millions of veterans to advance by attending technical schools, colleges, and universities. But because African Americans were largely excluded from White educational institutions at the time, they were unable—even though eligible—to take advantage of this government benefit to the same degree as Whites. The impact can still be seen in the educational deficit for people of color who could not parlay this benefit into educational credentials that would enable them to get technical, professional, and managerial jobs in the expanding post–World War II economy. Racial segregation thus prohibited many Black veterans and their families from using the very educational benefits they had earned through military service (Katznelson 2005). For White Americans, this policy enabled a huge expansion of the middle class, not only because of their own hard work but because of their ability to take advantage of government policies. This advantage also accrued to White Americans through various policies that, in the postwar period, allowed them to purchase homes and that created now flourishing suburbs—neighborhoods from which Black Americans were excluded via lending practices and other forms of housing discrimination. Still in cities, a small middle-class group of African Americans provided retail and professional services to their racial minority communities. By the time the civil rights legislation of the

1960s and beyond granted people of color access to formerly forbidden places (such as schools and neighborhoods), patterns of racial inequality were already cemented. White Americans were then many steps ahead because of their getting a foothold in an expanding set of opportunities early on.

Groups who have had advantages in the past are also able to pass them on to their children—creating the possibility of upward class mobility—while other groups are still having to catch up. Melvin Oliver and Thomas Shapiro (1995) discuss the **sedimentation of inequality:** The longer groups have been near the bottom of the social class hierarchy, the harder it is to advance. And, as the nation's economy shifts from an industrial to a postindustrial economy, deficits from the past create steeper roads to climb in the here and now. Thus, while earlier a person could get a decent job, perhaps with good job benefits, with only a high school education, now an advanced degree is essential for most forms of lucrative employment.

Thus, racial minorities (like everyone else) have to be equipped to compete in new ways, but past practices complicate their struggle. In the new economy, one's lifestyle and options come not just from earnings, but from whatever other assets and wealth—even if modest—that one has. Young people whose families can pay for their college education without the family or the student amassing great debt will emerge from college with a more secure economic lifestyle than those who graduate with large debts. Although the privilege on which this achievement rests may be invisible (or taken for granted) by some, the fact that some families can do this and others not reflects not just the family's hard work but also the policies of the past that have enabled them to secure these assets. You can see from this example that securing political and civil rights earlier in the nation's history means that some groups have accumulated advantages, while others struggle to maintain a secure foothold.

Melvin Oliver and Thomas Shapiro (1995) in their book *Black Wealth/White Wealth* introduced the importance of wealth to the study of racial inequality, meaning the assets that people have—homes, cars, stocks and bonds—once their debt is subtracted. Civil rights laws have enhanced access to work environments and narrowed the gap in earnings between White and Black Americans. By examining wealth, Oliver and Shapiro show how the past privileges of White Americans still produce patterns of inequality. In their article included here ("Wealth Inequality Trends"), Oliver and Shapiro update their findings to show how the wealth gap has increased since the 1990s—a time when inequality in the United States is at an all-time high.

For some, these economic trends mean holding onto—or trying to achieve—middle-class status. For others, faced with unemployment or very low-wage work, poverty is the result. Alan Jenkins ("Inequality, Race, and Remedy") demonstrates that poverty is still a part of the landscape both because of the legacy of racial inequality and because of the ineffectiveness of current social policies. Rather than blaming the poor for their plight, social policies could address the barriers that limit options for many people of color.

Poverty, even among people of color, is often invisible—at least to those with more privilege. The major hurricanes along the Gulf Coast in 2005 revealed to the wider public what people in poor communities around the nation intimately knew: They were not participating in the American Dream. Avis A. Jones-DeWeever and Heidi Hartmann ("Abandoned Before the Storms: The Glaring Disaster of Gender, Race, and Class Disparities in the Gulf") show how the intersection of class, race, *and* gender made women of color in this region vulnerable both before, during, and after this natural disaster. Many students might think that people are poor because they are not working, but Jones-DeWeever and Hartmann show that African American women in the Gulf region are in the labor market, but far too many have low earnings, because women are the majority of the nation's minimum-wage earners.

New Orleans is not unique; many urban communities have experienced a decline of industrial jobs, pushing people into low-wage service work. However, the growing service economy is two-tiered. Many people of color with high levels of educational attainment are now able to enter the middle class. Thus, African American and other racial-ethnic communities are more economically diverse than in the past. Opportunities for advancement are shaped not only by race and class but also by one's generation. Different age groups face different opportunities, depending on the economic and social conditions of their times. The Baby Boomer generation (those born roughly between 1946 and 1967), gained opportunities during an expanding economy and as civil rights legislation opened new doors. Those born later now find deindustrialization reshaping the work world they encounter. Celeste Watkins ("A Tale of Two Classes: The Socio-Economic Divide among Black Americans Under 35") addresses the political and economic trends currently affecting the opportunities for Black Americans under 35.

Vilna Bashi Bobb and Averil Clarke ("Experiencing Success: Structuring the Perception of Opportunities for West Indians") also look at the differences between generations. In their article, they challenge the myth that culture is the key to explaining the success of West Indian newcomers. The authors compare the experiences of West Indian immigrants in New York City with the second generation, those born here or who came as young children. The experiences of the children of immigrants with racism create different perceptions from those of their parents, and they are less likely to sacrifice and to believe in education as a strategy for advancement. Like the Lessinger article in Part VI about South Asian immigrants, Bobb and Clarke's article highlights how the persistence of racism challenges the notion that working hard is all newcomers need to do to achieve the American Dream. We see that now, as in the past, power and social policies provide very different opportunities for racial and ethnic groups, perpetuating racial and ethnic social inequalities.

REFERENCES

Acuña, Rudolfo. 1981. *Occupied America: A History of Chicanos*. Second Edition. New York: Harper and Row.

Katznelson, Ira. 2005. *When Affirmative Action Was White*. New York: Norton.

Marks, Carole. 1989. *Farewell—We're Good and Gone: The Great Black Migration*. Bloomington: Indiana University Press.

Oliver, Melvin, and Thomas Shapiro. 1995. *Black Wealth/White Wealth*. New York: Routledge

Takaki, Ronald. 1989. *Strangers from a Different Shore*. Boston: Little, Brown and Company.

29

Wealth Inequality Trends

MELVIN OLIVER AND THOMAS M. SHAPIRO

Oliver and Shapiro show the importance of distinguishing between wealth and income. Although income gaps remain between White Americans and other racial-ethnic groups, wealth is even more significant in reproducing racial inequality because it cumulates over time and can be passed on to future generations. The wealth gap in the United States is also the result of specific social policies—both in the past and the present—that suppress the economic well-being of African Americans and other people of color.

The growth and dispersion of wealth continues a trend anchored in the economic prosperity of post–World War II America. Between 1995 and 2001, the median net worth of all American families increased 39 percent, and median net financial assets grew by 60 percent. The growth of pension accounts (IRAs, Keogh plans, 401(k) plans, the accumulated value of defined contribution pension plans, and other retirement accounts) and stock holdings seems to account for much of this wealth accumulation.

While wealth grew and spread to many American families, there was little action at the bottom of the wealth spectrum as the percent of families with zero or negative net worth only dropped from 18.5 to 17.6, and those with no financial assets fell from 28.7 to 25.5.

Wealth remains highly concentrated, especially financial wealth, which excludes home equity. In 2001, the richest 5 percent of American households controlled over 67 percent of the country's financial wealth; the bottom 60 percent had 8.8 percent; and the bottom 40 percent just 1 percent.

The context of wealth growth and inequality in the last decade situates our concern about racial inequality and the progress of American families, as indeed, the rich have gotten richer. The number of families with net worth of $10 million or more in 2001 quadrupled since 1980. A *New York Times* article even bemoaned how the super rich are leaving the mere rich far behind. These 338,400 hyper-rich families emerged as the biggest winners in the new global economy, as new technologies spurred by tax incentives evolved, as the stock market soared, and as top executives in the corporate world received astronomical pay.

In 1965, CEOs took home twenty-four times more in pay than what average workers earned. In 1980, CEO compensation was forty-two times the average

SOURCE: From *Black Wealth/White Wealth*, Melvin L. Oliver and Thomas M. Shapiro.
© 2006. Reprinted by permission of Routledge.

American worker's pay. The gap in pay between average workers and large company CEOs surpassed 300-to-1 in 2003. In addition, in 2004, the heads of America's 500 biggest companies received a whopping 54 percent pay increase. Our point here is neither to paint CEOs as poster children for greed nor to argue that everybody should receive equal pay; rather, it is to underscore the growing magnitude of pay inequity and to understand it as a source of inequality. By comparison, in Japan a typical executive makes eleven times what a typical worker brings home and in Britain, twenty-two times.

The wealthy were the biggest beneficiaries of tax policy during President [George W.] Bush's first term. In fact, the bulk of the 2001 tax cuts—53 percent—will go to the top 10 percent of taxpayers. The tax cut share of the top 0.1 percent will amount to a 15 percent slice of the total value of the tax cut pie. Another reason that the wealthiest fare much better is that the tax cuts over the past decade have sharply lowered tax rates on income from investments, such as capital gains, interest, and dividends. While there are many reasons for the continuing wealth inequality trend, government policy has clearly abetted, encouraged, and privileged the property, capital, and income of America's wealthiest families.

Politicians' rhetoric aside, the tax treatment of income from labor compared with earnings from investment reveals the kernel of the kind of ownership society advocated by the present Bush administration and others....

WHAT FACTS HAVE CHANGED?

In 1995 when *Black Wealth/White Wealth* was published, we presented data that was in many respects a new way of gauging the economic progress of black Americans vis-à-vis white Americans. Most commentators and analysts were familiar and comfortable with income comparisons that provided a window on whether there was growing or declining racial economic inequality. But the focus on wealth, "the net value of assets (e.g., ownership of stocks, money in the bank, real estate, business ownership, etc.) less debts" created a different gestalt or perspective on racial inequality.

This gestalt had two dimensions. The first is the conceptual distinction between income and assets. While income represents the flow of resources earned in a particular period, say a week, month, or year, assets are a stock of resources that are saved or invested. Income is mainly used for day-to-day necessities, while assets are special monies not normally used for food or clothing but are rather a "surplus resource available for improving life chances, providing further opportunities, securing prestige, passing status along to one's family," and securing economic security for present and future generations. The second dimension is the quantitative: To what extent is there parity between blacks and whites on assets? Do blacks have access to resources that they can use to plan for their future, to enable their children to obtain a quality education, to provide for the next generation's head start, and to secure their place in the political community? For these reasons, we focused on inequality in wealth as the sine qua non indicator of material well-being. Without sufficient assets, it is difficult to lay claim to economic security in American society.

The baseline indicator of racial wealth inequality is the black–white ratio of median net worth. To what degree are blacks approaching parity with whites in terms of net worth? The change in gestalt is amply demonstrated in comparisons of black–white median income ratios to black–white median net worth ratios. For example, the 1988 data reported on in *Black Wealth/White Wealth* showed that black families earned sixty-two cents for every dollar of median income that white families earned. However, when the comparisons shift to wealth, the figure showed a remarkably deeper and disturbing level of racial inequality. For every dollar of median net worth that whites controlled, African Americans controlled only eight cents! This markedly different indicator of inequality formed the basis of the analysis contained in *Black Wealth/White Wealth* as we attempted to describe its origins, its maintenance, and its continuing significance for the life chances of African American families and children.

How has this landmark indicator on racial inequality changed in the period since the publication of *Black Wealth/White Wealth?* Using the most recent data available, it appears, not unsurprisingly, that the level of racial wealth inequality has not changed but has shown a stubborn persistence that makes the data presented in 1995 more relevant than ever because the pattern we discerned suggests a firmly embedded racial stratification. The most optimistic analyses suggest that the black–white median net worth ratio is 0.10, that is, blacks have control of ten cents for every dollar of net worth that whites possess. However, the most pessimistic estimate indicates that the ratio is closer to seven cents on the dollar. This slim range demonstrates that the level of wealth inequality has not changed appreciably since the publication of *Black Wealth/White Wealth*....

To gain a flavor for what has happened to African American wealth since 1998, we take the liberty to present four "composite cases" that characterize some of the circumstances that have contributed to the persistence of racial wealth inequality during this period. These "composites" are based on our extensive academic and policy experience in trying to understand the dynamics of wealth accumulation in the United States over the past ten years, and our work with many organizations around the country that has led to a deep familiarity with hundreds of families who have struggled to secure a solid asset base. None of these stories represent actual people, but their circumstances are those that many face, with the consequences being that black wealth has stagnated relative to white wealth, and that ten years after our initial publication of the baseline data on racial wealth inequality, there has been a gnawing persistence of the basic trends.

A STRIVING BLACK MIDDLE CLASS

Cynthia and James Braddock are the epitome of the black middle class. Both are college graduates, have steady jobs (Cynthia as a public school teacher and James as a corporate middle manager in the human resources division of a Fortune 500 company). During the 1980s and 1990s they have saved steadily in their employer-sponsored 403(b) and 401(k) savings plans, converting part of it into a down payment on their first home. They provide their two children with fashionable clothes, extra tutoring, and social opportunities that enrich them both academically and culturally. Since the stock market bust of 1999 they have seen their savings eaten away at by diminishing stock returns and found their purchasing power eroded by paltry pay raises. Consequently, maintaining their standard of living is increasingly dependent on the five credit cards that they have, two of which have reached their maximum. Their net worth has increased thanks to the increasing value of their home, but the decline of their stock portfolio in their employment-sponsored savings program has not been replaced by higher earnings even though they make steady monthly contributions.

THE STRUGGLING WORKING CLASS

The lives of Kenneth and Barbara Jones characterize the struggling working class of black America. Married for ten years, they have managed in the last five years to both gain steady employment. Kenneth works for a local retail establishment that does not have a broad array of benefits programs. His wife Barbara is a teacher's aide who does have access to health care and an employer-sponsored savings program, which provides needed benefits for the whole family (one preschooler and one child in a local public elementary school). They have not been able to save much, but demonstrated a strong record of making their rent and utility payments on time that helped them qualify for an affordable mortgage sponsored by a major bank in their city. With no down payment, they were able to purchase a small home that costs about the same in terms of monthly costs as their rental unit. They depend on credit cards to make routine and emergency repairs and are thus in debt. However, with the value of their home increasing, they generally feel optimistic about their future and that of their children.

THE DISENFRANCHISED BLACK ELDERLY

Rosa Williams is a resident of an inner city community whose homes were among the first owned by African Americans in the mid-fifties in a formerly all-white neighborhood. She is representative of what is happening to too many African American elders who worked hard to pay off their homes and had hoped to live a life of dignity and respect in their final years in their own home. Instead, Mrs. Williams has lost her home as a consequence of a contract she entered into with a subprime lender who promised her a sufficient loan to not only make a major repair but to also have money left over that would serve as a cushion for other major expenses. Mrs. Williams thought that the loan terms were similar to charges at a conventional bank, and even though her only income derived from social security, she felt confident she could pay it back. However, she did not read the fine print that was written in complicated legal language that contained a balloon payment that required full payment of the loan after one year and restricted her ability to refinance. Thus, Mrs. Williams lost her home, her only major asset and shelter, and has had to move in with her daughter and three grandchildren. Stripped of her hard-earned equity, Mrs. Williams has lost what little wealth she has and must spend her golden years on the edge of economic insecurity.

THE LEGACY OF TANF

Donna Smith, head of a household that includes three school-aged kids, joined the labor market for the first time because of the stringent requirements associated with Temporary Assistance For Needy Families [TANF]. Working first in a temporary job, Donna landed a position as a housecleaner in a hotel where the wages do not lift her and her family out of poverty. In fact, Donna cannot make ends meet from paycheck to paycheck. Without a credit history and with only a recent employment record, Donna cannot obtain credit from conventional lenders or even major credit card suppliers. Her only recourse is to secure payday loans so that she can pay her electricity before her service is canceled or provide money for her daughter's field trip that the school requires. Consequently, she is falling further and further behind economically because the payday lender has already garnished her meager check, making it even more difficult to make ends meet.

Embedded in these composite stories are some of the contradictory facts and new dimensions of financial life in America that have affected the persistence of the black–white racial wealth gap. They include a strong economy of the 1990s that enabled greater savings, especially in employer-based savings programs, but which has petered out recently; a stock market bust that punished some of the newest entrants into the market most severely; increasing credit card debt; a growing trend of black home ownership complemented by growing sub-prime and predatory lending directed at minority communities; and growth in the working poor due to the influx of the TANF population into the labor market. This mix of factors weaves the mosaic underlying the story of the continuing racial wealth gap in the first decade of the twenty-first century.

THE STORY OF THE PERSISTENCE OF THE RACIAL WEALTH GAP

...We developed a sociology of wealth and racial inequality in *Black Wealth/White Wealth*, which situated the study of wealth among concerns with race, class, and social inequality. This theoretical framework elucidated the social context in which wealth generation occurs and demonstrated the unique and diverse social circumstances that blacks and whites face. Three concepts we developed provided a sociologically grounded approach to understand the racial wealth gap and highlighted how the opportunity structure disadvantages blacks and contributes to massive wealth inequalities between the races. The first concept, racialization of state policy, explores how state policy has impaired the ability of most black Americans to accumulate wealth from slavery throughout American history to contemporary institutional discrimination. The "economic detour" helps us understand the relatively low level of entrepreneurship among the small scale and segmentally niched businesses of black Americans, leading to an emphasis on consumer spending as the route to economic assimilation. The third concept—the sedimentation of racial inequality—explores how the cumulative effects of the past have seemingly cemented blacks to the bottom of America's economic hierarchy in regards to wealth.

These concepts do much to show how this differential opportunity structure developed and worked to produce black wealth disadvantages. It also builds a strong case that layering wealth deprivation generation after generation has been central in not only blacks' lack of wealth but also whites' privileged position in accumulating wealth. As we noted [in *Black Wealth/White Wealth*]:

> What is often not acknowledged is that the accumulation of wealth for some whites is intimately tied to the poverty of wealth for most blacks. Just as blacks have had "cumulative disadvantages," whites have had "cumulative advantages." Practically every circumstance of bias and discrimination against blacks has produced a circumstance and opportunity of positive gain for whites.

The past opportunity structure that denied blacks access or full participation in wealth-building activities serves as a powerful deterrent to current black ambitions for wealth. Without an inheritance that is built on generations of steady economic success, blacks, even when they have similar human capital and class position, lag far behind their white counterparts in their quest to accumulate a healthy nest egg of assets....

The Rise and Decline of a Tight Labor Market and a Bull Stock Market

Black Wealth/White Wealth documented wealth data that reflected a period in the American economy characterized by relatively high unemployment rates and a stagnant economy. However, this period was followed by one of the largest and longest economic expansions in the history of the United States. From its beginning in March 1991 to its ending in November 2001, the United States endured a record expansion....

For African Americans this was a period of tight labor markets that led to greater levels of labor force participation owing to the existence of greater demand for their participation in the economy.* Employers made "extra efforts to overcome the barriers created by skill and spatial mismatch" to reach out to

* Eds. note: A "tight labor market" refers to one in which there is relatively low unemployment and more plentiful jobs.

African American workers to fill their growing labor needs. Moreover, "employers may find discrimination more costly when the economy is strong and their usually preferred type of job candidate is fully employed elsewhere." In the throes of a heated and tight labor market, *Business Week* proclaimed, "With the economy continuing to expand and unemployment at its lowest point in 30 years, companies are snapping up minorities, women, seniors, and any-one else willing to work for a day's pay."

African Americans, however, did not wholly benefit from this extraordinary period in American history. Black joblessness continued to be a problem. The historical ratio of two-to-one black-to-white unemployment rates persisted with black men averaging 7.1 percent compared with 3 percent for white men, while black women averaged 6.8 compared with 3.2 percent for white women in the latter half of 1999. Nevertheless, those African Americans who were employed during this period saw real wage gains that could be translated into savings, investments, and an increase in net worth.

Another aspect of the expansion of the economy during the 1990s was the rapid rise in the stock market. Fueled by technology stocks and the growth of key stocks like Microsoft, Sun, Yahoo, and other new stock offerings in the technology sector, the stock market started to attract investments not only from high-income and high-wealth individuals, but also from an increasing number of middle-class families and even working-class families. This investment was facilitated by growing participation in employer-sponsored savings programs that enabled employees to make tax-deferred and/or matched contributions through payroll deductions. The ease of the transaction and the constant media and public interest in the high-flying stock market encouraged mass participation. The market rose steadily and rapidly. Beginning from a monthly average in 1992 in the low 400s, the Standard and Poor's 500 tripled in size by 1999. If one were lucky enough to purchase Microsoft or Yahoo early, then his or her gains would have been astronomical. For example, when Yahoo was first available as a public offering shares were sold for $1.24. By December 1999 Yahoo listed for $108.17. It was the desire for these kinds of returns that fueled an overheated market and led to the description of "irrational exuberance" concerning the frenzy for the "market."

African Americans, while constrained by resources, also entered into this frenzy. The decade of the 1990s was the breakthrough era for African American involvement in the stock market. Facilitated by employer savings plans and, for the first time, sought after by stock and brokerage firms, African Americans invested readily into the market. In 1996 blacks had a median value of $4,626 invested in stocks and mutual funds. At the height of the market, that value had almost doubled to $8,669. During this period African American stock market investors had closed the black–white ratio of stock market value from twenty-eight cents on the dollar to forty cents on the dollar. However, the market's plunge after 1999 sent African American portfolio values down to a median average of $3,050. This brought the black–white ratio of stock market value back in line with the 1996 level, eroding all the gains that the bull market had bestowed....

HOME OWNERSHIP, NEW MORTGAGE AND CREDIT MARKETS

... Home equity is the most important wealth component for average American families, and even though home ownership rates are lower, it is even more prominent in the wealth profiles of African American families. Although housing appreciation is very sensitive to many characteristics relating to a community's demographics and profile (which realtors euphemistically call "location, location, location"), overall home ownership has been a prime source of wealth accumulation for black families. For example, for the average black home owner, homes created $6,000 more wealth between 1996 and 2002. However, fundamentally racialized dynamics create and distribute housing wealth unevenly. The Federal Reserve Board kept interest rates at historically low levels for much of this period, and this fueled both demand and hastened converting home equity wealth into cash.

... The typical home owned by white families increased in value by $28,000 more than homes owned by blacks. Persistent residential segregation, especially in cities where most blacks live, explains this equity difference as a compelling index of bias that costs blacks dearly. This data point corroborates other recent research demonstrating that rising housing wealth depends upon a community's demographic characteristics, especially racial composition. One study concludes that homes lost at least 16 percent of their value when located in neighborhoods that are more than 10 percent black. Thus, a "segregation tax" visits black home owners by depressing home values and reducing home equity in highly segregated neighborhoods. Shapiro summarizes the case: "The only prudent conclusion from these studies is that residential segregation costs African American home owners enormous amounts of money by suppressing their home equity in comparison to that of white home owners. The inescapable corollary is that residential segregation benefits white home owners with greater home equity wealth accumulation."[1] Furthermore, most African American families rent housing and thus are not positioned to accumulate housing wealth, mainly because of affordability, credit, and access issues....

In 1995, access to credit for minorities was a major issue. Financial institutions responded both to criticisms regarding credit discrimination and to the newly discovered buying capacity of minorities. Increasing numbers of African American and Hispanic families gained access to credit cards throughout the 1990s: 45 percent of African American and 43 percent of Hispanic families held credit cards in 1992 and by 2001 nearly 60 percent of African American and 53 percent of Hispanic families held credit cards. The irony here is that as access to credit broadened under terms highly favorable to lenders, debt became rampant and millions of families became ensnared in a debt vice.

Credit card debt nearly tripled from $238 billion in 1989 to $692 billion in 2001. These figures represent family reliance on financing consumption through debt, especially expensive credit in the form of credit cards and department store charge cards. During the 1990s, the average American family experienced a

53 percent increase in credit card debt—the average family's card debt rising from $2,697 to $4,126. Credit card debt among low-income families increased 184 percent. Even high-income families became more dependent on credit cards: There was 28 percent more debt in 2001 than in 1989. The main sources of credit card debt include spiraling health care costs, lower employer coverage of health insurance, and rising housing costs amid stagnating or declining wages after 2000 and increasingly unsteady employment for many. This suggests strongly that the increasing debt is not the result of frivolous or conspicuous spending or lack of budgetary discipline; instead, deferring payment to make ends meet is becoming the American way for many to finance daily life in the new economy.

Given that a period of rising income did not lift the African American standard of living, and given the context of overall rising family debt, an examination of the racial component of credit card debt furthers our understanding of the contemporary processes associated with the continuation of the economic detour and the further sedimentation of inequality. The average credit card debt of African Americans increased 22 percent between 1992 and 2001, when it reached an average of nearly $3,000. Hispanic credit card debt mirrored blacks by rising 20 percent in the same period to $3,691. As we know, the average white credit card debt was higher, reaching $4,381 in 2001. One of the most salient facts involves the magnitude and depth of African American reliance on debt. Among those holding credit cards with balances, nearly one in five African Americans earning less than $50,000 spend at least 40 percent of their income paying debt service. In other words, in every 8-hour working day these families labor 3.2 hours to pay off consumer debt. Even though black families carry smaller monthly balances, a higher percentage of their financial resources goes toward servicing debt.

The median net worth of African American families at the end of 2002 was $5,988, essentially the same as it was in 1988.[2] Again, it is not as if nothing happened since we wrote *Black Wealth/White Wealth;* indeed, African American fortunes expanded with good times and contracted with recessions and the bursting of the stock market bubble. In the last decade, the high point of African American (and Hispanic) wealth accumulation was 1999, when it registered $8,774, just before the bursting of the stock market bubble in early 2000. Between 1999 and 2002, African American wealth declined from $8,774 to $5,988 wiping out more than a decade's worth of financial gains....

THE DARK SIDE OF HOME OWNERSHIP

Subprime lending is targeted to prospective homebuyers with blemished credit histories or with high levels of debt who otherwise would not qualify for conventional mortgage loans. A legitimate niche for these kinds of loans brings home ownership within the grasp of millions of families. These loan products are essential in expanding home ownership rates. In return for these riskier investments, financial institutions charge borrowers higher interest rates, often requiring higher processing and closing fees, and often containing special loan conditions like prepayment penalties, balloon payments, and adjustable interest rates.

The subprime market expanded greatly in the last decade as part of new, aggressive marketing strategies among financial institutions hungrily eyeing rising home ownership and seeing promising new markets. Moreover, the mortgage finance system in the United States became well integrated into global capital markets, which offer an ever-growing array of financial products, including subprime loans. Subprime loan originations grew fifteen-fold, from $35 billion to $530 billion between 1994 and 2004. Reflecting the increasing importance of subprime loans to the financial industry, the subprime share of mortgage loans has seen a parallel meteoric rise from less than 4 percent in 1995 to representing about 17 percent of mortgage loans in 2004.

Loan terms like prepayment penalties and balloon payments increase the risk of mortgage foreclosure in subprime home loans, even after controlling for the borrower's credit score, loan terms, and varying economic conditions. One study from the Center for Community Capitalism demonstrates that subprime prepayment penalties and balloon payments place Americans at substantially greater risk of losing their homes.

A key finding is that subprime home loans with prepayment penalties with terms of three years or longer faced 20 percent greater odds of entering foreclosure than loans without prepayment penalties. Prepayment restrictions mean that home owners are stuck with loan terms, unable to refinance, to obtain lower rates, to weather financial difficulties, or to take advantage of lower interest rates. Another important finding shows that, subprime home loans with balloon payments, where a single lump sum payment many times the regular payment amount is due at the end of the loan term, face 46 percent greater odds of entering foreclosure than loans without such a term. In addition, borrowers whose subprime loans include interest rates that fluctuate face 49 percent greater odds of entering foreclosure than borrowers with fixed rate subprime mortgages. Under these terms, home borrowers are more likely to lose their homes....

Delinquency (falling behind in mortgage payments) and losing one's home through foreclosure are hitting vulnerable neighborhoods hardest. Concentrated foreclosures can negatively affect the surrounding neighborhoods, threatening to undo community building and revitalization efforts achieved through decades of collaborative public–private partnerships, community organizing, and local policy efforts.

BLACK CONSUMERS AND THE URBAN MARKET

Our discussion of the "economic detour" stressed that blacks were the only group that came to America and found insurmountable barriers to establishing sustainable business. This has led to their being "forced into the role of the consumer." At the eve of the twenty-first century, blacks have come to be identified in corporate America as one of the most important market niches for consumer goods.

Euphemistically termed the "urban market," black communities and youth are bombarded with and urged to consume a range of commodities that are designed to meet their perceived "cultural tastes" and "lifestyle." These goods range from fashionable clothes to the latest athletic shoes, from cigarettes and specialized alcoholic beverages to high-tech stereo and audio equipment. With a combined purchasing power of $372 billion in 2002, major retail and consumer services had found the "urban market" to be a motherlode of opportunities. As a consequence, African Americans are targeted through advertisement that are designed to reach them specifically through television, radio, magazines, or billboards. For example, magazine advertisers spend over $300 million dollars in 2002 in the top twelve advertising categories (e.g., toiletries and cosmetics, apparel, auto, etc.) directly targeted towards African Americans. Their advertising dollars bring strong returns. For example:

- African Americans account for more than 30 percent of industry spending in the $4 billion hair market.
- Black consumers spend more on telephone services than any other consumer group. Their expenditures in this category totals $918 per capita annually, or 8.1 percent more than the average.
- According to a 2001 study by Cotton Incorporated, African Americans consumers will spend an average of $1,427 on clothing per year for themselves—$458 more than the average consumer.
- The average African American family spends 30 percent more on weekly groceries than the U.S. population at large.

But the evidence is not that this is conspicuous consumption, but rather it is the consequences of living in communities and cities where the costs of good and services, especially for the poor, are higher. Data consistently show that the "poor pay more" for essential goods and services.

...Moreover, low-income families have less access to mainstream financial institutions and thus are much more likely to need the services of a check casher, typically a storefront operation offering check-cashing services and payday loans. Unusually high interest rates and fees distinguish these alternative financial services, which more low-income families rely upon because low-income and minority communities are not where mainstream banks look for customers or locate branches....

Many low-income families live paycheck-to-paycheck. When a pothole in the road blows out a tire, a child needs to visit the emergency room for medical treatment, or the family budget simply comes up short of the next paycheck, emergency loans seem to offer a temporary fix. Short-term loans come with a high price, especially when they come from payday lenders. These cash advances or borrowing against the next paycheck invariably lead to further financial crisis....

The bottom line is that the racial wealth gap worsened during the last decade. Thus, even though we could make an argument that African American achievements on the job and in schools were improving, an escalating racial wealth gap reversed these accomplishments.

REFERENCES

1. Thomas Shapiro. 2004. *The Hidden Cost of Being African American* (New York: Oxford University Press).
2. This section uses Survey of Income Program Participation (SIPP) data from 2002 as reported in the Pew Hispanic Center Report.

DISCUSSION QUESTIONS

1. In these days of increasing inequality, what advantages does an analysis of wealth add to a study of racial stratification?
2. Most Americans accumulate wealth by purchasing a home. Why is this path problematic for many African Americans?

30

Inequality, Race and Remedy

ALAN JENKINS

Despite the American ideal that one's social origin should not erect barriers to social and economic opportunity, the reality persists that race is a strong predictor of poverty. Here Jenkins discusses the rise in poverty rates since 2000, as well as the role of race in producing poverty. He also reviews some of the efforts that can be made to address the persistence of poverty in the United States.

Our nation, at its best, pursues the ideal that what we look like and where we come from should not determine the benefits, burdens, or responsibilities that we bear in our society. Because we believe that all people are created equal in terms of rights, dignity, and the potential to achieve great things, we see inequality based on race, gender, and other social characteristics as not only unfortunate but unjust. The value of equality, democratic voice, physical and economic security, social mobility, a shared sense of responsibility for one another, and a chance to start over after misfortune or missteps—what many Americans call redemption—are the moral pillars of the American ideal of opportunity.

Many Americans of goodwill who want to reduce poverty believe that race is no longer relevant to understanding the problem or to fashioning solutions for it. This view often reflects compassion as well as pragmatism. But we cannot solve the problem of poverty—or, indeed, be the country that we aspire to be—unless we honestly unravel the complex and continuing connection between poverty and race.

Since our country's inception, race-based barriers have hindered the fulfillment of our shared values and many of these barriers persist today. Experience shows, moreover, that reductions in poverty do not reliably reduce racial inequality, nor do they inevitably reach low-income people of color. Rising economic tides do not reliably lift all boats.

In 2000, after a decade of remarkable economic prosperity, the poverty rate among African Americans and Latinos taken together was still 2.6 times greater than that for white Americans. This disparity was stunning, yet it was the smallest difference in poverty rates between whites and others in more than three decades. And from 2001 to 2003, as the economy slowed, poverty rates for most communities of color increased more dramatically than they did for whites, widening the racial poverty gap. From 2004 to 2005, while the overall number

SOURCE: From *The American Prospect*, April 22, 2007.

of poor Americans declined by almost 1 million, to 37 million, poverty rates for most communities of color actually increased. Reductions in poverty do not inevitably close racial poverty gaps, nor do they reach all ethnic communities equally.

Poor people of color are also increasingly more likely than whites to find themselves living in high-poverty neighborhoods with limited resources and limited options. An analysis by The Opportunity Agenda and the Poverty & Race Research Action Council found that while the percentage of Americans of all races living in high-poverty neighborhoods (those with 30 percent or more residents living in poverty) declined between 1960 and 2000, the racial gap grew considerably. Low-income Latino families were three times as likely as low-income white families to live in these neighborhoods in 1960, but 5.7 times as likely in 2000. Low-income blacks were 3.8 times more likely than poor whites to live in high-poverty neighborhoods in 1960, but 7.3 times more likely in 2000.

These numbers are troubling not because living among poor people is somehow harmful in itself, but because concentrated high-poverty communities are far more likely to be cut off from quality schools, housing, health care, affordable consumer credit, and other pathways out of poverty. And African Americans and Latinos are increasingly more likely than whites to live in those communities. Today, low-income blacks are more than three times as likely as poor whites to be in "deep poverty"—meaning below half the poverty line—while poor Latinos are more than twice as likely.

THE PERSISTENCE OF DISCRIMINATION

Modern and historical forces combine to keep many communities of color disconnected from networks of economic opportunity and upward mobility. Among those forces is persistent racial discrimination that, while subtler than in past decades, continues to deny opportunity to millions of Americans. Decent employment and housing are milestones on the road out of poverty. Yet these are areas in which racial discrimination stubbornly persists. While the open hostility and "Whites Only" signs of the Jim Crow era have largely disappeared, research shows that identically qualified candidates for jobs and housing enjoy significantly different opportunities depending on their race.

In one study, researchers submitted identical résumés by mail for more than 1,300 job openings in Boston and Chicago, giving each "applicant" either a distinctively "white-sounding" or "black-sounding" name—for instance, "Brendan Baker" versus "Jamal Jones." Résumés with white-sounding names were 50 percent more likely than those with black-sounding names to receive callbacks from employers. Similar research in California found that Asian American and, especially, Arab American résumés received the least-favorable treatment compared to other groups. In recent studies in Milwaukee and New York City, meanwhile, live "tester pairs" with comparable qualifications but of differing races tested not only the effect of race on job prospects but also the impact of an apparent criminal record. In Milwaukee, whites reporting a criminal record were more likely to receive a callback from employers than were blacks without a criminal record. In New York, Latinos and African Americans without criminal records received fewer callbacks than did similarly situated whites, and at rates comparable to whites with a criminal record.

Similar patterns hamper the access of people of color to quality housing near good schools and jobs. Research by the U.S. Department of Housing and Urban Development (HUD) shows that people of color receive less information from real-estate agents, are shown fewer units, and are frequently steered away from predominantly white neighborhoods. In addition to identifying barriers facing African Americans and Latinos, this research found significant levels of discrimination against Asian Americans, and that Native American renters may face the highest discrimination rates (up to 29 percent) of all.

This kind of discrimination is largely invisible to its victims, who do not know that they have received inaccurate information or been steered away from desirable neighborhoods and jobs. But its influence on the perpetuation of poverty is nonetheless powerful.

THE PRESENT LEGACY OF PAST DISCRIMINATION

These modern discriminatory practices often combine with historical patterns. In New Orleans, for example, as in many other cities, low-income African Americans were intentionally concentrated in segregated, low-lying neighborhoods and public-housing developments at least into the 1960s. In 2005, when Hurricane Katrina struck and the levees broke, black neighborhoods were most at risk of devastation. And when HUD announced that it would close habitable public-housing developments in New Orleans rather than clean and reopen them, it was African Americans who were primarily prevented from returning home and rebuilding. This and other failures to rebuild and invest have exacerbated poverty—already at high levels—among these New Orleanians.

In the case of Native Americans, a quarter of whom are poor, our government continues to play a more flagrant role in thwarting pathways out of poverty. Unlike other racial and ethnic groups, most Native Americans are members of sovereign tribal nations with a recognized status under our Constitution. High levels of Native American poverty derive not only from a history of wars, forced relocations, and broken treaties by the United States but also from ongoing breaches of trust—like our government's failure to account for tens of billions of dollars that it was obligated to hold in trust for Native American individuals and families. After more than a decade of litigation, and multiple findings of governmental wrongdoing, the United States is trying to settle these cases for a tiny fraction of what it owes.

The trust-fund cases, of course, are just the latest in a string of broken promises by our government. But focusing as they do on dollars and cents, they offer an important window into the economic status that Native American communities and tribes might enjoy today if the U.S. government lived up to its legal and moral obligations.

Meanwhile, the growing diversity spurred by new immigrant communities adds to the complexity of contemporary poverty. Asian American communities, for example, are culturally, linguistically, and geographically diverse, and they span a particularly broad socioeconomic spectrum.

Census figures from 2000 show that while one-third of Asian American families have annual incomes of $75,000 or more, one-fifth have incomes of less than $25,000. While the Asian American poverty rate mirrored that of the country as a whole, Southeast Asian communities reflected far higher levels. Hmong men experienced the highest poverty level (40.3 percent) of any racial group in the nation.

RACE AND PUBLIC ATTITUDES

Americans' complex attitudes and emotions about race are crucial to understanding the public discourse about poverty and the public's will to address it. Researchers such as Martin Gilens and Herman Gray have repeatedly found that the mainstream media depict poor people as people of color—primarily African Americans—at rates far higher than their actual representation in the population. And that depiction, the research finds, interacts with societal biases to erode support for antipoverty programs that could reach all poor people.

Gilens found, for instance, that while blacks represented only 29 percent of poor Americans at the time he did his research, 65 percent of poor Americans shown on television news were black. In a more detailed analysis of TV newsmagazines in particular, Gilens found a generally unflattering framing of the poor, but the presentation of poor African Americans was more negative still. The most "sympathetic" subgroups of the poor—such as the working poor and the elderly—were underrepresented on these shows, while unemployed working-age adults were overrepresented. And those disparities were greater for African Americans than for others, creating an even more unflattering (and inaccurate) picture of the black poor.

Gray similarly found that poor African Americans were depicted as especially dysfunctional and undeserving of assistance, with an emphasis on violence, poor choices, and dependency. As Gray notes, "The black underclass appears as a menace and a source of social disorganization in news accounts of black urban crime, gang violence, drug use, teenage pregnancy, riots, homelessness, and general aimlessness. In news accounts .... poor blacks (and Hispanics) signify a social menace that must be contained."

Research also shows that Americans are more likely to blame the plight of poverty on poor people themselves, and less likely to support antipoverty efforts, when they perceive that the people needing help are black. These racial effects are especially pronounced when the poor person in the story is a black single mother. In one study, more than twice the number of respondents supported individual solutions (like the one that says poor people "should get a job") over societal solutions (such as increased education or social services) when the single mother was black.

This research should not be surprising. Ronald Reagan, among others, effectively used the "racialized" mental image of the African American "welfare queen" to undermine support for antipoverty efforts. And the media face of welfare recipients has long been a black one, despite the fact that African Americans have represented a minority of the welfare population. But this research also makes clear that unpacking and disputing racial stereotypes is important to rebuilding a shared sense of responsibility for reducing poverty in all of our communities.

REMOVING RACIAL BARRIERS

We cannot hope to address poverty in a meaningful or lasting way without addressing race-based barriers to opportunity. The most effective solutions will take on these challenges together.

That means, for example, job-training programs that prepare low-income workers for a globalized economy, combined with antidiscrimination enforcement that ensures equal access to those programs and the jobs to which they lead. Similarly, strengthening the right to organize is important in helping low-wage workers to move out of poverty, but it must be combined with civil-rights efforts that root out the racial exclusion that has sometimes infected union locals. And it means combining comprehensive immigration reform that offers newcomers a pathway to citizenship with living wages and labor protections that root out exploitation and discourage racial hierarchy.

Another crucial step is reducing financial barriers to college by increasing the share of need-based grants over student loans and better coordinating private-sector scholarship aid—for example, funds for federal Pell Grants should be at least double current levels. But colleges should also retain the flexibility to consider racial and socioeconomic background as two factors among many, in order to promote a diverse student body (as well as diverse workers and leaders once these students graduate). And Congress should pass the DREAM Act, which would clear the path to a college degree and legal immigration status for many undocumented students who've shown academic promise and the desire to contribute to our country.

Lack of access to affordable, quality health care is a major stress on low-income families, contributing to half of the nation's personal bankruptcies. Guaranteed health care for all is critical, and it must be combined with protections against poor quality and unequal access that, research shows, affect people of color irrespective of their insurance status.

Finally, we must begin planning for opportunity in the way we design metropolitan regions, transportation systems, housing, hospitals, and schools. That means, for example, creating incentives for mixed-income neighborhoods that are well-publicized and truly open to people of all races and backgrounds.

A particularly promising approach involves requiring an "opportunity impact statement" when public funds are to be used for development projects. The statement would explain, for example, whether a new highway will connect low-income communities to good jobs and schools or serve only affluent communities. It would detail where and how job opportunities would flow from the project and whether different communities would share the burden of environmental and other effects (rather than having the project reinforce traditional patterns of inequality). It would measure not only a project's expected effect on poverty but on opportunity for all.

When we think about race and poverty in terms of the shared values and linked fate of our people, our approach to politics as well as policy begins to change. Instead of balancing a list of constituencies and identity groups, our task becomes one of moving forward together as a diverse but cohesive society, addressing through unity the forces that have historically divided us.

DISCUSSION QUESTIONS

1. What have been the trends in national poverty rates in recent years? How are they related to race?
2. What changes in social policy does Jenkins argue are essential in addressing the connection between race and poverty?

31

Abandoned Before the Storms

The Glaring Disaster of Gender, Race, and Class Disparities in the Gulf

AVIS A. JONES-DEWEEVER AND HEIDI HARTMANN

Most now know that race and poverty were a fundamental aspect of the devastation caused by Hurricane Katrina in New Orleans and along the Gulf Coast. Less examined in the public response to Katrina is the relationship between race, class, and gender—a connection explored here by Jones-DeWeever and Hartmann's examination of class and poverty in New Orleans.

As Hurricanes Katrina and Rita ravaged the Gulf Coast region, the impact of the storms made clear the lingering implications of America's persistent divides. For a time, at least, the outcome of race and class disadvantage in America became crystal clear and garnered long overdue attention. Yet glaring gender disparities throughout the affected areas remained largely unexamined. This [article] fills that void by exploring how the multiple disadvantages faced by the women of the Gulf both increased their vulnerability in a time of crisis and, unless proactively addressed, remain an impediment to their ability to rebuild their lives long after the storms.

Long before the devastation of the hurricanes, many women within the Gulf region were already living at the bottom. Women in the affected states were the poorest women in the nation, among the least likely to have access to health insurance, and, despite high work participation rates, the most likely to find themselves stuck in a cycle of low-wage work. Most disadvantaged were women of color, who often faced quite limited opportunities and outcomes, especially with respect to employment and earnings, educational attainment, and ultimately, the likelihood of living in poverty. To tease out the plight of the women in the region prior to the devastation that compounded their challenges and to catch a glimpse of the opportunities awaiting them in their relocated homes, we use employment and poverty data from federal government sources as well as indicators developed by the Institute for Women's Policy Research (IWPR) through our *Status of Women and the States* series to uncover the multiple disadvantages faced by women

SOURCE: From Chester Hartman and Gregory D. Squires (Eds.) 2006, *There Is No Such Thing as a Natural Disaster: Race, Class, and Hurricane Katrina.* Reprinted by permission of Routledge.

in the city of New Orleans, its broader Metropolitan Statistical Area (MSA), and the MSAs of Biloxi–Gulfport–Pascagoula, Mississippi, and Beaumont–Port Arthur, Texas.

LIVING AT THE BOTTOM

The images splashed across CNN in the wake of Katrina exposed to many the disastrous effects of poverty in America. Yet, poverty had long been a festering problem in the United States, growing for four years in a row and leaving the poor further and further behind. By the time Katrina made it to shore, more than 36 million people across the nation were poor and a large proportion (42.3%) lived in deep poverty, with incomes below 50% of the poverty threshold (U.S. Census Bureau 2005). Even for the "average" poor person, the struggle to make ends meet was intense by historical standards. By 2004, the average income of those living in poverty had dropped further below the poverty line than at any time since that statistic was first recorded back in 1975 (Center on Budget and Policy Priorities 2005). America's poor had truly hit bottom in the modern age.

For most Americans, imagining the struggles of the poor in this country is quite difficult, if not disturbing. Poverty is typically tucked away, either confined to an urban enclave avoided by those who aren't within its boundaries by accident of birth, or dispersed broadly, on a lonely country road far away from neighbors jobs, and, in many respects, opportunity. In the lives of most Americans, one's only brush with poverty is the occasional discomfort felt due to the outstretched hand of a stranger on the street who claims to be homeless or the annual newscasts from the local food bank come Thanksgiving or Christmas. In this land of opportunity, poverty is hidden from view, like a messy closet one hopes goes undiscovered by an important houseguest. Although we all know it's there—somewhere—for most, poverty goes unseen and unacknowledged, as does its implications for everyday life, and ultimately, survival—that is, unless you're the one who is poor.

For America's poor, poverty is the one reality that all too often cannot be escaped. Especially among the deeply entrenched urban poor, poverty doesn't go on vacation. It doesn't provide an occasional break. Instead, it remains disturbingly consistent and touches lives in every way imaginable. On any one day in 2004, anywhere between 614,000 to 854,000 households in America had at least one family member who went hungry because there was not enough money to ensure that everyone could eat (Nord, Andrews, and Carlson 2005). Millions more lived amid overcrowded or substandard housing conditions or couldn't rely on consistent housing at all as the escalating price of keeping a roof over one's head continued to far outpace the value of wages, as it had done for the past 40 years (National Low Income Housing Coalition 2004). For many, being poor meant being trapped in failing school districts, unable to assure a quality education and thus, a way out for their children. It meant being relegated to a place where no one else wants to live, isolated, and disjointed from the rest of society, and having a higher likelihood than others to being exposed to environmental dangers (U.S. Environmental Protection Agency 1993). In New Orleans, that meant being trapped on low ground. In the real lives of real people, poverty is having to decide at the end of the month which is most important—electricity, running water, gas, or any of the other pressing expenses that must be paid. Poverty demands frequent juggling because there is simply not enough to fulfill every need. That is the reality of poverty in America, a reality that has become increasingly difficult to escape. And in an era of restructuring, outsourcing, and low, stagnant wages, poverty, America's dirty little secret, not only lives on, but thrives.

As the prevalence and depth of poverty has grown over the years, the crumbling value of the minimum wage has made finding a way out next to impossible. After adjusting for inflation, the value of the minimum wage is at its second lowest point since 1955, ensuring that working, and working very hard, will not be enough to guarantee an above-poverty-level existence (Bernstein and Shapiro 2005). A single mother of two, for example, working 40 hours per week, 52 weeks a year, earns only $10,712 before taxes—a full $5,000 below the poverty threshold (Boushey 2005). Even with the Earned Income Tax Credit, which maxes out at roughly $4,000, her family income still falls below the poverty line. As a result, she falls squarely within the ranks of the working poor (Shulman 2003; Shipler 2004).

The above-cited example is not atypical. The face of the working poor in America is, more often than not, the face of a woman. It is women who make up the majority (60%) of America's minimum-wage workers and 90% of those who remain stuck in low-wage work throughout their prime earning years (Lovell 2004; Rose and Hartmann 2004). And ironically, those women who lived in the metropolitan areas impacted by Hurricanes Katrina and Rita were especially likely to encounter this fate. The women of the Gulf participated in the labor force at about the same rate as women in the nation overall, and, with the lone exception of those who resided in the Beaumont-Port Arthur, Texas area, women in this region were less likely than women nationwide to work part-time or for only part of the year. Yet, despite their strong attachment to the labor force, in each geographic area examined, their median annual earnings trailed those of women nationally, with the women in the city of New Orleans falling furthest behind (Jones-DeWeever et al. 2006).

Unemployment, too, was especially prevalent in the Gulf region. Prior to Katrina, the city of New Orleans was particularly hard hit, with unemployment rates over 50% higher than that of the nation as a whole. Still, for many, even the advantage of holding a job meant little in terms of ensuring a life free of poverty. In each affected area examined, more than two in five of those who were poor experienced poverty despite being employed, and within the affected Metropolitan Statistical Area (MSA) of Biloxi–Gulfport–Pascagoula in Mississippi, more than half of those who lived below the poverty line (52.7%) worked in 2004, far exceeding the national rate of 45.3%.

Taken together, the evidence paints a picture of a hard-working people. Yet a strong work ethic failed to insulate them from the likelihood of being poor. Overall, poverty in the region was more commonplace than was the case nationwide. And, as is typical throughout this nation and around the world, women were considerably more likely to be poor than men, with women in New Orleans most at risk of poverty in the Gulf region. Fully one quarter (25.9%)

of the women who made New Orleans their home lived in poverty, well outpacing the national poverty rate (14.5%) for women (Gault et al. 2005).

For those who faced life beyond their working years, poverty was all too common. Like younger women, older women were much more likely to face poverty in their golden years than their male counterparts. With the exception of Biloxi–Gulfport–Pascagoula, in every one of the four areas examined, the chance that women aged 65 years or older would live in poverty in this region exceeded the national rate and roughly doubled the rate of poverty experienced by elderly men. Also, like their younger counterparts, those who lived within the city of New Orleans were at greatest risk of poverty, as nearly a quarter (24.3%) of the Crescent City's older women lived below the poverty line (Gault et al. 2005).

Women in this region, then, both young and old, and particularly those who lived in New Orleans, faced a remarkably high likelihood of being poor. Most at risk were those who lived up to the responsibility of raising children on their own. In both the city of New Orleans and its broader Metropolitan Statistical Area, the percentage of female-headed families falling below the federal poverty line exceeded the national average, topping off at roughly 40%. … And as Katrina loomed, it was these families who faced the impending crisis with the fewest resources at their disposal to make it out alive and to keep their families together as chaos emerged.

Even though poverty in the Katrina- and Rita-impacted areas was significant and clearly outpaced the nation as a whole, the high level of poverty experienced in the Gulf region is consistent with broader poverty proliferation throughout the South. As compiled and ranked by IWPR's *Status of Women in the States* (2004), all eight states in the South Central Region rank among the bottom third of all 50 states and the District of Columbia for the percent of women living above the poverty line. Mississippi ranks dead last, followed closely by Louisiana (ranked 47th) and Texas (ranked 44th). Kentucky (ranked 36th), the highest-ranked state in the region, still falls significantly below the mid-point for the nation as a whole (Werschkul and Williams 2004).

Not only are women in the Gulf region the most likely to be poor, they are also at high risk of being without health insurance. Texas ranked worst in the nation in terms of women's health insurance coverage, with Louisiana and Mississippi also falling near the bottom, ranking 49th and 43rd, respectively (Werschkul and Williams 2004).

So, while no state in the nation has escaped poverty's reach, and although most recently growth in poverty has been starkest in the Midwest (U.S. Census Bureau 2005), the South remains especially plagued by this problem. This southern legacy not only has a gendered dynamic, but a racial one as well.

EXAMINING THE SIGNIFICANCE OF RACE

Lost on no one during the days immediately following the breach of the levees was the high concentration of African Americans left to fend for themselves in the middle of a ravaged city. But long before the storm, New Orleans, not unlike other urban areas across the nation, was in many ways a city sharply divided by race and class, as well as gender. The women of New Orleans and in the Gulf region generally lived very different lives across the racial divide. For example, despite being home to three Historically Black Universities (Dillard University, Southern University at New Orleans, and Xavier University of Louisiana), college degree attainment among African-American women in the city of New Orleans only slightly outpaced the national rate (18.9% vs. 17.5%). In the broader New Orleans MSA, degree attainment by African-American women fell slightly below the national norm (16.4% vs. 17.5%). White women in the city of New Orleans, however, like women in urban areas more generally, had very high college graduation rates. They greatly exceeded their counterparts nationally in college degree attainment (50.6% vs. 27.8%) and more than doubled the rate of degree attainment achieved by African-American women (50.6% vs. 18.9%). Only within the Beaumont–Port Arthur area, where degree attainment is relatively low for all groups, did the difference in degree attainment rates between African-American and white women drop below ten percentage points.

The educational differences experienced by women across the racial divide most certainly carried over to the labor market. Within the managerial/professional fields, for example, white women's representation more than doubled that of African-American women in the city of New Orleans (65.7% of white women had managerial or professional jobs vs. only 27.2% of African-American women). White women also outpaced African-American women within the broader New Orleans metropolitan area, 45.9% vs. 28.6%, for a difference of more than 15 percentage points, significantly outpacing the national divide of 40% vs. 31%.

Similarly, white women in the city of New Orleans earned a higher proportion of the white male salary (80.2%) than the national average (71.7%) and nearly doubled the proportion of white men's earnings achieved by African-American women (80.2% vs. 43.9%). In fact, across each of the four hurricane-impacted areas examined that had available data, African-American women's earnings were less than half those of white men (43.9%) and well below the national ratio of African-American women's earnings to white men's earnings (62.7%). Although still lagging behind white men in each area examined, in no area did white women fail to exceed three-fifths of the earnings of white men.

Looking statewide, women's earnings were especially low among those who lived in areas that bore the brunt of Hurricane Katrina. In fact, of the 43 states with sample sizes large enough to provide a reliable measure of African-American women's earnings, Louisiana ranked worst in the nation with full-time annual earnings of only $19,400. Mississippi followed closely, ranking next to last with median earnings of $19,900 for African-American women. White women also garnered comparatively low earnings in those states. Although exceeding African-American women's earnings, compared to white women in other states, white women's median earnings in Mississippi ($25,700) and Louisiana ($26,500) put them in the bottom third in the nation and well below the $30,900 median earnings for white women nationally. In contrast, for both groups, women's earnings in Texas well outpaced those of their counterparts in Louisiana and

Mississippi and placed them within the top third of earnings nationally. Still, the earnings of African-American women in Texas failed to exceed the average earnings of African-American women nationwide, and the earnings of Hispanic women in Texas fell within the bottom third of all states (Werschkul and Williams 2004)....

Also consistent with national trends, poverty in the Gulf region is highly correlated with race and gender. While overall women are consistently more likely to be poor than men, both throughout the affected areas as well as the nation as a whole, African-American women are especially vulnerable to poverty. Hardest hit were those who lived within the Biloxi–Gulfport–Pascagoula area and the city of New Orleans, where roughly a third lived below the federal poverty line (32.5% and 32.3%). As was the case for African-American women, white women's poverty was most pronounced in the Biloxi–Gulfport–Paseagoula area, where 13.7% were poor; within New Orleans city and its broader MSA, however, white women's poverty was less than the national average of 10%.

What a sad irony that one of the worst disasters in our nation's history occurred in an area populated by some of the nation's most disadvantaged citizens. Both African-American and white women in Mississippi and Louisiana had many fewer resources at their disposal than their counterparts nationwide. African-American women especially found themselves at the bottom. They were the lowest earners across both race and sex in Mississippi and Louisiana and, consequently, were the least likely to be equipped with the necessary resources to take themselves and their families out of harm's way. As a result, they largely stayed to face the storm on their own; and for those who survived, life will never be the same again.

LIFE AFTER THE STORMS

One can only imagine the trauma of losing everything save the clothes on one's back, being separated from family, friends, and community, and suddenly forced to begin life anew in a random, unfamiliar place a long way—geographically, socially, and culturally—from home. This is the challenge facing many of those who have been displaced following the devastation brought by Katrina and Rita. But given the probability that those most likely to relocate were also the ones who faced the most severe challenges at home of poverty, limited educational background, and all the implications of a paycheck-to-paycheck existence, what might the relocated settlers now expect in the way of employment and other opportunities while awaiting the possibility of a newly rebuilt city? Unfortunately, our analysis of the data suggests very little in terms of improved economic opportunities.... In their new homes, many evacuees will still be faced with high poverty rates, particularly among women and female-headed households, and, for African-American women, low median earnings for year-round, full-time work. In some areas, such as Jackson, Mississippi; Baton Rouge, Louisiana; Little Rock, Arkansas;

Mobile, Alabama; and Atlanta, Georgia, poverty rates are even higher and/or earnings are even lower than in those environments left behind.

A few areas, though, suggest an environment that provides a more solid foundation for evacuees to rebuild their lives. Places like Charlotte, North Carolina; Richmond–Petersburg, Virginia; and Nashville, Tennessee boast comparatively lower poverty rates for women and female-headed families or have significantly higher earnings for African-American women. These positive conditions offer a glimmer of hope for improved opportunities to build a better and brighter future to those who have chosen to make their relocation permanent and to start the task of building a new life, livelihood, and home in new surroundings and without the social networks that had once defined a community.

CREATING A BETTER FUTURE FOR THE WOMEN OF THE GULF

Life will never be the same for those who lost loved ones, material assets, family mementos, and the one place in the world they called home. The best that can be done for them is to ensure that for all, even the most disadvantaged, we use this crisis as a springboard to creating a future replete with opportunities for a better tomorrow. To fulfill that commitment we must ensure that the particular needs of women are not overlooked in the rebuilding process. Acknowledging and addressing the needs of women ultimately amounts to fulfilling the needs of children, families, and entire communities. The following would be a good start:

Invest in education and training to increase women's earning potential and foster opportunities to participate in the rebuilding process—Increased education and job-training efforts in the Gulf Coast region as well as those areas where many have relocated could provide the best foundation possible for many evacuees to be able to begin the long process of building a stable and secure future for themselves and their families. This investment could help break the cycle of living paycheck-to-paycheck and provide the first real opportunity many have had to acquire employment that moves them above the poverty line. Particularly key for women is access to nontraditional training, such as in the trades (e.g., carpentry, electrical, etc.). Training in the trades often requires apprenticeships that provide the opportunity to earn while learning and would ultimately expose women to a skill and a profession that would offer much greater income potential than that associated with most traditionally female work. Also, an expansion of the Workforce Investment Act, the legislation meant to provide training, could serve as a key avenue toward increasing women's representation in living-wage jobs (Gault et al. 2005).

Expand earning opportunities through better wages and fair hiring practices—Increasing the federal minimum wage is a long overdue necessity. Last raised in 1997, today's wage rate falls woefully short of providing adequate income for families struggling to keep up with the increasing costs associated with housing, transportation, childcare, and energy. To give workers a fair chance at escaping the cycle of poverty, they must first be paid fairly. Working for nearly a decade without receiving a wage increase while the cost of living continues to climb falls far short of a fair deal for American workers. And since women make up the bulk of those who work at and near minimum wage, it is their earning power that is disproportionately suppressed. To offer a real chance for workers to escape poverty, the minimum wage must be increased.* But in addition to fair wages, fair hiring practices must be assured as well. Contractors should, whenever possible, be local or be required to hire local workers so that former residents of those areas most devastated have the opportunity to get back on their feet financially while rebuilding their lives and their communities. And finally, given the disproportionate race/class/gender dynamics of the post-Katrina impacts, now is the time to strengthen affirmative action requirements instead of dropping them entirely.

Stop the clock on TANF benefits for the victims of the storms—As early as January, 2002, 21,396 families in the state of Louisiana had reached the state or federal lifetime limit for TANF benefits. While the state did provide a number of extensions for a variety of reasons, including domestic violence, the lack of childcare or other support services, or recent job loss, over half (11,138) of those families had their TANF cases closed immediately (Bloom et al. 2002). Congress' post-Katrina action of providing extra assistance to states affected by Katrina or to states hosting evacuees does nothing to address the needs of those who had reached their limit prior to the storm or for legal immigrants affected by the storm. These oversights must be rectified. The clock must be stopped for those who have lost everything and must start over either in the affected areas or in an unfamiliar place, lacking the social networks that could help in rebuilding their lives.

Respect the needs of women and communities in the planning and rebuilding process and engage the black middle class—Community is much more than physical structures. A community is, at its core, its people. In rebuilding New Orleans, the needs and concerns of its original inhabitants should be sought and acted upon. A careful balance must be achieved that respects pre-existing community culture and social networks while also seeking to create a better-served and better-maintained environment. Making sure that the voices of neighborhood leaders, including women's, civic, and religious group representatives, are sought, heard, and acted upon is a key component of living up to this responsibility. Also, making sure women are included in every phase of the rebuilding process helps ensure that their needs will not be overlooked and that they too will benefit from a rebuilding economy. From the urban and regional planners to the workers on the construction sites, the inclusion of women is key to making sure that the communities that are rebuilt and the workforce that is developed will be better and stronger than ever.

Also key to the successful reemergence of New Orleans and its surrounding communities is the needed inclusion of the skills, talents, and cultural competency

* Editors' note: The federal minimum wage was finally raised in 2007 with additional raises planned for 2008 and 2009, after this article was published.

of the Black middle class. Home to three Historically Black Colleges or Universities, New Orleans has long been a city that has attracted and retained a strong African-American professional community. Drawing on this community now is key, not only for meeting the immediate and future skills needs of the region, but also as a means of doing so through a lens of cultural intelligence.

Make the provision of childcare a priority—Working families need access to affordable, quality childcare in order to be able to seek and maintain employment. This need is especially pronounced among those who are trying to rebuild their working lives in unfamiliar surroundings, as well as for those returning home and trying to start anew by helping to rebuild their damaged communities. Government subsidies could help by funding professional development and training for evacuees and returnees seeking childcare careers while simultaneously meeting the needs of those seeking quality, consistent care. Additionally, entrepreneurship support could be used to fund the establishment of centrally located, high-quality childcare centers and to enable some providers to initiate home-based childcare businesses. Childcare assistance for low-income families had fallen far short of the broader need well before Katrina came to shore. This tragedy most certainly has only increased that need. Now is the time to do a better job of supporting struggling families while fulfilling our unspoken commitment to the well-being of America's most helpless citizens, its infants and young children.

Fulfill the promise of attacking the poverty problem—When President Bush addressed the nation in September of 2005, he stated strongly that "we have a duty to confront this poverty with bold action" (Office of the Press Secretary 2005), yet the wheels of action have not only turned exceedingly slowly, one could argue that they have actually gone in reverse. Since that day, Congress has cut Medicaid services to poor children and the elderly in need of long-term care; it has expanded TANF work requirements without adequately funding the increased childcare burden working families will undoubtedly face; and it has reduced child support enforcement dollars, which will ultimately negatively impact low-income single-parent families. All told, through his 2006 and 2007 budgets submitted to Congress, the President has proposed some $435 billion in cuts to programs meant to aid America's neediest citizens. Counter to his own call for bold action to confront the poverty problem, the President has instead moved boldly in the other direction, pushing forward with his plan to extend $648 billion in tax cuts to America's millionaires over the next 10 years, dwarfing the $150 billion currently committed to relief and reconstruction efforts in Katrina-impacted areas (Aron-Dine and Friedman 2006; Weinstein, Beeson, and Warnhoff 2006). What of the commitment to the poor? What of the commitment to the people of New Orleans?

The President was right when he called for bold action to confront poverty. Now he has the responsibility of living up to those words. To do so would mean a dramatic change in course from the current policy direction. A bold anti-poverty strategy would require, among other things, a commitment to affordable housing, expansion of the Earned Income Tax Credit, broad-scale investments in public education and adult education, and increased health care and childcare availability. At the core, these and other anti-poverty policy goals would require a shift in priorities on a national scale. Such a shift is possible. If we have learned nothing else, the Katrina tragedy has made painfully clear the very real human and fiscal costs associated with years of neglect and indifference. We simply cannot afford to go back to business as usual.

CONCLUSION

Ultimately, it is our responsibility to learn from the heartbreak of seeing mothers begging for milk for their dehydrated infants, from seeing children and the elderly stranded on rooftops pleading for help, and seeing the dead left without dignity and the dying suffering for days in their own waste. Lest we forget, these are the images of Katrina that should stay with us forever as a reminder of what can happen when we turn a blind eye to the implications of a society divided. We can and must do better. The women, children, and families of the Gulf deserve nothing less.

REFERENCES

Aron-Dine, Aviva, and Joel Friedman. 2006. *The Skewed Benefits of the Tax Cuts, 2007–2016: If the Tax Cuts are Extended, Millionaires Will Receive More Than $600 Billion over the Next Decade.* Washington, D.C.: Center on Budget and Policy Priorities.

Bernstein, Jared, and Isaac Shapiro. 2005. *Unhappy Anniversary: Federal Minimum Wage Remains Unchanged for Eighth Straight Year, Falls to 56-Year. Low Relative to the Average Wage.* Washington, D.C.: Center on Budget and Policy Priorities and the Economic Policy Institute.

Bloom, Dan, Mary Farrell, and Barbara Pink with Diana Adams-Ciardullo. 2002. *Welfare Time Limits: State Policies, Implementation, and Effects on Families.* New York: Manpower Demonstration Research Corporation.

Boushey, Heather. 2005. *The Structure of Poverty in the U.S.* Paper presented at the Congressional. Black Caucus Foundation Summit Poverty, Race and Policy: Strategic Advancement of a Poverty Reduction Agenda.

Caiazza, Amy, April Shaw, and Misha Werschkul. 2004. *Women's Economic Status in the States: Wide Disparities by Race, Ethnicity, and Region.* Washington, D.C.: Institute for Women's Policy Research.

Center on Budget and Policy Priorities. 2005. *Economic Recovery Failed to Benefit Much of the Population in 2004.* Washington, D.C.: Center on Budget and Policy Priorities.

Gault, Barbara, Heidi Hartmann, Avis Jones-DeWeever, Misha Wershkul, and Erica Williams. 2005. *The Women of New Orleans and the Gulf Coast: Multiple Disadvantages and Key Assets for Recovery, Part I: Poverty, Race, Gender, and Class.* Washington, D.C.: Institute for Women's Policy Research.

Jones-DeWeever, Avis, Olga Sorokina, Erica Williams, Heidi Hartmann, and Barbara Gault. 2006. *The Women of New Orleans and the Gulf Coast: Multiple Disadvantages and Key Assets for Recovery, Part II: Employment and Earnings, Gender, Race, and Class.* Washington, D.C.: Institute for Women's Policy Research.

Lovell, Vicky. 2004. *No Time to Be Sick: Why Everyone Suffers When Workers Don't Have Paid Sick Leave.* Washington, D.C.: Institute for Women's Policy Research.

National Low Income Housing Coalition. 2004. *America's Neighbors: The Affordable Housing Crisis and the People it Affects.* Washington, D.C.: National Low Income Housing Coalition.

Nord, Mark, Margaret Andrews, and Steven Carlson. 2005. *Household Food Security in the United States,* 2004. Washington, D.C.: USDA Economic Research Service. http://www.ers.usda.gov/Publications/err11/

Office of the Press Secretary, White House. 2005. *President Discusses Hurricane Relief in Address to the Nation.* http://www.whitehouse.gov/news/releases/2005/09/20050915-8.html

Rose, Stephen J., and Heidi Hartmann. 2004. *Still A Man's Labor Market: The Long-Term Earnings Gap.* Washington, D.C.: Institute for Women's Policy Research.

Shipler, David. 2004. *The Working Poor: Invisible in America.* New York: Knopf.

Shulman, Beth. 2003. *The Betrayal of Work: How Low-Wage Jobs Fail 30 Million Americans.* New York: New Press.

U.S. Census Bureau. 2005. *Income, Poverty, and Health Insurance Coverage in the United States: 2004.* Washington, D.C.: U. S. Government Printing Office.

U.S. Environmental Protection Agency. 1993. *Environmental Equity: Reducing Risk for All Communities. Volume I: Workgroup Report to the Administrator.* Washington, D.C.: U.S. Environmental Protection Agency. http://www.epa.gov/compliance/resources/publications/ej/reducing_risk_com_vol1.pdf

Weinstein, Deborah, Jennifer Beeson, and Steve Wamhoff. 2006. *Guide to the FY 2007 Federal Budget: What's at Stake for Human Needs.* Washington, D.C.: Coalition on Human Needs.

Werschkul, Misha, and Erica Williams. 2004. *The Status of Women and the States.* Washington, D.C.: Institute for Women's Policy Research.

DISCUSSION QUESTIONS

1. Were a disaster to strike your community, which groups would be most vulnerable? How would race, gender, and class affect the vulnerability of different groups?
2. Find some current news reports on the recovery status of New Orleans and the Gulf Coast. How has race affected the recovery effort—up until now?

32

A Tale of Two Classes

The Socio-Economic Divide among Black Americans Under 35

CELESTE M. WATKINS

By examining generational differences among African Americans, Watkins shows that class differences exist within racial groups, but, more importantly, such differences stem from the particular social structural conditions that groups experience at different points in time.

For one segment of the generation of African Americans under 35, the current era is the best of times. Credentialed with college as well as graduate- and professional-school degrees, skilled in the use of the latest technological wizardry, possessed of the "right" jobs with the "right" potential, and supported by competitive salaries, they have benefited enormously from the expanded opportunity American society now offers more and more African Americans—and they readily express the self-confidence such success breeds.

But what is the apt characterization of the current era for the other segment of African Americans under 35?

For this group—who are not credentialed with higher education degrees, and thus, who are not a part of the heady swirl of the white-collar world; who may not have even high school diplomas, or, if they do, have neither the education nor the training that would make those diplomas meaningful; who are consigned now to the low-wage service jobs of the economy, if they have a job at all—the right words or phrases are far more difficult to select.

For the latter group, is the current era a new beginning, a climb out of the material and psychological swamp of continual un- and underemployment? The sharp, quick decline in the black unemployment rate in the late 1990s, powered by many of the so-called hard-core unemployed taking the low-wage service jobs the booming economy offered, showed that most of those in this profoundly disparaged cohort know all about the value of work, want to work, and will whenever they are given a chance.

Will they now, in the period when the air is thick with job cutbacks among American companies and foreboding about a recession, still be given a chance?

SOURCE: From *The State of Black America 2001*, ed. by Lee Daniels (New York: National Urban League, 2001), pp. 67–81. Reprinted by permission.

Or will the American economy return to its old "tradition" of not just double-digit black unemployment, but high double-digit black unemployment?

And if that happens, what will Black America do?

This issue—the future of those in their generation who have had relatively little opportunity—is the most critical question under-35 African Americans face. For the ways in which blacks of different socioeconomic groups seek to deal with and help each other will, in large measure, determine the future economic, social, and political health of Black America itself....

It is not, of course, as if class divisions are something new to Black America. What are new, however, are their scope and the forces behind it. When William Julius Wilson argued this very point in his groundbreaking 1978 book *The Declining Significance of Race*, asserting that the life chances of individual African Americans have more to do with their economic class positions than their day-to-day encounters with racism, he faced a wave of criticisms.[1] ...

Years later, Wilson's argument is less controversial. In fact, it is taken as a widely agreed upon thesis by scholars and non-scholars alike. Social scientists point out the widening class gap within the African-American community as income inequality increases across all American households. They highlight the implications of such a gap in terms of varying access to education, jobs, and the networks that can lead to socioeconomic mobility. Others, particularly blacks themselves, notice the gap as well, often in the diversity of class presentation in public spaces and within African-American familial networks....

While the condition of poorer blacks has occupied social scientists for decades, increased attention is now being paid to the experiences of working-class, middle-class, and upper-class blacks. These more precise ways of defining the experiences of African-Americans have led to at least three conclusions. First, as Wilson argued, the opportunities and experiences of African Americans heavily depend on their class status. Second, within the black community, the gap is indeed widening between the "haves" and the "have-nots." Third, such a widening gap has major implications for the interclass relations of blacks as well as for the broader policy agenda of the African-American community.

THE DEVELOPMENT OF THE DIVIDE: THE EVOLUTION OF CLASS IN BLACK AMERICA

... Segregation encouraged the interaction of classes within the black community and served to offset some of the effects of the systematic economic and political disempowerment of blacks. Black business owners, doctors and other professionals serviced predominately, and often exclusively, black clientele. Poorer blacks often lived in the same or in close-by neighborhoods to blacks that were more affluent. While the income of many blacks was derived almost solely from black clientele, others challenged the hegemonic arrangements that kept them out of the industrial economy, eventually opening access for blacks to many of the jobs that provided economic stability and potential socioeconomic mobility.

By the mid-1900s that challenge included increasingly successful efforts to wrest their share of local political power out of the hands of predominately white political machines....

Following World War II, large structural changes in the economy had major implications for the American class structure, affecting the African-American community significantly. In fact, the largest changes in black mobility occurred in the 1950s and 1960s. Increases in African-American employment in the government and private sector, sparked by the passage of the Fair Employment Practices Act and the economic expansion after the war, sharply increased the size of the African-American working and middle classes. By the 1970s, some suburban communities opened to African Americans, allowing the more affluent members of predominately African-American neighborhoods to move. In addition, increased economic opportunity through the gains of the Civil Rights Movement and affirmative action programs gave many African Americans increased opportunities to solidify their positions in the American middle class.

At the same time, however, economic shifts and public policies in the 1970s and 1980s intensified the stressful economic situation of the inner-city communities where many lower and some working-class blacks remained. Hit hard by economic recessions and inflation, poverty rates for the nation's central cities rose from 12.7 percent to 19 percent between 1969 and 1985....

Throughout the 1990s, scholars and journalists paid increased attention to those at the bottom of the economic ladder in their writings. Terms such as "the underclass" and "ghetto poor" were used to describe those inhabiting the core of America's central cities who, for the most part, had been shut out of the technological and economic advances of the past decades.[2] This group had evolved as a result of current and historic discrimination, the dynamics of the low-wage labor market, the decentralization of businesses, middle- and working-class outmigration, un- and underemployment, and limited education and skills. These macro-structural changes were linked to the concentration of poverty in inner cities as well as the rise in black female-headed families, out-of-wedlock births, and crime. Expressing long-term spells of poverty and welfare receipt, this group was a small segment of the low-income African-American families residing in inner cities, yet received major attention by scholars and policy makers alike.

The result of all these forces and developments is a more variegated class structure within Black America, with each stratum facing similar but often distinct challenges. In 1998, 26 percent of African Americans (9.1 million) were living in poverty, compared with 8 percent of non-Hispanic whites (15.8 million).[3] The condition of the black lower class and ghetto poor persists, with scholars fine-tuning their understanding of and policy prescriptions for this group, composed of not only the chronically un- and underemployed but also the working poor....

Scholars and journalists have also raised concerns about the actual economic and political power of the black middle class and have highlighted that while the black-white income gap my be decreasing, the wealth gap remains sizable.[4] In addition, while African Americans have gained political power in central cities, the surrounding, predominantly-white suburban areas have continued to attract both businesses and residents, challenging the influence that these localities can wield. In 1999, while the median income of African-American households was $27,910, the highest ever recorded, it still significantly trailed the $44,366 median income of white non-Hispanic households.[5]

Yet, one cannot deny that, at the same time, African Americans have gained significant ground on both economic and political fronts.... More and more blacks have become a part of the nation's political life through public service in elected and appointed positions. In 1999, about 47 percent of African-American householders were homeowners.[6] ... In fact, about 6.1 percent of black households had annual incomes over $100,000 in 1999.[7]

THE WIDENING CLASS GAP: ISSUES AND COMPLICATIONS

... Is a widening class gap among African Americans necessarily problematic? Some would argue that because income inequality is rising across American households, the gap should not be a concern. In fact, some would argue that such inequality is simply further evidence that the trends for blacks are becoming more in line with the trends for the rest of society. Perhaps such a gap will encourage those African Americans on the bottom of the economic ladder to strive for mobility as they see more and more blacks taking their places on the higher rungs....

While these points are well taken, a widening class gap within Black America presents similar problems as rising income inequality crosses all American households. While class mobility may be a beneficial and favorable goal, when that mobility occurs only for those already on the higher rungs of the economic ladder, intractable inequality rises. Inequality of outcomes (in this case, income) on the surface may not be a problem. However, in American society, too often inequality of outcomes leads to and reinforces inequality of opportunities, creating not a class system but a caste-like one. As some climb to the top, others will be left on the bottom with very few opportunities for upward mobility and a future so bleak that deprivation, particularly in a land of wealth, could potentially lead to desperation.

African Americans under 35 must forcefully come to grips with this issue. The reason is that although the problem is not new, they have come of age in an American society in which they have opportunities to compose their racial identities at a time of largely implicit rather than explicit racial tension. While racism is still very much a part of this generation's consciousness and overall life experiences, its often subtle manifestations in the members' daily experiences has enabled some to create political consciousness that may not hinge entirely on racial solidarity. Many note the class gap and wrestle with it in their daily encounters with African Americans higher and lower on the socioeconomic ladder. Many even think about it when engaging in dialogue about the future leadership needs of Black America. How does one represent the interests of a social group whose issues and objectives seem increasingly diverse and at times, in opposition to each other? There is a lack of clarity on how best to respond to the class gap, both on a broader community level and in the day-to-day interactions of blacks across class lines. Two complications are central.

First, many African Americans want to live in socioeconomically diverse black neighborhoods. But their desire is often sapped by the instability that pervades neighborhoods with significant numbers of poor residents. The resulting internal conflict frequently ends with those better-off opting for less economically diverse communities that wholeheartedly encourage upward mobility.[8] Social scientists Katherine Newman, Elijah Anderson, and Mary Pattillo-McCoy all highlight these tensions in neighborhoods whose members represent not only varying class positions but also different ways of thinking about the appropriate values and norms for their communities. At the same time, those with limited opportunities for advancement are often frustrated by the perceived lack of ongoing support and resources afforded by more successful members of the community....

Second, because of their history in this country, African Americans have a shared desire for economic, social, and political equality with the rest of America. However, some blacks are closer than others to achieving this goal, raising the question of whether the means by which this goal is to be pursued should be the same throughout the black community.... Largely through racial integration, blacks, particularly those under 35, are faced with the fact that their experiences—professional, educational and cultural—are often vastly different from others sharing their racial, but not socioeconomic, class backgrounds. In fact, depending on their backgrounds, some blacks would argue that in certain aspects, they have more in common with some of their white counterparts than with blacks of a different socioeconomic class....

BRIDGING THE DIVIDE: STRATEGIES AND DIRECTIVES FOR THE UNDER 35 GENERATION

It is imperative to respond to the socioeconomic divide on both intra- and interracial levels. The former must take place largely through the interpersonal interactions of people of diverse socioeconomic groups. The latter must take place through a broader political agenda that is sensitive to the socioeconomic diversity within Black America.

... Increased class-based tension is certainly an understandable—indeed, unavoidable—by-product of rising income inequality within the black community. In his book *Harlemworld: Doing Race and Class in Contemporary Black America*, anthropologist John L. Jackson argues that it is problematic, however, to think of Black America as two overtly discrete worlds—one lower class, one middle class—that rarely, if ever, interact in meaningful ways. While black suburban flight has encouraged geographic distance between blacks of different classes, there are often opportunities through familial ties, churches and other institutions, and in shared public spaces, for blacks to interact across class lines, mitigating some of this social distance.[9]

In order to respond to potential tensions, at least three actions are necessary, particularly from the generation of 25–35 year old African Americans poised to become leaders in their communities. First, there must be a cross-pollination of resources through familial ties, churches and other community organizations, and opportunities created by individuals. Research suggests that often what encourages mobility, and the behaviors necessary for mobility, are social networks that are extensive and diverse....

Second, there must be a fundamental acceptance among black people of their own diversity and, to borrow from Jackson, of different ways of "doing blackness" within the African-American community. Questioning racial authenticity due to political views, socioeconomic status, or presentation of self is futile at best....

Third, it is imperative for African Americans to be involved in debates that shape the political and economic future of this country. Allowing a diversity of views encourages creative, achievable, and effective approaches to responding to what ultimately perpetuates much of the class gap in America—inequality of opportunity.... Power in America is largely gained through economic resources obtained through education, ownership, and networks. Many blacks have obtained, or are on their way to obtaining, this power on an individual level.... The ability to leverage power allows one to sit at the table when decisions facing the nation are made. As a large heterogeneity of social and political views of African Americans come to the forefront, it is possible to garner broad support for some initiatives among African Americans yet respect and appreciate those times when there is little opportunity to gather support simply on the basis of racial solidarity....

TOWARD A BROADER POLITICAL AGENDA

First, it is important to remember that the class gap within Black America is not just the black community's problem. It is America's problem. Just as affluent African Americans should be encouraged to consider their responsibilities toward less affluent blacks, it is also essential to challenge those forces in American society that created and sustain widening income inequality in America. After all, those large reservoirs of low-wage workers who are poorly educated were not created by black elites. These reservoirs are the product of decades of systematic racial subjugation—and that is a problem for which America as a whole bears responsibility.

Second, policies designed to increase and sustain labor force participation can greatly assist those on the bottom third of the socioeconomic class structure.... Policies that encourage us to make work pay through the Earned Income Tax Credit, living wages, and family friendly employment benefits assist all low-income families and greatly assist families in the bottom third of the black economic class structure....

Third, we must support policies that allow working- and middle-class Americans to secure assets and build wealth, which would greatly aid African Americans in these socioeconomic groups. Policy conversations often focus on the needs of lower-class and middle-class Americans. Working-class Americans then, find themselves attracted to demands for education and training, affordable health care, and community development attractive to low-income families as well as home ownership policies, entrepreneurship opportunities, and wealth building incentives attractive to the middle class. These kinds of policy initiatives have major implications for the economic stability and mobility of black households and many of them, because they appeal across racial lines, have great potential for garnering widespread political support.

Fourth, the transmission of wealth intergenerationally positions African Americans to solidify and improve upon their economic standing and to reduce wealth inequality within the American socioeconomic class structure.... Through initiatives designed to educate households on how to not only build but pass on wealth, African-Americans could solidify the economic gains made in the past several years and position themselves to eliminate the wealth gap between themselves and their white counterparts.

REFERENCES

1. Wilson, William Julius. 1978. *The Declining Significance of Race: Blacks & Changing American Institutions.* Chicago: University of Chicago Press.
2. Wilson, William Julius. 1987. *The Truly Disadvantaged: The Inner City, the Underclass, & Public Policy.* Chicago: University of Chicago Press. Jencks, Christopher & Paul Peterson. 1991. *The Urban Underclass.* Washington, DC: Brookings Institution. Gans, Herbert. 1995. *The War Against the Poor: The Underclass & Anti-Poverty Policy.* New York: Basic Books.
3. The poverty rate of blacks has declined from a high of 38 percent in 1969. U.S. Census Bureau. 2000. *The Black Population in the United States,* March 1999. Washington, DC: On-line data.
4. Oliver, Melvin L. & Thomas M. Shapiro. 1995. *Black Wealth/White Wealth: A New Perspective on Racial Inequality.* New York: Routledge. Conley, Dalton. 1999. *Being Black, Living in the Red: Race, Wealth, & Social Policy in America.* Berkeley: University of California Press.
5. U.S. Census Bureau. 2002. *Money Income in the United States: 1999.* Current Population Reports, P60-209. U.S. Government Printing Office. Washington, DC.
6. U.S. Census Bureau. 2001. *Report on Residential Vacancies & Homeownership.* Press Release. Washington, DC.
7. U.S. Census Bureau. 2000. *Money Income in the United States: 1999.* Current Population Reports, P60-209. U.S. Government Printing Office. Washington, DC.
8. Anderson, Elijah. 1990. *Streetwise: Race, Class, & Change in an Urban Community.* Chicago: University of Chicago Press. Anderson, Elijah. 1999. *Code of the Street: Decency, Violence, & the Moral Life of the Inner City.* New York: W.W. Norton. Newman, Katherine. 1999. *No Shame in My Game.* New York: Knopf & Russell

Sage. Pattillo-McCoy, Mary. 1999. *Black Picket Fences: Privilege and Peril Among the Black Middle Class*. Chicago: University of Chicago Press.

9. Jackson, John L. 2001. *Harlemworld: Doing Race & Class in Contemporary Black America*. Chicago: University of Chicago Press.

DISCUSSION QUESTIONS

1. What generational differences does Watkins see as she analyzes social class among Black Americans?
2. What policies and practices can help bridge the gap between the prospects of people of color under 35 who are from different social classes?

33

Experiencing Success

Structuring the Perception of Opportunities for West Indians

VILNA F. BASHI BOBB AND AVERIL Y. CLARKE

Social mobility *is defined by sociologists as the movement of a group (or individual) up or down in the class system. As shown here, the possibility of social mobility is conditioned by the opportunities provided to groups in such arenas as education, work, and other social institutions. For West Indians, as for other groups, education is key to mobility, but education is mediated by the impact of racism.*

The question of social mobility is a critical one for West Indian migrants in the United States and for the lives of their children. . . .

We want to look at what first- and second-generation West Indians believe about the possibilities and opportunities for "getting ahead" in American society. Research to date has focused on "objective" indicators of socioeconomic status attainment, but none has assessed what members of the different generations believe about this success or lack thereof in their own social experience.[1] Although writings on West Indians have shown that social experience is important in explaining first- and second-generation perceptions of the social structure of economic opportunity, few works have emphasized social experience directly, and even fewer have compared the social experience of the first generation with that of the second. Here we use respondents' own words to report what the immigrant and second generations understand success to be. . . .

This study is based on two sets of interviews with West Indian immigrants and U.S.-born children of immigrants from the West Indies. The data on the immigrant generation's perspectives on social mobility come from Vilna Bashi's 1996 interviews with forty-four black immigrants from St. Vincent and the Grenadines and Trinidad and Tobago who migrated to New York City between 1930 and 1980 (Bashi 1997). From 1998 to 1999, Averil Clarke (2001) interviewed fifty-five college-educated women of African American and West Indian ancestry to study how fertility and nuptiality decisions affected the ability

SOURCE: From *Islands in the City: West Indian Migration to New York*, ed. by Nancy Foner (Berkeley, CA: University of California Press, 2001), pp. 216–236. Copyright © 2001 by The Regents of the University of California. Reprinted by permission.

to earn a college degree. For the present [article], Clarke used data only for those women whose parents emigrated from the West Indies to New York, and supplemented their stories with fifteen additional interviews of native-born persons, all between the ages of fifteen and forty-five, whose parents migrated from these same islands to New York City....

EDUCATION AND SOCIAL MOBILITY FOR WEST INDIANS

The first-generation immigrants interviewed for this study strongly believed that education would generate returns in the labor market; they were also committed to the idea of getting an education themselves. Many came to the United States expressly to gain an education, some exclusively for education and others in order to work and obtain schooling simultaneously.[2] A well-known and common scheme—for women at least—was to enter the country under a visa that allowed them to start work as babysitters or housekeepers (employers in these niches typically filed the necessary papers under exceptions to the usual immigration laws), and then to go to school to learn the skills that would permit a career change (Bashi 1998)....

The second generation was just as committed to the belief in education as a tool necessary for advancement. Almost every second-generation respondent directly or indirectly mentioned education as key factor in what they had achieved (or failed to achieve) to date. Roy[3] told us that his professional success was due to "the ability to learn and understand what I do." ... Jonathan maintained that "some of the values" that he possessed were "due to [his parents] being immigrants." He continued, "My mom and my dad always explained to me that you had to work to get whatever you wanted. Education was one of the ways of getting something in this country. So they came here and worked hard and I believe they instilled those things into me."

Although second-generation West Indians agreed with their parents that education was the key to success or failure in the economic sector, some perceived limitations in the value of an education and felt that a singular focus on education was not likely to suffice for economic advancement. For example, although Nordrick did not attribute his career success to his record of educational achievement, he added that, when he discovered the fact that his educational credentials surpassed those of his white coworkers, he realized he was being discriminated against in terms of salary and promotions. Kyle explained that the major obstacles to his success were the structure of society and racial barriers. Consider his comments on how a degree from a black institution did not carry the same weight as one from a majority white institution when applying to graduate school.

> Like going to my university. The way this world in America views that is that I can have the same grades as a person in a University of Penn, and just because he or she went to Penn and I went to Virginia State, I would get overlooked with a quickness. I feel like to be coerced to go

> to a predominantly white school or some other school because of a reputation—you know, I feel the person should speak for himself not the reputation of the school.

... Much like their parents, second-generation West Indians continue to hail the value of an education for gaining economic success, and they often attribute failure to inadequate, misdirected, or inappropriate education. Yet the second-generation and long-resident immigrants also perceive that there are limits to the returns to education, and more than a few of them would be suspicious of claims that education is the only or even the main ingredient to success in their field.... These perceptions emerge from the U.S.-born generation's understanding of how the racial structure in the United States negatively affects black people who seek to achieve higher social or economic status.

SOCIAL EXPERIENCE AS A FACTOR IN UNDERSTANDING THE ROLE OF EDUCATION

What are the social experiences that underpin the attitudes we have described? First-generation immigrants bring with them extensive experience in the economic and education systems of "Third World" immigrant-sending countries. This includes but is not limited to the idea that "education was a privilege, not a right, in the West Indies," as Randall, a first-generation immigrant, put it. During colonialism the West Indian education system was modeled after the British education system, which, albeit in neocolonial fashion, still exists today. In the West Indies it is not the case that all students can advance as far as they like. At various points one must take examinations in order to move to the next level, and there are only a very limited number of slots available for advancement. Thus, not only must you take qualifying examinations, but it is your placement among all exam-takers that determines whether you will win a coveted slot that allows you to further your education.

Sufficient experience in this context generates investment in the idea that lack of educational opportunity limits one's economic mobility (since it literally does so)....

Immigrant West Indians encourage their second-generation children to adopt the West Indian immigrant worldview in an environment where, despite severe inequities, education appears to be free and available to all. However, in the United States, education does not always translate into attractive social positions. In this context, black immigrant parents' attempt to pass on their intense attachment to education is a "harder sell" to black children who know the "raw deal" of the black experience in this different racial structure. Education is the culturally accepted means for achieving social mobility, and our second-generation respondents (only two-thirds of whom held college degrees) generally agreed with this but saw race as a mediating factor. Moving our lens from the first to the second generation, we move from a generation that wholly believes in education as a means for mobility to one that does not. What needs to be explained is not why some young black people believe in the power of education, but why it is that many young blacks do not. We emphasize that this difference between the generations is due to structural limitations that restrict blacks in the United States and the fact that these limitations differ from those that affect blacks who live in other nations....

It is evident that West Indians have used geographic mobility to take advantage of opportunities for socioeconomic mobility for more than a century. "Strategic flexibility" is the term Carnegie (1987) has coined to describe the West Indian philosophy of organizing one's life in order to be ready to take advantage of whatever opportunities for work and migration may arise.... Emigration has long been a primary means of economic advancement, from the emancipation in 1834 of black slaves in the islands until today....

Our first-generation informants corroborated the existence of and adherence to the "strategic flexibility" philosophy, particularly as it relates to seeking education for job advancement. For example, one informant explained that when he first came to New York, he "did a little of everything," starting with jobs in carpentry and then moving on to spraypainting. "You've gotta have more than one skill," he said.... Randall told a similar story when asked if his work history consistently used one of the skills he had honed. "No, I did a couple of other jobs, I work other little places. Listen, let me tell you, any place that was offering me five or ten dollars more than what I was doing, I'm gone. Oh yes, that's the way." To take advantage of opportunities for advancement, some people moved, not only from one job to another, but across state borders....

Among other things, strategic flexibility depends upon an understanding of the world in which personal accommodation to economic and social systems is part and parcel of economic mobility. Struggling to gain an education becomes just one kind of adjustment or accommodation that one must make. Furthermore, education is bound up with the migration act itself....

With regard to socialization in the education system of the United States, second-generation immigrants experience a system in which all education is not equal (e.g., "black" schools are valued less than "white" schools), and this inequality affects their belief in the value of an education.... It is well known that primary education systems in the United States differ in quality by regional status within and across counties. The children of West Indian immigrants living in urban areas (as most do) are likely to attend a school with a predominantly black student body and inadequate funding and resources.... The American school system is not always helpful in educating young blacks, despite their high aspirations and hard work.

In sum, although both generations see education as the primary factor in social advancement, their differing social experiences lead them to interpret the importance of education somewhat differently. For the second generation, a lack of experience in the postcolonial Third World education system and economy has a dampening influence on their perceptions about the rewards associated with education. The investment in strategic flexibility driven by Third World–style limits to social and economic opportunity is another quality possessed by the first generation but lacking in the second.

While these experiences are enough to cause differences in the generations' perspectives on the rewards to education, we also believe that racial socialization has its own independent effect. The way in which black West Indian immigrants as a group are able to distance themselves economically and psychologically from racism in the United States leaves intact their beliefs about the rewards of education. In contrast, the second generation, as well as those who immigrated at young ages or have remained in the United States for long periods of time, becomes immersed in the American racial hierarchy and has no such distance. These West Indians have a less optimistic understanding of the meaning of blackness for their chances for advancement opportunities in the economic arena.

RACISM AND SOCIAL ADVANCEMENT FOR WEST INDIANS

... West Indians in the immigrant generation are shocked or surprised by racism in the United States. "We don't have that back home," as one person explained and many others expressed in similar words. West Indian migrants come to the United States with an idea of race developed in their Caribbean homelands. Immigrants described the West Indian racial structure as akin to a class system, where access to sources of human capital—particularly education—makes a significant difference in one's ability to achieve social mobility (Gopaul-McNichol 1993; Foner 1985; Dominguez 1975)....

Evidence suggests that there is an ideal typical trajectory for the adaptation of West Indian immigrants to racism in the United States (Bashi 1996). When they arrive, West Indians believe that, because they are committed to hard work and social advancement, and because they are foreigners, racist behavior will not be directed toward them. They soon learn that racism is directed toward all black people (i.e., racism is not a set of social behaviors applied by white Americans just toward black Americans, as many immigrants initially believe). Once they learn that racism indeed applies to them, West Indians develop coping strategies to ignore, avoid, or overlook it. Then they learn a second lesson—that living with racism is a demoralizing process, and that racism cannot be wholly ignored or avoided.

Immigrant responses to racism in the United States indicate an understanding of how racism has demoralized African Americans, though only one immigrant reported being demoralized herself. Although they saw themselves as targets of racism, West Indians still reported that they did not let racism bother them or get in the way of achieving their dreams.

> I lived in St. Vincent, where there is a good enclave of so-called white people, and we worked together but we didn't see black and white, we saw people. We worked together, socialized together.... So when coming to America, and they were calling you "black" ... You work in the hospital and they would spit on you and tell you don't touch them

> and tell you to take your black so-and-so away from them … After a while you learn to rise above it, like water off the duck's back.

Of course, racism did bother some immigrants enough that it became one reason for deciding to leave New York City and return home. Still, they did not seem to think that racism was a problem that hindered mobility.…

Second-generation respondents unanimously stated that racism should be an issue of great concern in our society. They readily recited incidents and situations that they believed were indicative of racism and its deleterious consequences for blacks. As Nordrick said in a matter-of-fact tone, "Racism is still a big problem. It's still nowhere near equal." … Nordrick stated that he was affected by racism early in his professional life.

> During my eight years of working construction management, just seeing numerous white bosses bring their children, nieces, sons, and daughters in, bring them into a position above me, making more money than me, and they had no formal education and I had been working there for years. There was basically no question for me to ask. So eventually I caught on and knew that I had to move on.

… While the U.S.-born and U.S.-raised respondents were unanimous in their perceptions of the problem of racism, there was more variance in the degrees to which they felt that racism affected them personally.… Carla declared that minorities face racism as they attempt to climb the rungs of the job ladder. However, when she spoke about racism's effect on her own life, Carla said that she herself had not been its victim.

In the same vein, Christopher, who mentioned racism in the military, said, "Me, no, I haven't been personally affected by racism. I rise above the top when it comes to that. I know how to deal with it, I guess. I don't let it really get to me."…

CONCLUSION

Both first- and second-generation West Indians believe that education is key to achieving social mobility. In a racialized society, however, race and ethnicity mediate the impact of education. Both generations experience racism, but they perceive its personal and societal impact differently.… Despite having experienced racism in the United States, immigrants believed that racism need not be an obstacle to mobility, but that it is something readily ignored or avoided. We believe that these differences can be explained largely by one's ability to invoke West Indian ethnic status, which is related to the degree that the social experience common to a foreign-born, foreign-raised West Indian is incorporated into one's socialization.

The first generation has the maximum opportunity to invoke West Indian ethnic identity—they have the social experiences of a foreign reference point, a network of coethnics with which to live and work, and a foreign accent—all of which translate into social distance from a racially hierarchical social system. The second generation is less strategically flexible, particularly in those areas most influenced by the racial system. Social experience variables, rather than immigrant status or cultural ethnicity, best explain differences between the generations. What characterizes the first generation is not just their foreign birth but that they have moved from a black-majority society to the white-majority United States and successfully remained, reaching employment and educational goals that most of those who stayed behind can only dream of. What characterizes the second generation is not just their American birth but that they have difficulty marshaling their West Indianness in a society that racializes black people with little regard to ethnicity. As Linda, a second-generation respondent, told us,

> We are affected by race 'cause a lot of people just see us as black Americans. We're treated, as far as I know, the same ways black Americans are treated because we're clumped in that category. But I think our own perception is different than that of black American because our parents had such a drive and a strive that other Americans did not have. We come from a totally different perspective of work ethic and a different perspective of, you know, their education.... So I think we're affected because we are clumped in those categories, but I think that our perception is different.

Since subsequent American-born generations will have less and less ability to invoke West Indian ethnicity, they may end up exposed to the oppressive aspects of American racial structure. In contrast, ethnic West Indians may be relieved of racism's worst effects if West Indian ethnicity continues to be recognized as a marker of difference. Thus, West Indians of both the first and the second generations may gain from investing in the cultural difference stereotype.

NOTES

1. By *social experience*, we mean immersion in a social context of sufficient duration to socialize a person or group so that they are fully familiar with the way that social context operates.
2. Over 95 percent of the first-generation respondents arrived in New York City well before 1980. They were here, then, before the 1986 crackdown on employers who hired foreigners without proper working papers. They were also here before New York City's near bankruptcy and fiscal crisis in 1975–76, which brought about restrictions in City University of New York admission policies for bachelor's degree programs. Thus, it was relatively easy for these immigrants to combine both work and schooling under the immigration, employment, and college admissions climates they faced in New York City between 1965 and 1976.
3. We invented pseudonyms to refer to our respondents. None of their real names are used in this article.

REFERENCES

Bashi, Vilna. 1996. "'We Don't Have That Back Home': West Indian Immigrant Perspectives on American Racism." Paper presented at the annual meeting of the American Sociological Association, New York, August.

———.1997. "Survival of the Knitted: The Social Networks of West Indian Immigrants." Ph.D. diss., University of Wisconsin, Madison.

———.1998. "Racist Ideology in Immigration Policy and its 'Effect' on Black Immigrant Social Networks." Paper presented at the annual meeting of the Social Science History Association, Chicago, November.

Carnegie, Charles V. 1987. "A Social Psychology of Caribbean Migration: Strategic Flexibility in the West Indies." In Barry Levine (ed.) *The Caribbean Exodus*. New York: Praeger.

Clarke, Averil. 2001. "Black Women, Silver Spoons: How African American Women Manage Social Mobility, Fertility, and Nuptiality Decisions." Ph.D. diss., University of Pennsylvania.

Dominguez, Virginia. 1975. *From Neighbor to Stranger: The Dilemma of Caribbean Peoples in the United States*. Occasional Papers #5. New Haven: Yale University, Antilles Research Program.

Foner, Nancy. 1978. *Jamaica Farewell: Jamaican Migrants in London*. Berkeley: University of California Press.

——.1985. "Race and Color: Jamaican Migrants in London and New York City." *International Migration Review* 19:708–727.

Gopaul-McNichol, Sharon-Ann. 1993. *Working with West Indian Families*. New York: The Guilford Press.

DISCUSSION QUESTIONS

1. Why do Bobb and Clarke think that social experiences are important in modifying one's values regarding education?
2. Define "strategic flexibility" and describe how immigrants use it to achieve social mobility.
3. How do members of the second generation differ from their immigrant parents in the way they are prepared to be successful in the United States?

Assimulation – Adapting
Talking Over

Part VII
Student Exercises

1. You are a college student, but are you the first generation, second, third, fourth, or even fifth generation in your family to attend college? Think about the concepts that Oliver and Shapiro discuss with regard to your own family history. What are the assets in your family? What has that meant in terms of your own life in terms of residence, education, leisure, and planning for your future? As you discuss your family history with classmates, what do you learn about the way that the past affects the present?
2. What is the minimum wage in your state? Once you find out this figure, try using the newspaper or Internet to identify jobs that pay the minimum wage. What would you earn in one month working 40 hours a week? What is the nature of the work—does it involve manual labor or social skills? Do you get health care and other benefits with that job? Turning again to newspapers or the Internet, where would you find an apartment that would fit your budget? After paying rent, hopefully not more than 50 percent of your income, how would you budget for food, utilities, clothing, and transportation? We doubt you will have money for leisure, so find a nice park to take a walk or other free events in your community. Where would you live and how does this scenario compare with your current lifestyle?

PART VIII

INTRODUCTION

Institutional Segregation and Inequality

The United States has made progress toward reducing inequality by changing key laws that excluded people from entering the country, becoming citizens and voting, and accessing work opportunities and education. However, even after the passage of key civil rights legislation and important Supreme Court decisions, we still have institutionalized racism; that is, power and privileges based on race are still a part of most areas of social life. The history of discrimination—both how race was a factor in the labor market, in education, and in other spheres of social life and the legacy of the ideas that allegedly justified those arrangements—still plagues us. Institutionalized racism persists in the patterns of segregation in most areas of society. People of different races are socially and spatially separated from each other. Segregation means that people's daily lives are structurally very different; such differences are linked to opportunities. Differences in power and resources also mean variation in a group's or individual's abilities to secure housing, provide for their families, advance educationally, protect themselves from exploitation, as well as generally enjoy the fruits of society.

In its 1954 *Brown v. The Board of Education of Topeka, Kansas* decision, the Supreme Court declared segregation unconstitutional. Brown included four public education cases woven together that contested the 1898 *Plessy v. Ferguson* Supreme Court decision. In that case, the Court had recognized segregation as constitutional because it assumed that communities could provide separate but equal facilities in schools, public transportation, and other spheres of life. In the South, legalized segregation—separate facilities sanctioned by *de jure segregation*—shaped very different lives for people based on race. The system that resulted was referred to as Jim Crow. In the North, people were also socially and spatially separated, but rather than the separation being legally enforced, the system was made up of common practices, or *de facto segregation* (resulting from economic or social factors rather than law).

Both systems caused harm, just as our new post–civil rights segregation is detrimental to the lives of both disadvantaged and advantaged persons. At this time, when the nation includes a great deal of racial and ethnic diversity, we could share experiences and shape a vision that promotes equality. Instead, the shape of our institutions means that struggle and hardship characterize many people's lives, while more privileged groups know little about those lives other than what the media project. Part VIII explores the segregated nature of many contemporary social institutions and what that means for various segments of the population. Here scholars document discriminatory treatment on the part of private corporations and different levels of government, so that employment options, social policies, and educational opportunities all reflect patterns of exclusion or differences in treatment. Segregation is also supported by the actions of private individuals, whose choices about residence and educational options reflect their own racial privileges.

Racial oppression in the United States was originally grounded in economic exploitation, so scholars study employment patterns to see how, and the degree to which, race still shapes access to jobs. In Section A, Cedric Herring in "Is Job Discrimination Dead?" finds strong documentation for practices that are continuing to limit employment options for Black Americans in working- and middle-class jobs. The U.S. political economy is shifting from industrial jobs to service-based jobs, a process called *economic restructuring*. Economic restructuring is shaping the work opportunities of different groups and in different ways. For some, it means finding no work; for others, the character of work has changed. Whether one can find a good job is, as the popular adage goes, often a matter of "who you know." And, as Deirdre A. Royster shows ("Race and The Invisible Hand: How White Networks Exclude Black Men from Blue-Collar Jobs"), racism trumps the "invisible hand" of the market by shaping the networks and connections that Black men have, leaving them less likely than White men to find employment, even in working-class jobs. Her research provides a rich account of the processes by which discrimination operates against Black men, as well as other groups who do not have access to White networks.

The visible hand of racism that Royster identifies also works in patterning other employment practices. Marta Tienda and Haya Stier ("The Wages of Race: Color and Employment Opportunity in Chicago's Inner City") show how continuing discrimination results in unemployment for Black, Mexican, and Puerto Rican workers. Their analysis of a single city and their review of different explanations of unemployment show solid evidence of the continuing impact of racial discrimination in employment.

Patterns of segregation and inequality are also found in the work of new immigrants. Pierrette Hondagneu-Sotelo ("Families on the Frontier") examines how immigrant women, especially those from Latin America, have found a racialized and gendered niche in private homes, hotels, nursing homes, and hospitals. Many women send their wages back home to support families in their nation of origin. New *transnational families*, in which family members live in two countries, are different from immigrant families of the past. Hondagneu-Sotelo notes that, earlier, men would cross borders for industrial and agricultural work to support their families in home countries, but now it is most frequently women who cross borders.

The economic rewards promised by civil rights legislation and improved educational attainment have not materialized for all people of color, including new immigrant groups. These articles on employment help explain why communities of color are increasingly stratified, with some people doing well and others struggling to escape poverty. Patterns of segregation and inequality in government or state policy also directly influence family life and health for racial and ethnic groups. Section B, "Families, Communities, and Welfare," opens with a discussion of the impact of racism on Black families and communities. Joe R. Feagin and Karyn D. McKinney ("The Family and Community Costs of Racism") show that racism is not just an abstract force. It does real harm to people and the places they live. Yet, people of color actively resist such psychic and physical assaults, working as they can to build families and communities for social support and, at times, overt resistance to the forces of racism.

Racial segregation disrupts not only families and communities but also relationships that people might otherwise have. Segregation can distort people's ideas about each other and make them more susceptible to accepting the racial stereotypes present in areas such as popular culture, with consequences for both White people and people of color. With racial segregation, there is less sharing of culture, histories, ideas, and caring across racial lines. One recent study shows that, among other things, friendships are affected by segregation. A team of sociologists studied cross-race friendships in several high schools around the nation and found that cross-race friendships increase when school populations are more diverse (Quillian and Campbell 2003).

One of the hallmarks of a free society is that people are able to freely associate with others—as peers, friends, neighbors, lovers, marriage partners, or in any other relationship. But this has not always been the case in the United States. Historically, laws in thirty states prohibited White people from marrying someone of a different race. These *antimiscegenation* laws prohibited so-called "race mixing." For example, the state of California passed a law in 1880 prohibiting any White person from marrying a "negro, mulatto, or Mongolian." This law was designed to prevent marriages between White people and Chinese immigrants (Takaki 1989). Specific laws against intermarriage varied from state to state. Most southern states prohibited White people from marrying Negroes, while some western states, like California, were also anti-Asian. Laws did not prohibit non-White groups from marrying each other, however. Thus in Mississippi the Chinese married so-called Negroes, although neither group was allowed to marry White people.

In order to enforce antimiscegenation laws, states had to devise ways to define race. States varied in this practice as well. "Alabama and Arkansas defined anyone with one-drop of 'Negro' blood as Black; Florida had a one-eighth rule," and other states varied in their racial definitions (Lopez 1996, p. 118). If someone wanted to marry a person of another race, they had to do so in a state that did not prohibit the union. However, they risked having their marriage denied if they moved to a state where such arrangements were illegal.

Not until 1967 were such laws declared unconstitutional by the decision in a U.S Supreme Court case, *Loving v. Virginia*—a case taken to the court by an interracial couple, Mildred Jeter (a Black woman) and Richard Loving (a White man), who had been married in the District of Columbia in 1958. When they returned to their home in Virginia after marrying, they were indicted and charged with violating Virginia's law banning interracial marriage. They were convicted and sentenced to one year in jail—a term they never served because the judge suspended the sentence on the condition that they leave Virginia. They returned five years later to appeal the decision. The Supreme Court decided the case in 1967 based on the argument that laws against intermarriage violated the 14th Amendment, which states: "No State shall make or enforce any laws which shall abridge the privileges or immunities of citizens of the United States; nor shall any State deprive any person of life, liberty, or property, without due process of law; nor deny to any person within its jurisdiction the equal protection of the laws" (U.S. Constitution, Amendment 14, Section 1).

Now interracial marriage is legal, although it is still relatively rare as discussed by Zhenchao Qian ("Breaking the 'Last Taboo': Interracial Marriage in America"). Qian also shows how public opinion about interracial dating and marriage has changed—and how it has not. Yet these trends of cross race interactions are also part of major racial division between families.

Race, as well as other dimensions of inequality, like social class, influences not only a family's resources, but how others in the society perceive them. Segregation means that rather than understanding the experiences of families different from our own, we might have images of them from the media. [As we could see from Part III of this volume, such representations can be distortions.] Empirical research, that is the work of scholars who conduct studies from an objective perspective, can provide us with access to how people see their lives and opportunities. Today there is often much negative press about single mothers, showing us the poverty and hardships they face—often attributing their problems to their own actions. Too frequently these are images of Black women, but the population of single mothers is really more diverse than presented in the media. However, what the media does not tell us is how the single mothers view their own lives.

Kathryn Edin and Maria Kefalas ("How Motherhood Changed My Life") provide an alternative view based on interviews with low-income Black, White, and Latina single mothers. Often blamed for causing their own poverty through having children at a young age, these young women view motherhood as an identity that encourages, rather than discourages, their desires to succeed. Edin and Kefalas' study shows us how these women want to achieve on behalf of their children and embrace a concept of motherhood that is enabling, not disabling. As you can see, research can prove a window into communities we are segregated from and therefore dependent upon the media to inform us about. Yet, research shows us that we have to think critically about the images of other groups, as well as our own, that are projected in the media.

In addition to the media teaching us about other groups, it also informs us about social policies that are suppose to address critical problems, but as we can see such images can be problematic because they are not developed by the people involved, but rather by people with power. Many citizens might think that social policies are protective, but they can be experienced very differently by the people directly affected by these polices. In particular, rather than protecting people of color, many social policies harm communities. The next two chapters explore this issue with regard to families. They draw our attention to how negative stereotypes—what Patricia Hill Collins (2000) calls *controlling images*, ones that separate racial "minority" women from "majority" women. Such stereotypes of poor and minority women as inferior to White women mean that social policies are based on myths rather than the social contexts of people's lives. As a result, they can cause more poverty and hardship. Many families have been historically neglected by government agencies, and current trends indicate that agencies intervene in their lives in negative ways—a burden that more privileged families seldom face. We see this in the Personal Responsibility and Work Opportunities Reconciliation Act (PRWORA) of 1996 or what is known as "welfare reform." Those in more privileged families may think that policies are helping "the disadvantaged," but in reality few social policies are addressing the critical issues in poor people's their lives. Rather than a path out of poverty women can find themselves trapped by social polices that make it hard for them to care for their families.

Dorothy Roberts ("Child Welfare as a Racial Justice Issue") argues that the greater likelihood of children of color to be in foster care is evidence of racial inequality within state policies. As the clients in the child welfare systems around the nation changed from being a majority of White children to a majority of children of color, the remedy also changed—from providing services and keeping children at home to removing them from homes and placing them in foster care. Negative ideas about people of color still found in the media and other segments of the society foster the type of mistreatment that Roberts addresses. Beth Ritchie also examines this practice in "The Social Construction of the Immoral Black Mother," in which she shows that the real needs of these women are not being addressed because the policies relevant to them are shaped by myths and misinformation. When she interviewed Black women in low-income communities, she found that they faced barriers to mothering that social policies neither recognized nor helped to remove.

Patterns of segregation shape what people experience and what people know about the world. In 1968, Congress passed fair housing legislation to address discrimination in housing options. As we enter the twenty-first century, however, our housing is still not integrated. In Section C of Part VIII, we see that *residential segregation* isolates people from one another, with particular consequences for our nation's schools. John E. Farley and Gregory D. Squires ("Fences and Neighbors: Segregation in the 21st Century" detail the patterns of residential segregation that characterize our cities and surrounding areas. Moreover, *hypersegregation*—that is, the concentration of people of color in poor, urban areas—concentrates poverty in certain communities, thus also shielding those in more advantaged areas from understanding the current realities of race and class. One place where this plays out is in our nation's schools.

As part of an effort to provide opportunities in the face of residential segregation, efforts have been made to desegregate and integrate public schools. Now those trends are reversing. *Resegregation* in public schools is a consequence of both residential segregation and the ending of court-ordered desegregation plans. Not only are majority Black and Latino schools likely to receive fewer resources, but the social isolation thus engendered is also a factor in the development of individuals and groups. Linda Darling-Hammond ("The Color Line in American Education") shows how current attempts to reform education have left many behind. Schools attended mostly by poor and minority students are not being helped by current federal policies.

What do these patterns of segregation of neighborhoods and schools mean for young people coming of age? Jonathan Kozol, a long-time advocate of better education for poor and minority children, poignantly describes some of the devastating impacts of poor schools for the worldviews of young, poor, Black children. In his essay ("The Shame of the Nation") he uncovers some of the ironies of an educational system that purports to value diversity, while denying so many students of color a rich opportunity to learn.

Heidi Lasley Barajas and Jennifer L. Pierce ("The Significance of Race and Gender in School Success among Latinas and Latinos in College") also explore gender differences via interviews with Latina and Latino college students. Gendered experiences play a role in the strategies they employ to adjust to college and the challenge of mobility. For men in their study, sports reinforced a sense of manhood as an individual achievement, but the women employ more collective relationships to get and give support to each other. Thus, we see that segregation limits opportunities for young people of color, particularly urban youth and those who succeed academically have to scale various social barriers. Understanding how people cope in educational institutions can help us shape institutions that do not demand as much sacrifice by some groups.

The denial of justice to people of color began with the Constitution. After decades of struggle, there has been some progress in securing civil rights, but mistreatment of people of color is common. There is racial profiling on the highways, on city streets, in airports and in other public places. Police and other law enforcement officials can be a hostile presence in minority communities. In our final section on the examination of segregation and inequality, we look at how *social control*—the practices employed to push individuals and groups to conform to the rules of the society—often conflicts with social justice. In a democratic society, ideally all individuals should be treated fairly by the law, law enforcement officials, and the courts. However, race still plays a role in the treatment of individuals in the criminal justice system.

Thus, Eduardo Bonilla-Silva ("Keeping Them in Their Place") uses empirical studies to argue that Black people have a unique relationship with law enforcement. Because of the segregation of Black Americans, surveillance and differential treatment is invisible to many White Americans, who are often taught through the media to fear minority men as potential criminals. This is evidenced in popular culture where, as Tricia Rose ("Hidden Politics: Discursive and Institutional Policing of Rap Music") points out, young, minority people are socially controlled—often through police force—even while attending rap and hip-hop concerts. As a society we are granting more power to the criminal justice system, even though this is a poor substitute for other social reforms such as education.

As noted above, the separation of the different groups of people means that only a few actually see and question these new practices.

Christina Swarns ("The Uneven Scales of Capital Justice") discusses how race and social class play a role in capital criminal cases, particular where the death penalty is involved. The U.S. Supreme Court declared the death penalty unconstitutional in 1972, but it has been reinstated in many states across the nation since 1976. Swarm examines the biases that still exist at various levels of the criminal justice system. She explores why unchecked power influences who is likely to be prosecuted, found guilty, and sentenced to death. In addition to identifying the bias in the system, we can see how patterns of segregation originating in some areas reverberate in the courts and prisons. This is also demonstrated in an important study by Devah Pager ("The Mark of a Criminal Record"). Pager shows, by experimentally matching Black and White men, each with criminal records, that although having a past criminal record disadvantages both groups for subsequent employment, the effect for Black men is greater.

Ideologies as well as spatial borders, are still used to separate people. Fear of the "other" has jeopardized citizenship rights and social justice for racial groups at different points in history and up to the present. After the bombing of Pearl Harbor by the Japanese and the U.S. entrance into World War II, President Franklin Roosevelt issued Executive Order 9066 that called for the internment of Japanese immigrants and Japanese Americans living on the West Coast for security reasons. Roger Daniels ("Detaining Minority Citizens, Then and Now") uses that history as a vantage point from which to examine recent trends, including the experiences of many Arab Americans, who, after the tragedies of September 11, 2001, found themselves racialized and suspect. Understanding the history and the implications of policies that are supposed to make some of us feel safe is important, because what we need is to preserve social justice for all citizens.

Perhaps the challenge for this new century is to dismantle both the current racial segregation and the ideologies that target people of color as problems. Such injustices keep us from working together to make this nation a land of equality and social justice for all. In our final part, Part IX, we will look at the work that members of excluded groups—such as African Americans, American Indians, and Asian Americans—did to change the legal opportunity structure. Here we might find lessons on how to work to change the racial landscape.

REFERENCES

Collins, Patricia Hill. 2000. *Black Feminist Thought: Knowledge, Consciousness, and Empowerment*, 2nd ed. New York: Routledge.

34

Is Job Discrimination Dead?

CEDRIC HERRING

This article debunks the idea that racial discrimination is a thing of the past. Current research finds strong evidence of the persistence of discrimination, despite opinions that it has disappeared or weakened.

In November 1996, Texaco settled a case for $176 million with African-American employees who charged that the company systematically denied them promotions. Texaco originally vowed to fight the charges. When irrefutable evidence surfaced, however, Texaco changed its position. The New York Times released a tape recording of several Texaco executives referring to black employees as "niggers" and "black jelly beans" who would stay stuck at the bottom of the bag. Texaco also ultimately acknowledged that they used two promotion lists—a public one that included the names of blacks and a secret one that excluded all black employee names. The $176 million settlement was at the time the largest amount ever awarded in a discrimination suit.

Much has changed in American race relations over the past 50 years. In the old days, job discrimination against African Americans was clear, pervasive and undeniable. There were "white jobs" for which blacks need not apply, and there were "Negro jobs" in which no self-respecting white person would be found. No laws prohibited racial discrimination in employment. Indeed, in several states laws required separation of blacks and whites in virtually every public realm. Not only was racial discrimination the reality of the day, but also many whites supported the idea that job discrimination against blacks was appropriate. In 1944, 55 percent of whites admitted to interviewers that they thought whites should receive preference over blacks in access to jobs, compared with only 3 percent who offered such opinions in 1972.

Many blatant forms of racism have disappeared. Civil rights laws make overt and covert acts of discrimination illegal. Also, fewer Americans admit to traditional racist beliefs than ever before. Such changes have inspired many scholars and social commentators to herald the "end of racism" and to declare that we have created a color-blind society. They point to declines in prejudice, growth in the proportion of blacks who hold positions of responsibility, a closing of the earnings gap between young blacks and young whites and other evidence of "racial progress."

SOURCE: From *Contexts* 1(2): 13–18. Copyright © 2002, American Sociological Association. All rights reserved. Reprinted by permission.

However, racial discrimination in employment is still widespread; it has just gone underground and become more sophisticated. Many citizens, especially whites who have never experienced such treatment, find it hard to believe that such discriminatory behavior by employers exists. Indeed, 75 percent of whites in a 1994 survey said that whites were likely to lose a job to a less-qualified black. Nevertheless, clear and convincing evidence of discriminatory patterns against black job seekers exists.

In addition to the landmark Texaco case, other corporate giants have made the dishonor roll in recent years. In 2000, a court ordered Ford Motor Company to pay $9 million to victims of sexual and racial harassment. Ford also agreed to pay $3.8 million to settle another suit with the U.S. Labor Department involving discrimination in hiring women and minorities at seven of the company's plants. Similarly in 1999, Boeing agreed to pay $82 million to end racially based pay disparities at its plants. In April 2000, Amtrak paid $16 million to settle a race discrimination lawsuit that alleged Amtrak had discriminated against black employees in hiring, promotion, discipline and training. And in November 2000, the Coca-Cola Company settled a federal lawsuit brought by black employees for more than $190 million. These employees accused Coca-Cola of erecting a corporate hierarchy in which black employees were clustered at the bottom of the pay scale, averaging $26,000 a year less than white workers.

The list of companies engaged in discrimination against black workers is long and includes many pillars of American industry, not just marginal or maverick firms. Yet when incidents of discrimination come into public view, many of us are still mystified and hard-pressed for explanations. This is so, in part, because discrimination has become so illegitimate that companies expend millions of dollars to conceal it. They have managed to discriminate without using the blatant racism of the old days. While still common, job discrimination against blacks has become more elusive and less apparent.

HOW COMMON?

Most whites think that discriminatory acts are rare and sensationalized by a few high-profile cases and that the nation is well on its way to becoming a colorblind society. According to a 2001 Gallup survey, nearly 7 in 10 whites (69 percent) said that blacks are treated "the same as whites" in their local communities. The numbers, however, tell a different story. Annually, the federal government receives about 80,000 complaints of employment discrimination, and another 60,000 cases are filed with state and local fair employment practices commissions. One recent study found that about 60 percent of blacks reported racial barriers in their workplace in the last year, and a 1997 Gallup survey found that one in five reported workplace discrimination in the previous month.

The results of "social audits" suggest that the actual frequency of job discrimination against blacks is even higher than blacks themselves realize. Audit studies test for discrimination by sending white and minority "job seekers" with comparable resumes and skills to the same hiring firms to apply for the same job.

The differential treatment they receive provides a measure of discrimination. These audits consistently find that employers are less likely to interview or offer jobs to minority applicants. For example, studies by the Fair Employment Practices Commission of Washington, D.C., found that blacks face discrimination in one out of every five job interviews and that they are denied job offers 20 percent of the time. A similar study by the Urban Institute matched equally qualified white and black testers who applied for the same jobs in Chicago. About 38 percent of the time, white applicants advanced further in the hiring process than equally qualified blacks. Similarly, a General Accounting Office audit study uncovered significant discrimination against black and Latino testers. In comparison with whites, black and Latino candidates with equal credentials received 25 percent fewer job interviews and 34 percent fewer job offers.

These audit studies suggest that present-day discrimination is more sophisticated than in the old days. For example, discriminating employers do not explicitly deny jobs to blacks; rather, they use the different phases of the hiring process to discriminate in ways that are difficult to detect. In particular, when comparable resumes of black and white testers are sent to firms, discriminatory firms systematically call whites first and repeatedly until they exhaust their list of white applicants before they approach their black prospects. They offer whites jobs on the spot but tell blacks that they will give them a call back in a few weeks. These mechanisms mean that white applicants go through the hiring process before any qualified blacks are even considered.

Discriminatory employers also offer higher salaries and higher-status positions to white applicants. For example, audit studies have documented that discriminatory employment agencies often note race in the files of black applicants and steer them away from desirable and lucrative positions. A Fair Employment Practices Commission study found that these agencies, which control much of the applicant flow into white-collar jobs, discriminate against black applicants more than 60 percent of the time.

Surprisingly, many employers are willing to detail (in confidence to researchers) how they discriminate against black job seekers. Some admit refusing to consider any black applicants. Many others admit to engaging in recruitment practices that artificially reduce the number of black applicants who know about and apply for entry-level jobs in their firms. One effective way is to avoid ads in mainstream newspapers. In one Chicago study, more than 40 percent of the employers from firms within the city did not advertise their entry-level job openings in mainstream newspapers. Instead, they advertised job vacancies in neighborhood or ethnic newspapers that targeted particular groups, mainly Hispanics or white East European immigrants. For the employer who wants to avoid blacks, this strategy can be quite effective when employment ads are written in languages other than English, or when the circulation of such newspapers is through channels that usually do not reach many blacks.

Employers described recruiting young workers largely from Catholic schools or schools in white areas. Besides avoiding public schools, these employers also avoided recruiting from job-training, welfare and state employment service programs. Consequently, some job-training programs have had unanticipated negative effects on the incomes and employment prospects of their African-American enrollees. For instance, research on the effect of such training programs on the earnings and employability of black inner-city residents found that those who participated in various job-training programs earned less per month and had higher unemployment rates than their counterparts who had not participated in such programs.

WHO SUFFERS?

Generally, no black person is immune from discriminatory treatment. A few factors make some even more vulnerable to discrimination than others. In particular, research has shown that African Americans with dark complexions are likelier to report discrimination—one-half do—than those with lighter complexions. Job discrimination is also associated with education in a peculiar fashion: Those blacks with more education report more discrimination. For example, in a Los Angeles study, more than 80 percent of black workers with college degrees and more than 90 percent of those with graduate-level educations reported facing workplace discrimination. Black immigrants are more likely than nonimmigrants to report discrimination experiences, residents of smaller communities report more than those of larger ones, and younger African Americans report more than older ones. Rates of job discrimination are lower among those who are married than among those who are not wed. Research also shows that some employment characteristics also appear to make a difference: African Americans who are hired through personal contacts report discrimination less often, as do those who work in the manufacturing sector and those who work for larger firms.

Discrimination exacts a financial cost. African Americans interviewed in the General Social Survey in 1991 who reported discrimination in the prior year earned $6,200 less than those who reported none. (In addition, blacks earn $3,800 less than whites because of differences in educational attainment, occupation, age and other factors.) A one-time survey cannot determine whether experiences of discrimination lead to low income or whether low income leads to feeling discriminated against. Multivariate research based on data from the Census Bureau, which controls for education and other wage-related factors, shows that the white-black wage gap (i.e., "the cost of being black") has continued to be more than 10 percent—about the same as in the mid-1970s. Moreover, research looking at the effects of discrimination over the life course suggests a cumulative effect of discrimination on wages such that the earnings gap between young blacks and whites becomes greater as both groups age.

HOW CAN THERE BE DISCRIMINATION?

Many economists who study employment suggest that job discrimination against blacks cannot (long) exist in a rational market economy because jobs are allocated based on ability and earnings maximization. Discrimination, they argue, cannot play a major role in the rational employer's efforts to hire the most productive worker at the lowest price. If employers bypass productive workers to satisfy their racism, competitors will hire these workers at lower-than-market wages and offer their goods and services at lower prices, undercutting discriminatory employers. When presented with evidence that discrimination does occur, many economists point to discriminators' market monopoly: Some firms, they argue, are shielded from competition and that allows them to act on their "taste for discrimination." These economists, however, do not explain why employers would prefer to discriminate in the first place. Other economists suggest that employers may rationally rely on "statistical discrimination." Lacking sufficient information about would-be employees, employers use presumed "average" productivity characteristics of the groups to which the potential employees belong to predict who will make the best workers. In other words, stereotypes about black workers (on average) being worse than whites make it "justifiable" for employers to bypass qualified black individuals. In these ways, those economists who acknowledge racial discrimination explain it as a "rational" response to imperfect information and imperfect markets.

In contrast, most sociologists point to prejudice and group conflict over scarce resources as reasons for job discrimination. For example, racial groups create and preserve their identities and advantages by reserving opportunities for their own members. Racially based labor queues and differential terms of employment allow members to allocate work according to criteria that have little to do with productivity or earnings maximization. Those who discriminate against blacks often use negative stereotypes to rationalize their behavior after the fact, which, in turn, reinforces racism, negative stereotypes and caricatures of blacks.

In particular, labor market segregation theory suggests that the U.S. labor market is divided into two fundamentally different sectors: (1) the primary sector and (2) the secondary sector. The primary sector is composed of jobs that offer job security, work rules that define job responsibilities and duties, upward mobility, and higher incomes and earnings. These jobs allow incumbents to accumulate skills that lead to progressively more responsibility and higher pay. In contrast, secondary sector jobs tend to be low-paying, dead-end jobs with few benefits, arbitrary work rules and pay structures that are not related to job tenure. Workers in such jobs have less motivation to develop attachments to their firms or to perform their jobs well. Thus, it is mostly workers who cannot gain employment in the primary sector who work in the secondary sector. Race discrimination—sometimes by employers but at times by restrictive unions and professional associations fearful that the inclusion of blacks may drive down their overall wages or prestige—plays a role in determining who gets access to jobs in the primary sector. As a consequence, African Americans are locked out of jobs in the primary labor market, where they would receive higher pay and better treatment, and they tend to be crowded into the secondary sector. And these disparities compound over time as primary sector workers enhance their skills and advance while secondary sector workers stay mired in dead-end jobs.

An alternative sociological explanation of African-American disadvantage in the U.S. labor market is what can be referred to as "structural discrimination." In this view, African Americans are denied access to good jobs through practices that appear to be race-neutral but that work to the detriment of African Americans. Examples of such seemingly race-neutral practices include seniority rules, employers' plant location decisions, policy makers' public transit decisions, funding of public education, economic recessions and immigration and trade policies.

In the seniority rules example, if blacks are hired later than whites because they are later in the employers' employment queue (for whatever reason), operating strictly by traditional seniority rules will ensure greater job security and higher pay to whites than to African Americans. Such rules virtually guarantee that blacks, who were the last hired, will be the "first fired" and the worst paid. The more general point is that employers do not have to be prejudiced in implementing their seniority rules for the rules to have the effects of structural discrimination on African Americans. Unequal outcomes are built into the rules themselves.

These same dynamics apply when (1) companies decide to locate away from urban areas with high concentrations of black residents; (2) policy makers decide to build public transit that provides easy access from the suburbs to central city job sites but not from the inner city to central city job sites or to suburban job sites; (3) public education is funded through local property tax revenues that may be lower in inner-city communities where property values are depressed and higher in suburban areas where property values are higher and where tax revenues are supplemented by corporations that have fled the inner city; (4) policy makers attempt to blunt the effects of inflation and high interest rates by allowing unemployment rates to climb, especially when they climb more rapidly in African-American communities; and (5) policy makers negotiate immigration and trade agreements that may lead to lower employer costs but may also lead to a reduction in the number of jobs available to African Americans in the industries affected by such agreements. Again, in none of these cases do decision makers need to be racially prejudiced for their decisions to have disproportionately negative effects on the job prospects or life chances of African Americans.

WHAT CAN BE DONE?

Employment discrimination, overt or covert, is against the law, yet it clearly happens. Discrimination still damages the lives of African Americans. Therefore, policies designed to reduce discrimination should be strengthened and expanded rather than reduced or eliminated, as has recently occurred. Light must be shed on the practice, and heat must be applied to those who engage in it. Some modest steps can be taken to reduce the incidence and costs of racial discrimination:

Conduct More Social Audits of Employers in Various Industries of Varying Sizes and Locations

In 2000, the courts upheld the right of testers (working with the Legal Assistance Foundation of Chicago) to sue discriminatory employers. Expanded use of evidence from social audits in lawsuits against discriminatory employers provides more information about discriminatory processes, arms black applicants more effectively and provides greater deterrence to would-be discriminators who do not want to be exposed. Even when prevention is not successful, documentation from social audits makes it easier to prosecute illegal discrimination. As in the Texaco case, it has often been through exposure and successful litigation that discriminatory employers mended their ways.

Restrict Government Funding to and Public Contracts with Firms that Have Records of Repeated Discrimination Against Black Applicants and Black Employees

The government needs to ensure that discriminatory employers do not use taxpayer money to carry out their unfair treatment of African Americans. Firms that continue discriminating against blacks should have their funding and their reputations linked to their performance. Also, as lawsuits over this issue proliferate, defense of such practices becomes an expensive proposition. Again, those found guilty of such activities should have to rely on their own resources and not receive additional allocations from the state. Such monetary deterrence may act as a reminder that racial discrimination is costly.

Redouble Affirmative Action Efforts

Affirmative action consists of activities undertaken specifically to identify, recruit, promote or retain qualified members of disadvantaged minority groups to overcome the results of past discrimination and to deter discriminatory practices in the present. It presumes that simply removing existing impediments is not sufficient for changing the relative positions of various groups. In addition, it is based on the premise that to truly affect unequal distribution of life chances, employers must take specific steps to remedy the consequences of discrimination.

Speak Out When Episodes of Discrimination Occur

It is fairly clear that much discrimination against African Americans goes unreported because it occurs behind closed doors and in surreptitious ways. Often, it is only when some (white) insider provides irrefutable evidence that such incidents come to light. It is incumbent upon white Americans to do their part to help stamp out this malignancy.

Now that racial discrimination in employment is illegal, stamping it out should be eminently easier to accomplish. The irony is that because job discrimination against blacks has been driven underground, many people are willing to declare victory and thereby let this scourge continue to flourish in its camouflaged state. If we truly want to move toward a color-blind society, however, we must punish such hurtful discriminatory behaviors when they occur, and we should reward efforts by employers who seek to diversify their workforce by eliminating racial discrimination. This is precisely what happened in the landmark Texaco case, as well as the recent Coca-Cola settlement. In both cases, job discrimination against African Americans was driven above ground, made costly to those who practiced it and offset by policies that attempted to level the playing field.

DISCUSSION QUESTIONS

1. Why does Herring think that the 1964 Civil Rights Act has not ended employment discrimination?
2. What is an audit study? How is this tool used to investigate discriminatory practices?
3. What is the impact of new discriminatory practices on minorities?

35

Race and the Invisible Hand

How White Networks Exclude Black Men from Blue-Collar Jobs

DEIRDRE A. ROYSTER

The "invisible hand" refers to a meritocracy, where what you know shapes employment options. By comparing the experiences of Black and White working-class men who graduated from the same vocational high school, Royster shows how race can shape Black men's access to job networks—making who you know a critical factor in their economic well-being.

In the late 1970s, [African American sociologist Williams Julius Wilson] published an extremely influential book, *The Declining Significance of Race.* In this book, Wilson argued that race was becoming less and less important in predicting the economic possibilities for well-educated African Americans. In other words, the black-led Civil Rights movement had been successful in removing many of the barriers that made it difficult, if not impossible, for well-trained blacks to gain access to appropriate educational and occupational opportunities. Wilson argued that this new pattern of much greater (but not perfect) access was unprecedented in the racial stratification system in the United States and that it would result in significant and lasting gains for African American families with significant educational attainment.

Recent research on the black middle class has only partially supported Wilson's optimistic prognosis. While blacks did experience significant educational and occupational gains during the 1970s, their upward trajectory appears to have tapered off in the 1980s and 1990s. Moreover, some blacks have found themselves tracked into minority-oriented community relations positions within the professional and managerial occupational sphere. Even more troubling are data indicating that the proportion of blacks who attend and graduate from college appears to be shrinking, with the inevitable result that fewer blacks will have the credentials and skills necessary to get the better jobs in the growing technical and professional occupational categories. Despite real concerns about the stability of the

SOURCE: From Deirdre A. Royster, *Race and the Invisible Hand: How White Networks Exclude Black Men from Blue-Collar Jobs* (Berkeley, CA: University of California Press, 2003), pp. 18–23, 58–59, 180–189. Copyright © 2003 by The Regents of the University of California. Reprinted by permission.

black middle class and some glitches in the workings of the professional labor market, no one doubts that a substantial portion of the black population now enjoys access to middle-class opportunities and amenities—including decent homes, educational facilities, public services, and most importantly, jobs—commensurate with their substantial education and job experience, or in economic terms, their endowment of human capital.

While other scholars were investigating his theories about the black middle class, Wilson became distressed about the pessimistic prospects of blacks who were both poorly educated and increasingly isolated in urban ghettos with high rates of poverty and unemployment. His main concern was that changing labor demands that increase opportunities for highly skilled workers have the potential of making unskilled black labor obsolete. According to Wilson's next two books, *The Truly Disadvantaged* and *When Work Disappears*, this group's inability to gain access to mobility-enhancing educational opportunities is exacerbated by the further problems of a deficiency of useful employment contacts, lack of reliable transportation, crowded and substandard housing options, a growing sense of frustration, and an image among urban employers that blacks are undesirable workers, not to mention the loss of manufacturing and other blue-collar jobs. These factors, and a host of others, contribute to the extraordinarily difficult and unique problems faced by the poorest inner-city blacks in attempting to advance economically. Wilson and hundreds of other scholars—even those who disagree with certain aspects of his thesis—argue that this group needs special assistance in order to overcome the obstacles they face.

If Wilson intended the *Declining Significance of Race* thesis (and its Underclass corollary) to apply mainly to well educated blacks and ghetto residents, then Wilson only explained the life chances of at most 30 to 40 percent of the black population.[1] The rest of the black population neither resides in socially and geographically isolated ghettoes, nor holds significant human capital, in the form of college degrees or professional work experience. Looking at five-year cohorts beginning at the turn of the century, Mare found that the cohort born between 1946 and 1950 reached a record high when 13 percent of its members managed to earn bachelor's degrees. Recent cohorts born during the Civil Rights era (1960s) have not reached the 13 percent high mark set by the first cohort to benefit from Civil Rights era victories. As a result, today the total percentage of African Americans age 25 and over who have four or more years of college is just under 14 percent.[2] According to demographer Reynolds Farley, while college attendance rates for white males (age 18–24) have rebounded from dips in the 1970s back to about 40 percent, black male rates have remained constant at about 30 percent since the 1960s. Figures like these suggest that Civil Rights era "victories" have not resulted in increasing percentages of blacks gaining access to college training. Instead, most blacks today attempt to establish careers with only modest educational credentials, just as earlier cohorts did. Thus the vast majority of blacks are neither extremely poor nor particularly well educated; most blacks would be considered lower-middle- or working-class and modestly educated. That is, most blacks (75 percent) lack bachelor's degrees but hold high school diplomas or GEDs; most blacks (92 percent) are working rather than unemployed; and most (79 percent) work at jobs that are lower-white- or blue-collar rather than professional.[3] Given that modestly educated blacks make up the bulk of the black population, it is surprising that more attention has not been devoted to explicating the factors that influence their life chances.

Wilson's focus on the extremes within the black population, though understandable, points to a troubling underspecification in his thesis: it is unclear whether Wilson sees individuals with modest educational credentials—high school diplomas, GEDs, associate's degrees, or some college or other post-secondary training, but not the bachelor's degree—as cobeneficiaries of civil rights victories alongside more affluent blacks. The logic of his thesis implies that as long as they do not reside in socially and geographically isolated communities filled with poor and unemployed residents, from which industrial jobs have departed, then modestly educated blacks, like highly educated blacks, ought to do about as well as their white counterparts....

Because he argues that *past* racial discrimination created the ghetto poor, or underclass, while macro-economic changes—and not current racial discrimination—explain their current economic plight, Wilson's perspective implies that white attempts to exclude blacks are probably of little significance today. In addition, Wilson offers a geographic, rather than racial, explanation for whites' labor market advantages when he argues that because most poor whites live outside urban centers, they do not suffer the same sort of structural dislocation or labor obsolescence as black ghetto residents. If Wilson's reasoning holds, there is no reason to expect parity among the poorest blacks and whites in the United States without significant government intervention. Despite a conspicuous silence regarding the prospects for parity among modestly educated blacks and whites, Wilson's corpus of research and theory offers the most race and class integrative market approach available. First, Wilson specifies how supply and demand mechanisms work differently for blacks depending on their class status. Specifically, he argues that there is now a permanent and thriving pool (labor supply) of educationally competitive middle-class blacks, while simultaneously arguing that changes in the job structure in inner cities have disrupted the employment opportunities (labor demand) for poorer blacks. Second, Wilson argues that contemporary racial disparity results, by and large, from the structural difficulties faced by poor blacks rather than racial privileges enjoyed by (or racial discrimination practiced by) poor or more affluent whites. One of the questions guiding this study is whether the life chances of modestly educated whites and blacks are becoming more similar, as with blacks and whites who are well-educated, or more divergent, as with blacks and whites on the bottom....

The fifty young men interviewed for this study may have been, in some ways, atypical. For example, none of them had dropped out of school, and all were extremely polite and articulate. I suspect that these men were among the easiest to contact because of high residential stability and well-maintained friendship networks. Their phone numbers had remained the same or they had kept in touch with friends since graduating from high school two to three years earlier. Because of these factors, I may have tapped into a sample of men who were more likely to be success stories than most would have been. Researchers call this sampling dilemma creaming, because the sample may reflect those who were most likely to rise to the top or be seen as the cream of the crop, rather than those of average or mixed potential. In this study, however, it may have been an advantage to have "creamed," since I wanted to compare black and white men with as much potential for success as possible. Moreover, men who are personable and who have stable residences and friendship networks might be most able to tap into institutional and personal contacts in their job searches—one of my main research queries....

While I don't think there were idiosyncratic differences among the black and white men I studied or between the men I found and their same-race peers that I didn't find, some of the positive attributes of my sample suggest that my findings may generalize only to young men who generally play by the rules. Of course, it isn't all that clear what proportion of young working-class males (black or white) try to play by the rules—maybe the vast majority try to do so. Nor is it clear by what full set of criteria my subjects did, in fact, play by the rules. My sample includes men who had brushes with the law as well as some who might be considered "goody two-shoes." In other words I'm not sure that the specific men with whom I spoke are atypical among working-class men, but I am willing to acknowledge that they may be. Perhaps what is most important to remember about the sample, who seemed to me to be pretty ordinary, "All-American" men, is my contention that this set of men ought to have similar levels of success in the blue-collar labor market—if, that is, we have finally reached a time when race doesn't matter....

One narrative, the achievement ideology, asserts that formal training, demonstrated ability, and appropriate personal traits will assure employment access and career mobility. The second narrative, the contacts ideology, emphasizes personal ties and affiliations as a mechanism for employment referrals, access, and mobility.... The achievement ideology has persistently dominated American understanding of occupational success, even though everyone, it seems, is willing to admit that "who you know" is at least as important as "what you know" in gaining access to opportunities in American society. All of the men in this study, for example, said that contacts were very important in establishing young men like themselves in careers. One offered a more nuanced explanation: "It's not [just] who you know, it's how they know you." That is, it is not simply knowing the right people that matters; it is sharing the right sort of bonds with the right people that influences what those people would be willing to do to assist you.

The black and white men in this study had more achievements in common than contacts. They were trained in the same school, in many of the same trades, and by the same instructors. They had formal access to the same job listing services and work-study programs. Instructors and students alike agreed, and records confirm, that in terms of vocational skills and performance, the blacks and whites in this sample were among the stronger students....

Black Glendale graduates trail behind their white peers. They are employed less often in the skilled trades, especially within the fields for which they have been trained; they earn less per hour; they hold lower status positions; they receive fewer promotions; and they experience more and longer periods of unemployment. No set of educational, skill, performance, or personal characteristics unique to either the black or white students differentiates them in ways that would explain the unequal outcomes.... Only their racial status and the way it situates them in racially exclusive networks during the school-work transition process adequately explain their divergent paths from seemingly equal beginnings.

In this study of one variant of the "who you know" versus "what you know" conundrum, it is manifestly and perpetually evident that racial dynamics are a key arbiter of employment outcome. Yet challenging the power of the achievement ideology in American society requires a careful exposition of how factors such as race throw a wrench into the presumption of meritocracy. In addition, the contacts ideology must be uniquely construed to take into account the significance of racially determined patterns of affiliation within a class, in this case the working class.

RACE, AN ARBITER OF EMPLOYMENT NETWORKS

Researchers have long argued that black males lack access to the types of personal contacts that white males appear to have in abundance.[4] I would argue that it's more than not having the right contacts. In terms of social networks, black men are at a disadvantage in terms of configuration, content, and operation, a disadvantage that is exacerbated in sectors with long traditions of racial exclusion, such as the blue-collar trades. Even when blacks and whites have access to some of the same connections, as in this study, care must be taken to examine exactly what transpires. For example, I noted that black and white males were assisted differently by the same white male teachers. If I had only asked students whether they considered their shop teachers contacts on which they could rely, equal numbers of black and white males would have answered affirmatively. But this would have told us nothing about *how* white male teachers *chose to know and help* their black male students. The teachers chose to verbally encourage black students, while providing more active assistance to white students. I discovered a munificent flow of various forms of assistance, including vacancy information, referrals, direct job recruitment, formal and informal training, vouching behaviors, and leniency in supervision. For white students, this practice, which repeated neighborhood and community patterns within school walls, served to convert institutional ties (as teachers) into personal ones (as friends) that are intended to and do endure well beyond high school....

Even without teachers consciously discriminating, significant employment information and assistance remained racially privatized within this public school context. In that white male teachers provided a parallel or shadow transition system for white students that was not equally available to black students, segregated networks still governed the school-work transition at Glendale even though classrooms had long been desegregated.

The implications for black men are devastating. Despite having unprecedented access to the same preparatory institution as their white peers, black males could not effectively use the institutional connection to establish successful trade entry. Moreover, segregation in multiple social arenas, beyond schools, all but precluded the possibility of network overlaps among working-class black and white men. As a result, black men sought employment using a truncated, resource-impoverished network consisting of strong ties to other blacks (family, friends, and school officials) who like themselves lacked efficacious ties to employment.

Beyond school, matters were even worse. Without being aware of it, white males' descriptions of their experiences revealed a pattern of intergenerational intraracial assistance networks among young and older white men that assured even the worst young troublemaker a solid place within the blue-collar fold. The white men I studied were not in any way rugged individualists; rather they survived and thrived in rich, racially exclusive networks.

For the white men, neighborhood taverns, restaurants, and bars served as informal job placement centers where busboys were recruited to union apprentice programs, pizza delivery boys learned to be refrigeration specialists, and dishwashers studying drafting could work alongside master electricians then switch back to drafting if they wished. I learned of opportunities that kept coming, even when young men weren't particularly deserving. One young man had been able to hold onto his job after verbally abusing his boss. Another got a job installing burglar alarms after meeting the vice president of the company at a cookout—without ever having to reveal his prison record, which included a conviction for burglary.

Again and again, the white men I spoke with described opportunities that had landed in their laps, not as the result of outstanding achievements or personal characteristics, but rather as the result of the assistance of older white neighbors, brothers, family friends, teachers, uncles, fathers, and sometimes mothers, aunts, and girlfriends (and their families), all of whom overlooked the men's flaws. It never seemed to matter that the men were not A students, that they occasionally got into legal trouble, that they lied about work experiences from time to time, or that they engaged in horseplay on the job. All of this was expected, brushed off as typical "boys will be boys" behavior, and it was sometimes the source of laughter at the dining room table. In other words, there were no significant costs for white men associated with being young and inexperienced, somewhat immature, and undisciplined.

The sympathetic pleasure I felt at hearing stories of easy survival among working-class white men in an era of deindustrialization was only offset by the depressing stories I heard from the twenty-five black men. Their early employment experiences were dismal in comparison, providing a stark and disturbing contrast. Whereas white men can be thought of as the second-chance kids, black men's opportunities were so fragile that most could not have recovered from even the relatively insignificant mishaps that white men reported in passing.

Black men were rarely able to stay in the trades they studied, and they were far less likely than white men to start in one trade and later switch to a different one, landing on their feet. Once out of the skilled trade sphere, they sank to the low-skill service sector, usually retail or food services. The black men had numerous experiences of discrimination at the hands of older white male supervisors, who did not offer to help them and frequently denigrated them, using familiar racial epithets. The young black men I spoke with also had to be careful when using older black social contacts. More than one man indicated to me that, when being interviewed by a white person, the wisest course of action is to behave as if you don't know anyone who works at the plant, even if a current worker told you about the opening. These young black men, who had been on the labor market between two and three years, were becoming discouraged. While they had not yet left the labor force altogether, many (with the help of parents) had invested time and resources in training programs or college courses that they and their families hoped would open up new opportunities in or beyond the blue-collar skilled labor market. Many of the men had begun to lose the skills they had learned in high school; others, particularly those who'd had a spell or two of unemployment, showed signs of depression.

My systematic examination of the experiences of these fifty matched young men leads me to conclude that the blue-collar labor market does not function as a market in the classic sense. No pool of workers presents itself, offering sets of skills and work values that determine who gets matched with the most and least desirable opportunities. Rather, older men who recruit, hire, and fire young workers choose those with whom they are comfortable or familiar. Visible hands trump the "invisible hand"—and norms of racial exclusivity passed down from generation to generation in American cities continue to inhibit black men's entry into the better skilled jobs in the blue-collar sector.

Claims of meritocratic sorting in the blue-collar sector are simply false; equally false are claims that young black men are inadequately educated, inherently hostile, or too uninterested in hard work or skill mastery to be desirable workers. These sorts of claims seek to locate working-class black men's employment difficulties in the men's alleged deficits—bad attitudes, shiftlessness, poor skills—rather than in the structures and procedures of worker selection that are typically under the direct control of older white men whose preferences, by custom, do not reflect meritocratic criteria.

Few, if any, political pressures, laws, or policies provide sufficient incentives or sanctions to prevent such employers from arbitrarily excluding black workers or hiring them only for menial jobs for which they are vastly overqualified. Moreover, in recent years, affirmative action policies that required that government contracts occasionally be awarded to black-owned firms or white-owned firms that consistently hire black workers have come under attack—eroding the paltry incentives for inclusion set forth during the Civil Rights era, nearly forty years ago. Indeed, there is far less pressure today than in the past for white-owned firms to hire black working men. And given persistent patterns of segregation—equivalent to an American apartheid, according to leading sociologists—there remain few incentives for white men to adopt young black men into informal, neighborhood-generated networks. As a result, occupational apartheid reigns in the sector that has always held the greatest potential for upward mobility, or just basic security, for modestly educated Americans.

IDEOLOGY AND THE DEFENSE OF RACIALIZED EMPLOYMENT NETWORKS

The public perception of the causes of black men's labor difficulties—namely, that the men themselves are to be blamed—contrasts with my findings. And my research is consistent with that of hundreds of social scientists who have demonstrated state-supported and informal patterns of racial exclusion in housing, education, labor markets, and even investment opportunities. Racism continues to limit the life chances of modestly educated black men....

WHITE PRIVILEGE, BLACK ACCOMMODATION

How, then, do black males, if they wish to earn a living in the surviving trades, negotiate training and employment opportunities in which networks of gatekeepers remain committed to maintaining white privilege? The present research suggests that the options are few: either accommodation to the parameters of a racialized system or failure in establishing a successful trade career. The interviews revealed that forms of black accommodation begin early, as when young men avoided training in trades of interest because they were known to hold little promise for integration and advancement. For those who made such discoveries later, accommodation took the form of disengaging from specific trades, such as electrical construction, and pursuing whatever jobs became available. For some, the disengaging process involved the claim that they were never really committed to the original trade field, but I suspect that such claims merely served to soften the blow of almost inevitable career failure. For the determined, accommodation required suppressing anger at racially motivated insults and biased employment decisions in the majority-white trade settings. If this strategy wore thin, two difficult accommodations remained. The first involved finding a work setting—not necessarily within one's trade—in which the workplace culture was, if not actively receptive to black inclusion, at least neutral. The second involved finding ways to work in white-dominated fields without having to work beside whites.

A word needs to be said about a particularly troubling accommodative behavior adopted by the black men: not actively and persistently pursuing offers of assistance. It is not clear to what extent the black men were fully cognizant of the extent and potency of whites' informal networks or of the cultural norms governing their operation. But, while it is evident that the older white men who were network gatekeepers did not extend the same access and support to the black men, the black men may also have been less proactive in pursuing older white men who might have assisted them.

Generally, the white men in the study appear to have more actively followed up on offers of assistance. And although their careers developed much more smoothly than those of the black men, they were certainly not without the difficulties of not being hired, workplace dissatisfaction, competing vocational interests, and unemployment. Nevertheless, they returned, sometimes repeatedly, to contacts for further assistance. Certainly, demonstrating proactivity toward a typically racially exclusive white network would be especially problematic for black men.

Undoubtedly, black men's exclusion from white personal settings where easy informal contact is facilitated, like neighborhoods and family, contributes to black men's reluctance to pursue whites for assistance. In addition, black men's lack of personal familiarity with normative expectations among whites probably hampers their efforts to imitate their white peers' more forward network behaviors. Furthermore, any efforts by blacks to engage in such behaviors might not be similarly regarded as appropriate, and might instead be interpreted as aggressive, "uppity," or indicative of a feeling of entitlement. Finally, black men's early experiences of racial exclusion, bias, and hostility in the school and the workplace inform not only their assessment of employment prospects, but also their actual employment strategies. Given these complicated contingencies, perhaps the somewhat hesitant responses of black men are, on the whole, not unreasonable.

CONCLUSION

Black men have paid a great price for exclusion from blue-collar trades and the networks that supply those trades, but they have not paid it alone. The pain of black men's unemployment and underemployment spreads across black communities in a ripple effect. Less able to contribute financially to the care of children and parents, or to combine resources with black women or assist other men with work entry and "learning the ropes" on the job, black men withdraw from the support structures that they need and that they are needed to support emotionally as well as economically. The enduring power of segregated networks in the blue-collar trades is as responsible as segregated neighborhoods for the existence of extremely poor and isolated black communities and of the disproportionately black and male prison population—in fact, more so. While many black families live in stable communities that are mostly, if not entirely, black, the inability to find remunerative jobs that do not require expensive college training makes living decently anywhere extremely difficult. And the loss of manufacturing jobs cannot account for black men's underemployment in the remaining blue-collar fields—especially construction, auto mechanics, plumbing, computer repair, and carpentry....

My findings demonstrate that, without governmental initiatives that provide strong incentives for inclusion, white tradesmen will have no reason to open their networks to men of color. As a result, the work trajectories of white and black men who start out on an equal footing will continue to diverge into skilled and unskilled work paths because of business-as-usual patterns of exclusion. Although there are few precedents for intervening in the private sector, there are strong precedents for intervening in the public sector, where the tax dollars of majority and minority citizens must not be redistributed in ways that condone customs of exclusion....

Without the government taking a lead, the young black men I studied—who played by the rules—are unlikely to ever reach their potential as skilled workers or to take their places as blue-collar entrepreneurs, as so many of their white peers are poised to do. This tragedy could have been averted. My hope is that it will be averted in the next generation.

REFERENCES

1. Haywood Horton, Beverlyn Lundy, Cedric Herring, and Melvin E. Thomas, "Lost in the Storm: The Sociology of the Black Working Class, 1850 to 1990," *American Sociological Review* 65, no.1 (2000): 128–137.
2. Robert Mare, "Changes in Educational Attainment and School Enrollment," in *State of the Union*, ed. Reynolds Farley (New York: Russell Sage Foundation, 1995); Nancy Folbre, *The Field Guide to the U.S. Economy* (New York: New Press, 1999).
3. Ibid.
4. Richard Freeman and Harry J. Holzer, *The Black Youth Employment Crisis* (Chicago: University of Chicago Press, 1986); William Julius Wilson, *The Truly Disadvantaged* (Chicago: University of Chicago Press, 1987); Paul Osterman, *Getting Started: The Youth Labor Market* (Cambridge: Massachusetts Institute of Technology Press, 1980).

DISCUSSION QUESTIONS

1. Why is economic advancement difficult for inner-city Black men with modest levels of education?
2. What are the differences in the way networks operated for Black men and White men in Royster's study? What are the implications of those outcomes for their futures?
3. Rather than just let market forces work as they will, why does Royster think government intervention is important?

36

The Wages of Race

Color and Employment Opportunity in Chicago's Inner City

MARTA TIENDA AND HAYA STIER

This article examines different arguments used to explain wage differences between African American, Hispanic, and White workers. Using data from the city of Chicago, the authors show the continuing effect of discrimination against Black, Mexican, and Puerto Rican workers.

As the second largest U.S. city at the turn of the [20th] century, Chicago was a booming industrial center on its way to becoming a major distribution center as well. Rapid industrial growth meant job growth—employment opportunities—which served as a magnet for southern and eastern European immigrants and northbound Blacks seeking better destinies than their southern origins had yielded. The dimension's of the "Great Migration" which relocated thousands of unskilled laborers from southern plantations to northern ghettos have been amply documented....

As the end of the [20th] century approaches, Chicago is a different city—a place where opportunity has been dimmed as industrial restructuring eliminated thousands of semiskilled and unskilled jobs that paid family wages. Like other northern cities (that is, Detroit, Cleveland, Philadelphia, New York, and St. Louis), Chicago's industrial decline is mirrored in its demographic decline; the city has lost nearly 1 million people since mid-century. During the 1980s Chicago fell from second to third largest city, surpassed by Los Angeles. Yet Chicago remains a destination for thousands of immigrants who continue to diversify the ethnic terrain. Contemporary immigrants destined for Chicago and other major cities originate from Mexico, Central America, Korea, and other parts of Asia rather than Europe....

Against this backdrop, it is unsurprising that Chicago was a core city for the Civil Rights movement, for Jessie Jackson's, Operation PUSH, and for William. J. Wilson's (1987) *"Truly Disadvantaged."* That poverty and joblessness has become more concentrated and devastating for, Chicago's Black population

SOURCE: From *Great Divides: Readings in Social Inequality in the United States* (3rd ed.). Edited by Thomas M. Shapiro (New York: McGraw Hill 2005).

suggests that race, matters as much at the close of the century as it did at the beginning. This study is an investigation of just that question—how does color influence job opportunities in the city of "Big Shoulders?" Accordingly, in this chapter we examine differences in labor force activity among parents residing in economically and ethnically diverse neighborhoods. Our general aim is to identify and evaluate the circumstances that sustain inequities in employment opportunities within a single labor market....

OPPORTUNITIES AND ETHNIC INEQUALITY

A dominant structural explanation for the rise of inner-city joblessness focuses on the decline of well-paying manufacturing jobs in old industrial centers (see Kasarda 1985: Wilson 1987). Wilson proposed that "social isolation" from mainstream work norms and behavior sustains chronic labor market inactivity in ghetto poverty neighborhoods. Yet this explanation does not resolve the puzzle of large race and ethnic differences in employment behavior among inner-city populations residing in a single labor market.

"Queuing theory" provides another structural perspective of ethnic differentials in employment (Hodge 1973). Simply stated, when job opportunities requiring low to moderate skills (education) decline, ethnic, employment competition intensifies for two reasons. First, employers can be more "choosy" about which workers to hire, allowing their preferences and prejudices to play themselves out differently from a situation of labor scarcity. Second, the changing ethnic composition of neighborhoods resulting from differential migration of minority workers can reconfigure the extent and nature of job competition along color lines (Hodge 1973; Tienda, Donato, and Cordero-Guzmán 1992).

Several studies have documented that immigrants participate in the labor force at a higher rate than natives (Borjas and Tienda 1987), but researchers do not agree whether immigrants displace native minority workers. Aggregate econometric evidence indicates that the labor market impacts of recent immigrants are benign, but studies of specific firms and communities reveal intense competition between immigrants and native minorities (Tienda and Liang 1993). At issue for the contemporary debates about pervasive inner-city joblessness is whether native minorities both perceive competition from immigrants and experience lower employment rates than comparably or less-skilled immigrants.

Finally, discrimination against people of color could also give rise to race and ethnic differences in joblessness even in the absence of skill differences because it implies unequal treatment of entire groups. For example, a recent study showed that the employment returns to education came to depend increasingly on race and ethnicity between 1960 and 1985, particularly among women with less than a high school degree (Tienda, Donato, and Cordero-Guzmán 1992). However, this study did not consider whether higher minority joblessness was associated with group differences in willingness to work.

INDIVIDUAL DIFFERENCES AND ETHNIC INEQUALITY

The human capital explanation for the disadvantaged labor market standing of minority workers is essentially one that focuses on workers' differences in skills. Assuming that education has uniform exchange value, then the highest labor force participation rates should correspond to groups with the most education. The skill deficit explanation of rising inner-city joblessness also has been tied to the "spatial mismatch hypothesis" (Holzer 1991; Kasarda 1985). Simply put, the rapid decline of unskilled and semi-skilled jobs from old industrial centers has produced a major imbalance between the supply of workers available and the demand for a highly educated workforce. Stated as a variant of the spatial mismatch hypothesis, the skill deficit interpretation maintains that increased inequities between White and non-White workers reflect the failure of minorities to keep pace with the rising educational requisites of new jobs.

An alternative individualistic explanation for pervasive inner-city joblessness suggests that weak labor force attachment is coupled with a preference for welfare (Murray 1984). Although several critics have challenged this interpretation, few studies have investigated the willingness of inner-city dwellers to work (Tienda and Stier 1991; Van Haitsma 1989). One variant of the weak attachment explanation is that jobless inner-city residents require higher wage rates than employers are willing to pay, given their skills and prior employment experience. Reservation wage refers to the minimum hourly pay required by workers in order to accept a job offer. Applied to race and ethnic differences in labor force activity, this implies that jobless minority workers demand higher compensation as a condition of accepting a job than similarly skilled nonminority workers. This argument has been summoned to explain the anomaly of high minority unemployment coupled with high rates of labor force activity among undocumented workers. However, few studies have examined empirically whether people of color actually demand higher wages than their skill characteristics warrant....

MAKING SENSE OF ETHNIC INEQUALITY: VOICES FROM THE INNER CITY

Robert Hodge's (1973) conception of the labor market as an ethnic queue is helpful in understanding race and ethnic differences in employment statuses and provides some insight into the emergence of wage inequities. The basic principle of queuing is that in the process of matching workers to jobs, the most desirable workers are hired first and the least desirable workers are hired last. Thus, when labor demand shrinks, increases in joblessness will be greatest for workers at the bottom than for workers ranked higher in the queue (Hodge 1973; Lieberson 1980; Reskin and Roos 1990). In a totally meritocratic society, skills would be the primary basis for determining individuals' rank in a hiring queue; in practice, gender, race, and national origin are also decisive. Distortions in rankings based purely on skills result because employers have imperfect information about prospective workers productivity, organizational procedures often protect and reward others with seniority, and employers discriminate against entire groups of workers (Hodge 1973).

We believe that the worsened labor market position of Black and Puerto Rican inner-city workers results partly from displacement through competition with new sources of labor. Responses to the open-ended interviews from the Social Opportunity Survey led support to this interpretation and reinforce our contention that the significance of color increases when the supply of jobs at a given skill level contracts and the supply of immigrant workers increases. These responses provide richly textured substance about how opportunity depends on race and national origin.

Jack's comments in response to a question about who gets ahead are especially illuminating about how color shapes opportunity in Chicago. Jack is married, Black, and forty-four years old. He has five kids to support. He lives in an extremely poor, all-Black neighborhood and has never been on welfare, though he lost his house when the steel mill shut down. To the question on who gets ahead, he responded:

> It's still the same old thing: Whites get ahead much quicker than Blacks. Still the same, that hasn't changed, 'cause on this job right now, I can see that. But there's nothing you can do about it if you want keep your job, you gotta just lay dead and try to make it.... I can say this much: a man gotta do what he gotta do. If he gotta work, he gotta work: it's as simple as that. (emphatically) *You got to work.* If you don't do that, you gonna rob and steal and I can't do that, 'cause what would I do in jail with five kids and a wife, you know? So I *have to work.*

When probed about who is least likely to get ahead, he replied:

> It would go between the Puerto Ricans and the Blacks. They, they catch hell too. 'Cause a lot of them work out there with me. I don't know why, I don't know: I wish I knew why. I don't know why it's like that. They're just saying "that's the way it is." It's like a set pattern.

A Mexican respondent was equally blunt in linking job opportunities to ethnicity. Says he: "The good jobs go to the gringos!"

Many respondents—too numerous to report—spontaneously answered questions about opportunity and access to good and bad jobs with references to race and ethnicity. Some acknowledged that color could be used to reserve slots for minorities, but there was general agreement that the best jobs go to Whites, and especially educated Whites, while the worst jobs go to Blacks and Latinos. "Someone has to do the dirty work," replied a White woman in response to several questions about employment opportunities. Having networks and connections was another recurrent theme in responses to questions about access to good and bad jobs, but no one clarified how one gets connections in the first place, except that the rich and White folks have them.

Belinda, a thirty-nine-year-old, married, Black woman with six kids who is on public aid, sees this quite clearly:

> Well, like, I've been here all my life, you know, and it's a White-ruled system, you know. If you're White, you've got a better chance. And I've personally experienced that as far as moving.... You know, I've went on jobs and seen a White person get hired and I was told there was no job. The job wasn't even, didn't even require skills, you know, things like that.

Several respondents, especially Blacks, lamented the better times they experienced in the past. Letonya, a forty-one-year-old mother of two, legally separated, and on strike from a certified nurse's aid job, articulated this position:

> Now they don't have the jobs that they had, even back in the sixties ... before, I don't think it's as many people are from other countries was over here like it is now, you know. And it seems to me they're sort of moving out the Americans (laughs). I mean taking up the jobs where the peoples, Americans used to have. 'Cause I had worked for quite a few in my life, I worked in a hospital until four years ago and they laid us off. And it was a lot of foreigners over here you know. And when they come over here, they work for little or nothing, you know what I'm saying.... *Americans understand that they couldn't live off with what little money they were paying.* (authors' emphasis)

Some would say Letonya had a higher reservation wage than immigrants, meaning that she demanded higher wages than other workers with similar characteristics. However, a more perceptive interpretation, particularly in light of the empirical evidence provided below showing that Blacks have lower reservation wages than similarly skilled Hispanics or Whites, is that Letonya understands what it takes to live in Chicago, and appropriately questions a decline in her economic status.

Letonya was not the only respondent who blamed her weakened labor market position on illegal immigrants. Several respondents implicated immigration as a major reason for the narrowing of economic opportunities among Chicago's native residents. In contrast to several econometric studies based on national data which show little competition between native minorities and recent immigrants, Chicago's inner-city parents perceived acute competition and job displacement from recent immigrants, and especially undocumented immigrants.

According to Sabrina, a forty-two-year-old widow in a visiting union, who has seven kids and a history of chronic unemployment, opportunity is lessened by the presence of immigrants:

> NO. They're bringing too many foreigners over here from other countries for any of us to get ahead or do what it's our right to do, and that's put us down.

Jane, a married mother of four who holds a GED and has been on welfare since 1972, thinks only the bad jobs go to illegals. Her response supports the view that undocumented migrants take the jobs U.S. natives refuse:

> Well, the bad jobs are usually for the poor Spanish people that come here illegally and they'll work for 2 dollars an hour ... because they're afraid, you know, that they're ... and they get mistreated and ... stuff. Wetbacks they call 'em, which I don't see why they do. Call 'em Spanish or whatever, but ....

Yet other respondents felt that unauthorized immigrants were taking both good and bad jobs. For example, Ann, a thirty-year-old mother of four who has never been on welfare and is currently married, doesn't believe there are enough good jobs for everybody:

> 'cause the illegal aliens ... they're taking all the jobs, 'cause its cheap labor. They can work for $3.00 an hour, whereas they're looking for $8 or $9 an hour.... I'm not prejudiced against any illegal alien, but they shouldn't come here, and be taking our jobs and what belong to us.

Susan, a disabled welfare mother of three, also agrees:

> I don't know, I might sound ... like I'm prejudice or something, but I think they should go back to where they belong and let the White people work and make a living.... The Mexicans should go to Mexico and the Puerto Ricans should go back to Puerto Rico and the Blacks to Africa.

These responses from Chicago's inner-city poor illustrate that competition for jobs is not simply a matter of Blacks versus Whites, or undocumented migrants versus native minorities. We were particularly struck by the responses of Puerto Ricans and Mexicans indicating job competition between them, intense discrimination against Puerto Ricans, and more generally, perceived rankings of Mexicans as workers preferred over Puerto Ricans. Dolores, a forty-two-year-old Puerto Rican welfare mother of four, has never been married and has been on public aid since 1969, following a highly unstable employment history between the ages of fifteen and twenty-four. She has lived in Chicago thirty-three years, and in her current neighborhood sixteen years. It is not lack of effort that keeps her jobless:

> I go down to ... I would love to work there, you make good money there. Every time I go they say they are not hiring. Every time I go I sign an application. They tell me I already have a lot of them there, but I say give me another.... I go down to the next factory where they make plastic flowers and apply there too. And to the next one and like that.... My daughters tell me they see help wanted signs, but when I go over they say no. My daughters tell me they want Mexicans and that I should just say I'm Mexican. Because they'll work for less and all that.

> But I say, "But they'll ask me for my papers!" But if you want the job, maybe you have to do that. Because I have some Mexican friends they go and get the job just like that. Because it's a $5 an hour job but they would only pay the Mexicans like $3.50.

When probed if she would work for $3.50, she replied:

> I wouldn't mind. I used to work for less, for eighty cents an hour. I wouldn't mind, but then when I go they say they aren't hiring. But the other ones get the jobs and then I don't want to go back, you know.

While argument about differences in the labor market standing of Mexicans and Puerto Ricans based on national data lend themselves quite readily to structural interpretations that emphasize the decline of blue-collar operative jobs as being responsible for the declining economic status of Puerto Ricans, this is only part of the story. In particular, it is difficult to argue that geography is a major factor in the differential labor force activity of Mexicans and Puerto Ricans living in poor Chicago neighborhoods. Also, a recent study provided strong evidence that the position of Puerto Ricaris in New York was undermined by recent immigrants (Tienda and Donato 1993). Therefore, we wish to emphasize the importance of intense discrimination as a factor responsible for the declining labor market status of Puerto Ricans.

EXPLAINING RACE AND ETHNIC INEQUALITY IN LABOR FORCE PARTICIPATION

Our theoretical discussion identified differences in skill and willingness to work as two plausible reasons for race and ethnic variation in labor force activity. If educational deficits were the primary reason for high joblessness among inner-city minorities, We would expect lower education levels in Chicago than the United States, and among ethnic groups, the lowest educational levels for Blacks. In fact, ... the educational characteristics of Chicago's Blacks and Whites are generally similar to those of their national counterparts. Hispanic parents were far more educationally disadvantaged than Black inner-city parents, averaging seven to ten years for Mexicans and Puerto Ricans, respectively. Yet the highest labor force activity rate (and lowest unemployment rate) corresponds to Mexican, not White or Black fathers.

Despite the similar mean levels of completed schooling among the Chicago and the national samples, the proportion of high school graduates revealed greater discrepancies. For example, between 55 and 58 percent of Chicago's Black parents were high school graduates, compared to about 80 percent of the Black parents nationally. Among Mexicans, 35 to 38 percent of Chicago parents reported having graduated from high school, compared to 43 to 63 percent of mothers and fathers nationally. The greatest discrepancy in high school graduation rates corresponds to Whites, as over 90 percent of the national sample compared to two-thirds of the inner-city sample claimed high school diplomas....

Skill, Color, and Labor Force Activity

A more succinct assessment of the relationship between ethnicity, education, and work activity is possible by examining simultaneously several determinants of labor force participation. For this we estimated a model that allowed us to evaluate the unique influence of individual, family, and neighborhood characteristics on the likelihood that respondents obtained a job. These results are illuminating and we summarize their lessons below. First, there are striking similarities in the determinants of labor force activity for both mother and fathers, despite some noteworthy differences. For example, marriage increases the probability of labor force participation for both men and women, and the presence of young children has no influence on labor force activity of either mothers or fathers. Also, disability lowers the odds of labor force participation for both mothers and fathers. However, the presence of other adults increases the probability that mothers will work, but not fathers. This result has a parallel based on national populations (Tienda and Glass 1985).

Second, two findings are particularly noteworthy for our hypothesis about the relative importance of color and skill in determining labor force outcomes. One is that high school graduation status significantly increased the labor market activity of inner-city mothers, but had no influence on the market activity of inner-city fathers. This indicates that education is a necessary, albeit insufficient condition for labor market success. Another key finding is that race and ethnic differences in labor force activity persist, even after standardizing our sample for differences in education, family status, and neighborhood characteristics. Specifically, Black fathers were significantly less likely, and Mexican fathers more likely, to participate in the labor force than their statistically equivalent White counterparts. Among mothers, significant ethnic differences were obtained only for Mexican-origin women, whose labor force activity exceeded that of White mothers with similar educational, family, and personal characteristics.

Finally, consistent with prior theorizing about the pervasiveness of joblessness in ghetto poverty areas, our results clearly show that labor force activity is significantly lower in neighborhoods with poverty rates in excess of 39 percent. It is unclear whether this result is obtained because residents of ghetto poverty areas refuse to work, or because they experience greater barriers obtaining jobs....

CONCLUSION

Our emphasis on discrimination is not intended to discount the importance of explanations of persisting race and ethnic differences in wages which emphasize the surplus of laborers willing to work at low wages. Although we have taken into account individual differences in education and experience, there probably remain unmeasured skill differences (English proficiency, for example) among the groups. Ultimately, shifts in both supply and demand curves set wages, hence the various individual and structural interpretations reviewed at the outset are pertinent for understanding persisting wage disparities along color lines (Holzer 1993). Our approach considered both supply and demand factors that have been summoned to explain increased labor market inequities along race and ethnic lines.

By the way of a tentative conclusion we propose that discrimination against people of color residing in the inner city, but especially Blacks, was heightened during the 1980s as the share of low-skill jobs declined. Even though Black jobless parents reported the lowest reservation wage rates and received the lowest wages among those with a job, employers are increasingly reluctant to hire them, particularly when alternative sources of immigrant labor are readily available. Stated differently, employers seem to prefer Mexican and White workers over Blacks, and to a somewhat lesser extent Puerto Ricans. Not only does color matter, but it has an exchange value that fluctuates according to general economic conditions and the availability of alternative sources of unskilled labor. But we want to be clear that, while the market may very well set prices, it is employers, not the market, who discriminate.

REFERENCES

Borjas, George J., and Marta Tienda. 1987. "The Economic Consequences of Immigration." *Science* 235 (February 6): 613–620.

Hodge, Robert W. 1973. "Toward a Theory of Racial Differences in Employment." *Social Forces* 52 (1): 16–31.

Holzer, Harry J. 1991. "The Spatial Mismatch Hypothesis: What has the evidence Shown?" *Urban Studies* 28 (1): 105–122.

———. 1993. "Black Employment Problems: New Evidence, Old Questions." Unpublished paper, Department of Economics, Michigan State University.

Kasarda, John P. 1985. Urban Change and Minority Opportunities." In *The New Urban Reality*, ed. Paul E. Peterson. Washington D.C.: Brookings Institution.

Lieberson, Stanley. 1980. *A Piece of the Pie: Blacks and White Immigrants Since 1880.* Berkeley and Los Angeles: University of California Press.

Murray, Charles. 1984. *Losing Ground.* New York: Basic Books.

Reskin, Barbara, and Patricia A. Roos. 1990. *Job Queues, Gender Queues: Explaining Women's Inroads into Male Occupations.* Philadelphia: Temple University Press.

Tienda, Marta, Katharine Donato, and Héctor Cordero-Guzmán. 1992. "Schooling. Color and Labor Force Activity of Women." *Social Forces* 71 (2): 365–395.

Tienda, Marta, and Jennifer Glass. 1985. "Household Structure and Labor Force Participation of Black, Hispanic and White Mothers." *Demography* 22 (3): 381–394.

Tienda, Marta, and Zai Liang. 1993. "Poverty and Immigration in Policy Perspective." In *Poverty and Public Policy*, ed. Sheldon H, Danziger, Gary D. Sandefur, and Daniel H. Weinberg. Cambridge: Harvard University Press.

Tienda, Marta, and Haya Stier. 1991. "Joblessness or Shiftlessness: Labor Force Activity in Chicago's Inner City." In *The Urban Underclass*, ed. Christopher Jencks and Paul Peterson. Washington D.C.: Brookings Institution.

Van Haitsma, Martha. 1989. "A Contextual Definition of the Underclass." *Focus* 12 (spring/summer): 27–31.

Wilson, William Julius. 1987. *The Truly Disadvantaged: The Inner City, the Underclass, and Public Policy*. Chicago: University of Chicago Press.

DISCUSSION QUESTIONS

1. What are the different explanations that have been offered to explain the high rate of Black and Hispanic joblessness?
2. What explanations of joblessness do the comments of those interviewed in this article support? What does this lead Tienda and Stier to conclude?

37

Families on the Frontier

From Braceros in the Fields to Braceras in the Home

PIERRETTE HONDAGNEU-SOTELO

The increased employment of women, especially White women, in the United States has created a growing demand for paid domestic work. Hondagneu-Sotelo's study of immigrant domestic workers shows how inequality between nations means that women who fill this need. Women (especially those from the Philippines, the Caribbean, Central America, and Mexico) have had to move for such work, often becoming separated from their own families.

Why are thousands of Central American and Mexican immigrant women living and working in California and other parts of the United States while their children and other family members remain in their countries of origin?... I argue that U.S. labor demand, immigration restrictions, and cultural transformations have encouraged the emergence of new transnational family forms among Central American and Mexican immigrant women. Postindustrial economies bring with them a labor demand for immigrant workers that is differently gendered from that typical of industrial or industrializing societies. In all post-industrial nations, we see an increase in demand for jobs dedicated to social reproduction, jobs typically coded as "women's jobs." In many of these countries, such jobs are filled by immigrant women from developing nations. Many of these women, because of occupational constraints—and, in some cases, specific restrictionist contract labor policies—must live and work apart from their families.

My discussion focuses on private paid domestic work, a job that in California is nearly always performed by Central American and Mexican immigrant women. Not formally negotiated labor contracts, but rather informal occupational constraints, as well as legal status, mandate the long-term spatial and temporal separation of these women from their families and children. For many Central American and Mexican women who work in the United States, new international divisions of social reproductive labor have brought about transnational family forms and new meanings of family and motherhood. In this respect,

SOURCE: From *Latinos: Remaking America*, ed. by Marcelo Suarez-Orozco and Mariela M. Paez (Berkeley, CA: University of California Press, 2002), pp. 259–266. Copyright © 2002 by The Regents of the University of California. Reprinted by permission.

the United States has entered a new era of dependency on braceras. Consequently, many Mexican, Salvadoran, and Guatemalan immigrant families look quite different from the images suggested by Latino familism.

This [article] is informed by an occupational study I conducted of over two hundred Mexican and Central American women who do paid domestic work in private homes in Los Angeles (Hondagneu-Sotelo 2001). Here, I focus not on the work but on the migration and family arrangements conditioned by the way paid domestic work is organized today in the United States. I begin by noting the ways in which demand for Mexican—and increasingly Central American—immigrant labor shifted in the twentieth century from a gendered labor demand favoring men to one characterized by robust labor demand for women in a diversity of jobs, including those devoted to commodified social reproduction. Commodified social reproduction refers to the purchase of all kinds of services needed for daily human upkeep, such as cleaning and caring work. The way these jobs are organized often mandates transnational family forms....

I ... note the parallels between family migration patterns prompted by the Bracero Program and long-term male sojourning, when many women sought to follow their husbands to the United States, and the situation today, when many children and youths are apparently traveling north unaccompanied by adults, in hopes of being reunited with their mothers. In the earlier era, men were recruited and wives struggled to migrate; in a minority of cases, Mexican immigrant husbands working in the United States brought their wives against the latters' will. Today, women are recruited for work, and increasingly, their children migrate north some ten to fifteen years after their mothers. Just as Mexican immigrant husbands and wives did not necessarily agree on migration strategies in the earlier era, we see conflicts among today's immigrant mothers in the United States and the children with whom they are being reunited. In this regard, we might suggest that the contention of family power in migration has shifted from gender to generation....

GENDERED LABOR DEMAND AND SOCIAL REPRODUCTION

Throughout the United States, a plethora of occupations today increasingly rely on the work performed by Latina and Asian immigrant women. Among these are jobs in downgraded manufacturing, jobs in retail, and a broad spectrum of service jobs in hotels, restaurants, hospitals, convalescent homes, office buildings, and private residences. In some cases, such as in the janitorial industry and in light manufacturing, jobs have been re-gendered and re-racialized so that jobs previously held by U.S.-born white or black men are now increasingly held by Latina immigrant women. Jobs in nursing and paid domestic work have long been regarded as "women's jobs," seen as natural outgrowths of essential notions of women as care providers. In the late twentieth-century United States, however, these jobs have entered the global marketplace, and immigrant women from developing nations around the globe are increasingly represented in them. In major metropolitan centers around the country, Filipina and Indian immigrant women make up a sizable proportion of HMO nursing staffs—a result due in no small part to deliberate recruitment efforts. Caribbean, Mexican, and Central American women increasingly predominate in low-wage service jobs, including paid domestic work.

This diverse gendered labor demand is quite a departure from patterns that prevailed in the western region of the United States only a few decades ago. The relatively dramatic transition from the explicit demand for Mexican and Asian immigrant *male* workers to demand that today includes women has its roots in a changing political economy. From the late nineteenth century until 1964, the period during which various contract labor programs were in place, the economies of the Southwest and the West relied on primary extractive industries. As is well known, Mexican, Chinese, Japanese, and Filipino immigrant workers, primarily men, were recruited for jobs in agriculture, mining, and railroads. These migrant workers were recruited and incorporated in ways that mandated their long-term separation from their families of origin.

As the twentieth century turned into the twenty-first, the United States was once again a nation of immigration. This time, however, immigrant labor is not involved in primary, extractive industry. Agribusiness continues to be a financial leader in the state of California, relying primarily on Mexican immigrant labor and increasingly on indigenous workers from Mexico, but only a fraction of Mexican immigrant workers are employed in agriculture. Labor demand is now extremely heterogeneous and is structurally embedded in the economy of California (Cornelius 1998). In the current period, which some commentators have termed "postindustrial," business and financial services, computer and other high-technology firms, and trade and retail prevail alongside manufacturing, construction, hotels, restaurants, and agriculture as the principal sources of demand for immigrant labor in the western region of the United States.

As the demand for immigrant women's labor has increased, more and more Mexican and (especially) Central American women have left their families and young children behind to seek employment in the United States. Women who work in the United States in order to maintain their families in their countries of origin constitute members of new transnational families, and because these arrangements are choices that the women make in the context of very limited options, they resemble apartheid-like exclusions. These women work in one nation-state but raise their children in another. Strikingly, no formalized temporary contract labor program mandates these separations. Rather, this pattern is related to the contemporary arrangements of social reproduction in the United States.

WHY THE EXPANSION IN PAID DOMESTIC WORK?

Who could have foreseen that as the twentieth century turned into the twenty-first, paid domestic work would become a growth occupation? Only a few decades ago, observers confidently predicted that this job would soon become obsolete, replaced by such labor-saving household devices as automatic dishwashers, disposable diapers, and microwave ovens and by consumer goods and services purchased outside the home, such as fast food and dry cleaning (Coser 1974). Instead, paid domestic work has expanded. Why?

The exponential growth in paid domestic work is due in large part to the increased employment of women, especially married women with children, to the underdeveloped nature of child care centers in the United States, and to patterns of U.S. income inequality and global inequalities. National and global trends have fueled this growing demand for paid domestic services. Increasing global competition and new communications technologies have led to work speedups in all sorts of jobs, and the much bemoaned "time bind" has hit professionals and managers particularly hard (Hochschild 1997). Meanwhile, normative middle-class ideals of child rearing have been elaborated (consider the proliferation of soccer, music lessons, and tutors). At the other end of the age spectrum, greater longevity among the elderly has prompted new demands for care work.

Several commentators, most notably Saskia Sassen, have commented on the expansion of jobs in personal services in the late twentieth century. Sassen located this trend in the rise of new "global cities," cities that serve as business and managerial command points in a new system of intricately connected nodes of global corporations. Unlike New York City, Los Angeles is not home to a slew of Fortune 500 companies, but in the 1990s it exhibited remarkable economic dynamism. Entrepreneurial endeavors proliferated and continued to drive the creation of jobs in business services, such as insurance, real estate, public relations, and so on. These industries, together with the high-tech and entertainment industries in Los Angeles, spawned many high-income managerial and professional jobs, and the occupants of these high-income positions require many personal services that are performed by low-wage immigrant workers. Sassen provides the quintessentially "New York" examples of dog walkers and cooks who prepare gourmet take-out food for penthouse dwellers. The Los Angeles counterparts might include gardeners and car valets, jobs filled primarily by Mexican and Central American immigrant men, and nannies and house cleaners, jobs filled by Mexican and Central American immigrant women. In fact, the numbers of domestic workers in private homes counted by the Bureau of the Census doubled from 1980 to 1990 (Waldinger 1996).

I favor an analysis that does not speak in terms of "personal services," which seems to imply services that are somehow private, individual rather than social, and are superfluous to the way society is organized. A feminist concept that was originally introduced to valorize the nonremunerated household work of women, *social reproduction* or alternately, *reproductive labor*, might be more usefully employed. Replacing *personal services* with *social reproduction* shifts the focus by underlining the objective of the work, the societal functions, and the impact on immigrant workers and their own families.

Social reproduction consists of those activities that are necessary to maintain human life, daily and intergenerationally. This includes how we take care of ourselves, our children and elderly, and our homes. Social reproduction encompasses the purchasing and preparation of food, shelter, and clothing; the routine daily upkeep of these, such as cooking, cleaning and laundering; the emotional care and support of children and adults; and the maintenance of family and community ties. The way a society organizes social reproduction has far-reaching consequences not only for individuals and families but also for macrohistorical processes (Laslett and Brenner 1989).

Many components of social reproduction have become commodified and outsourced in all kinds of new ways. Today, for example, not only can you purchase fast-food meals, but you can also purchase, through the Internet, the home delivery of customized lists of grocery items. Whereas mothers were once available to buy and wrap Christmas presents, pick up dry cleaning, shop for groceries and wait around for the plumber, today new businesses have sprung up to meet these demands—for a fee.

In this new milieu, private paid domestic work is just one example of the commodification of social reproduction. Of course, domestic workers and servants of all kinds have been cleaning and cooking for others and caring for other people's children for centuries, but there is today an increasing proliferation of these services among various class sectors and a new flexibility in how these services are purchased.

GLOBAL TRENDS IN PAID DOMESTIC WORK

Just as paid domestic work has expanded in the United States, so too it appears to have grown in many other postindustrial societies, in the "newly industrialized countries" (NICs) of Asia, in the oil-rich nations of the Middle East, in Canada, and in parts of Europe. In paid domestic work around the globe, Caribbean, Mexican, Central American, Peruvian, Sri Lankan, Indonesian, Eastern European, and Filipina women—the latter in disproportionately large numbers—predominate. Worldwide, paid domestic work continues its long legacy as a racialized and gendered occupation, but today, divisions of nation and citizenship are increasingly salient.

The inequality of nations is a key factor in the globalization of contemporary paid domestic work. This has led to three outcomes: (1) Around the globe, paid domestic work is increasingly performed by women who leave their own nations, their communities, and often their families of origin to do the work. (2) The occupation draws not only women from the poor socioeconomic classes, but also women who hail from nations that colonialism has made much poorer than those countries where they go to do domestic work. This explains why it is not unusual to find college-educated women from the middle class working in other countries as private domestic workers. (3) Largely because of the long, uninterrupted schedules of service required, domestic workers are not allowed to migrate as members of families.

Nations that "import" domestic workers from other countries do so using vastly different methods. Some countries have developed highly regulated, government-operated, contract labor programs that have institutionalized both the recruitment and the bonded servitude of migrant domestic workers. Canada and Hong Kong provide paradigmatic examples of this approach. Since 1981 the Canadian federal government has formally recruited thousands of women to work as live-in nannies/housekeepers for Canadian families. Most of these women came from Third World countries in the 1990s (the majority came from the Philippines, in the 1980s from the Caribbean), and once in Canada, they must remain in live-in domestic service for two years, until they obtain their landed immigrant status, the equivalent of the U.S. "green card." This reflects, as Bakan and Stasiulis (1997) have noted, a type of indentured servitude and a decline in the citizenship rights of foreign domestic workers, one that coincides with the racialization of the occupation. When Canadians recruited white British women for domestic work in the 1940s, they did so under far less controlling mechanisms than those applied to Caribbean and Filipina domestic workers. Today, foreign domestic workers in Canada may not quit their jobs or collectively organize to improve the conditions under which they work.

Similarly, since 1973 Hong Kong has relied on the formal recruitment of domestic workers, mostly Filipinas, to work on a full-time, live-in basis for Chinese families. Of the 150,000 foreign domestic workers in Hong Kong in 1995, 130,000 hailed from the Philippines, and smaller numbers were drawn from Thailand, Indonesia, India, Sri Lanka, and Nepal (Constable 1997, p. 3). Just as it is now rare to find African American women employed in private domestic work in Los Angeles, so too have Chinese women vanished from the occupation in Hong Kong. As Nicole Constable reveals in her detailed study, Filipina domestic workers in Hong Kong are controlled and disciplined by official employment agencies, employers, and strict government policies. Filipinas and other foreign domestic workers recruited to Hong Kong find themselves working primarily in live-in jobs and bound by two-year contracts that stipulate lists of job rules, regulations for bodily display and discipline (no lipstick, nail polish, or long hair, submission to pregnancy tests, etc.), task timetables, and the policing of personal privacy. Taiwan has adopted a similarly formal and restrictive government policy to regulate the incorporation of Filipina domestic workers (Lan 2000).

In this global context, the United States remains distinctive, because it takes more of a laissez-faire approach to the incorporation of immigrant women into paid domestic work. No formal government system or policy exists to legally contract foreign domestic workers in the United States. Although in the past, private employers in the United States were able to "sponsor" individual immigrant women who were working as domestics for their "green cards" using labor certification (sometimes these employers personally recruited them while vacationing or working in foreign countries), this route is unusual in Los Angeles today. Obtaining legal status through labor certification requires documentation that there is a shortage of labor to perform a particular, specialized occupation. In Los Angeles and in many parts of the country today, a shortage of domestic workers is increasingly difficult to prove. And it is apparently unnecessary, because the significant demand for domestic workers in the United States is largely filled not through formal channels of foreign recruitment but through informal recruitment from the growing number of Caribbean and Latina immigrant women who are *already* legally or illegally living in the United States. The Immigration and Naturalization Service, the federal agency charged with enforcement of illegal-migration laws, has historically served the interests of domestic employers and winked at the employment of undocumented immigrant women in private homes.

As we compare the hyperregulated employment systems in Hong Kong and Canada with the more laissez-faire system for domestic work in the United States, we find that although the methods of recruitment and hiring and the roles of the state in these processes are quite different, the consequences are similar. Both systems require the incorporation as workers of migrant women who can be separated from their families.

The requirements of live-in domestic jobs, in particular, virtually mandate this. Many immigrant women who work in live-in jobs find that they must be "on call" during all waking hours and often throughout the night, so there is no clear line between working and nonworking hours. The line between job space and private space is similarly blurred, and rules and regulations may extend around the clock. Some employers restrict the ability of their live-in employees to receive phone calls, entertain friends, attend evening ESL classes, or see boyfriends during the workweek. Other employers do not impose these sorts of restrictions, but because their homes are located in remote hillsides, suburban enclaves, or gated communities, live-in nannies/housekeepers are effectively restricted from participating in anything resembling social life, family life of their own, or public culture.

These domestic workers—the Filipinas working in Hong Kong or Taiwan, the Caribbean women working on the East Coast, and the Central American and Mexican immigrant women working in California constitute the new "braceras." They are literally "pairs of arms," disembodied and dislocated from their families and communities of origin, and yet they are not temporary sojourners.

REFERENCES

Bakan, Abigail B., and Daiva Stasiulis. 1997. "Foreign Domestic Worker Policy in Canada and the Social Boundaries of Modern Citizenship," In Abigail B. Bakan and Daiva Stasiulis (eds.), *Not One of the Family: Foreign Domestic Workers in Canada*, Toronto: University of Toronto Press, pp. 29–52.

Constable, Nicole. 1997. *Maid to Order in Hong Kong: Stories of Filipina Workers*. Ithaca and London: Cornell University Press.

Cornelius, Wayne. 1998. "The Structural Embeddedness of Demand for Mexican Immigrant Labor: New Evidence from California." In Marcelo M. Suárez-Orozco (ed.), *Crossings: Mexican Immigration in Interdisciplinary Perspectives*. Cambridge, MA: Harvard University, David Rockefeller Center for Latin American Studies, pp. 113–44.

Coser, Lewis. 1974. "Servants: The Obsolescence of an Occupational Role." *Social Forces* 52:31–40.

Hochschild, Arlie. 1997. *The Time Bind: When Work Becomes Home and Home Becomes Work*. New York: Metropolitan Books, Henry Holt.

Hondagneu-Sotelo, Pierrette. 1994. *Doméstica: Immigrant Workers and Their Employers.* Berkeley: University of California Press.

Lan, Pei-chia. 2000. "Global Divisions, Local Identities: Filipina Migrant Domestic Workers and Taiwanese Employers." Dissertation, Northwestern University.

Laslett, Barbara, and Johanna Brenner. 1989. "Gender and Social Reproduction: Historical Perspectives," *Annual Review of Sociology* 15:381–404.

Waldinger, Roger, and Mehdi Bozorgmehr. 1996. "The Making of a Multicultural Metropolis." In Roger Waldinger and Mehdi Bozorgmehr (eds.), *Ethnic Los Angeles.* New York: Russell Sage Foundation, pp. 3–37.

DISCUSSION QUESTIONS

1. What does Hondagneu-Sotelo mean by the term *transnational families*?
2. Define social reproductive work and explain why Hondagneu-Sotelo thinks this work is performed by immigrant women.
3. Social reproductive work is part of the industrial societies of the United States, Canada, and Hong Kong, but what are the different national strategies for securing workers?

38

The Family and Community Costs of Racism

JOE R. FEAGIN AND KARYN D.MCKINNEY

Racism takes a toll on people of color in multiple ways. Feagin and McKinney show how, even in the face of racist beliefs and actions that harm Black families and communities, those same institutions provide a defense against the harm that racism produces.

Most African Americans see their families and communities as primary defenses against the daily assaults of racism. It is in these families that most black children first learn how to cope with racism. Additionally; the local black community often operates as extended kin. As they become adults, most black Americans seek support from others in their communities when they face increasing numbers of racist incidents in a variety of societal settings, including places of employment. As our respondents often note, the damage of a racially hostile workplace does not end at the workplace door. A black individual's experience with racial animosity and mistreatment is personally painful at the moment it happens, and also can have a cumulative and negative impact on other individuals, on one's family, and on one's community.

African American families and communities are negatively affected in many ways by continuing white-on-black racism. The black family has faced physical, ideological, and material assaults from whites for nearly four centuries. Over these centuries, family and community have been closely linked for African Americans, and for that reason the various white assaults on black families often have a significant impact on the larger black communities....

THE LONG-TERM IDEOLOGICAL ASSAULT ON THE BLACK FAMILY

Most social science researchers view African American families as distinctive in certain characteristics when compared to similar white families. What is disputed is whether these differences are problematic. Rather than looking at African

SOURCE: From Joe E. Feagin and Karyn D. McKinney. *The Many Costs of Racism.* Copyright © 2003. Reprinted by permission of Rowan & Littlefield Publishing Group.

American families as simply *different* in certain ways from white families, most of the early literature and much recent literature addressing certain family forms found in black communities has described these family forms as more or less problematical or pathological....

Perhaps the best-known example of this pathological-family viewpoint is the 1965 report by Daniel Patrick Moynihan, *The Negro Family: The Case for National Action*. In this widely cited and influential government policy initiative, Moynihan writes about an allegedly distinctive pathology of black communities: "Obviously, not every instance of social pathology can be traced to the weakness of family structure.... Nonetheless, at the center of the tangle of pathology is the weakness of the family structure. Once or twice removed, it will be found to be the principal source of most of the aberrant, inadequate, or antisocial behavior, that did not establish, but now serves to perpetuate the cycle of poverty and deprivation."[1] In this report, white society is all but absolved of contemporary responsibility for poverty, deprivation, and family stresses in black communities. This labeling of urban black communities as a "tangle of pathology" is commonplace and has been copied to the present day. According to many advocates of this pathology, the typical black family is matriarchal in character, has weak kinship ties, has "illegitimate" children, lacks mainstream values, and is present-oriented rather than future-oriented. This group of factors is sometimes referred to as a "culture of poverty."[2] Thus defined, it can seem inevitable and somehow right to whites that African Americans are impoverished....

A primary problem with the "culture of poverty" notion is that it confuses cause and result. While admitting that slavery and racism *began* the problems for the black family and community, those who adhere to this perspective believe that certain contemporary black family and community problems persist primarily because of an unhealthy black subculture. They do not recognize or admit that the original cause of stress on the black family, institutionalized racial oppression is *still* the major and pervasive cause of much everyday stress for black families and communities today—and that the *results* of that stress include some of the adaptive characteristics often described in negative terms, or that are distorted in many analyses as supposed deviations from white family norms.

CONFRONTING AND COPING WITH RACISM: STRONG AND ADAPTIVE FAMILIES

Contrary to the common view of the pathology advocates, the African American family has long been characterized by strong kinship bonds and interaction. During slavery, one of the primary reasons that enslaved African Americans ran away was to return to their families. In fact, many who had escaped to freedom in the North again risked their lives to return South in the attempt to free their families. After more than two centuries of colonial and U.S. slavery, during Reconstruction many African American families that had been separated by slaveholders began to regroup and re-create themselves. Although most of these newly freed black parents were struggling to feed their children, they often added to their families' innumerable "fictive kin"—the extended family and non-kin children who were orphaned after slavery. Similar family restructuring, and the care of extended family and "fictive kin," were seen again during the Great Depression of the 1930s. These patterns of care and survival can still be seen in many African American families and communities today.

For several decades now, numerous scholars have shown that African American families and communities are remarkably adaptive and strong....

Contrary to the "culture of poverty" thesis and commonplace white stereotyping, most members of black families exhibit a strong orientation toward work. Many studies show that whites and blacks are similarly oriented in their attitudes toward work. Our data show that rather than working less hard than whites, most African Americans realize that they often must work *harder* than whites to achieve the same results. For example, a sheriff's deputy in one focus group put it this way: "And that's the same thing ... we were talking about on the energy. Burning so much energy trying to educate these people, that we qualify, you know? And I always said if you see a black doctor and a white doctor standing side by side, equal in status, that black man is *twice* as [good], because he had to work harder.... In every profession."

Another strength of the black family is its adaptability in regard to family roles. Contrary to the popular notion that the black family is usually "matriarchal," even when there is a husband present, scholars have found that African American families are often more egalitarian than are white families. Because they are not as patriarchal as the traditional white family, black families thus appear to many (especially white) outsiders as part of a "matriarchal" family system. Evidence of the adaptability of black families can also be seen when young women become mothers, in that the larger family is usually supportive. African American families are twice as likely as whites to take the children of premarital births into the homes of family or friends of the parents, and are *seven times less* likely than whites to put up these children for adoption.

Another adaptation to unfavorable circumstances is a strong achievement orientation.... Education is valued highly in the black family and community as one way to help overcome poverty and racism.

Another distinctive strength of African American families is their often strong religious orientation. This religious orientation contributes to cohesive black families, and has been used as a mechanism of survival for black communities from slavery through the 1960s civil rights movement to the present day....

BLACK FAMILIES: THE CHALLENGE OF DIFFICULT ECONOMIC CONDITIONS

Some of the most difficult problems faced by black families stem from outside economic forces. Some of these economic forces have, unintentionally, a differential impact on African Americans, while other economic problems are created by government or private sector policies that intentionally, or at least knowingly, have a negative impact on African Americans. In other work, Feagin has distinguished between direct and indirect institutionalized discrimination. Direct institutionalized discrimination is that discrimination intentionally built into organizations and institutions by whites in order to have a significant negative impact on people in subordinate racial groups, especially African Americans. Examples include the intentional exclusion, blatant or subtle, of African Americans from traditionally white jobs and neighborhoods, discrimination that can still be found in some sectors of U.S. society. Indirect institutionalized discrimination involves the differential and negative treatment by whites of subordinate racial groups without that treatment being intentional in the present moment. This is oppressive and damaging discrimination, nonetheless, because it carries into the present the impact of the often blatant discrimination of the past. For example, intentional discrimination in the past often creates limited resources for the descendants of those so discriminated against, and current generations of groups once severely subordinated have less inherited wealth and other socioeconomic resources than do current generations of whites. This means that, today, the average African American typically does not have the same economic and educational resources and, thus, opportunities as the average white American. For example,...the average black family today has about 60 percent of the income of the average white family—and only about a tenth of the average white family's wealth....

In our focus groups and in informal discussions with a number of other African American husbands and wives, we have found that many people explain in some detail how the stress of discrimination at work places an added burden on their relationships with their spouses. When husbands or wives are under great pressure and stress from the racist incidents in their predominantly white workplaces, their energy for, or willingness to interact with, their spouses when they come home is affected. They may wish to be secluded from family relationships, and perhaps just watch television for a long time in order to unwind for a while from the stress of dealing with whites in the workplace or other social settings.... Many African Americans are aware of how this impacts the family, and they try to develop counter-strategies. In other cases, even these veterans of contending with racism do not realize fully just how that racism has negatively affected their intimate family relationships. In these cases, white racism has yet other negative, often energy-draining, effects on African American families.

For the family, another obvious consequence of workplace discrimination occurs when an important breadwinner is fired unfairly from his or her job because of discrimination, and then is not able to provide adequately for the family. Thus, one black man quoted in an earlier research study had worked for ten years as a school employee whose job it was to deal with the community. Although he was hired for his interpersonal skills and life experiences, after ten years on the job his employer decided to put some formal screening credentials into place for such community contact positions. Under this newly imposed standard, although he had personally moved up into the middle-class, his working-class demeanor and speech were no longer valued, and he was fired. At the time of his interview, he was preparing to sue to win back his job and was working a temporary job that paid a very low wage. He had also separated from his wife. In his interview, he spoke of the costs of this experience on his children:

> I honestly believed and felt that I was put on this earth to help to work with kids, I really did. And I was good, I was damn good at my job to have, let's say, less education than probably anybody at that school.... I *am* angry. I'm very angry. Because what they did, they didn't do because I was not performing my job. They did [it] because I was a black man and I spoke out on what I believe and felt. If I hadda been one of their little henchmen to say "yes, no"...I would still be there.... They don't know what the hell they have put me through. I know I could have a job if it wasn't for them folks down there. They have shut off my livelihood. They don't know the suffering, not that I am absorbing, but my kids.... I don't know how many times I didn't have the money to pay the rent. They got food on their table when they come in, in the evening. Sometimes I don't have food here on my table for my kids.

Damage is done to African American families because of workplace mistreatment along racial or racial-class lines. This man cannot feed his children adequately. He also notes that he has had to take his daughter out of college, thereby jeopardizing her ability to be economically successful in the future. And he has separated from his wife. His family has been torn apart by the racist actions of whites in his workplace. We see here the powerful domino effects on families of racist actions in yet another U.S. workplace.

THE IMPACT OF RACISM: THE BLACK COMMUNITY

... The spin-off effects of animosity and mistreatment in employment settings can be seen in yet other areas of the lives of these African Americans. Another respondent in a focus group sadly noted the negative impact on participation in church activities:

> I have withdrawn from some of the things I was involved with at church that were very important to me, like dealing with the kids at church. Or we had an outreach ministry where we would go out into the low-income housing and we would share about our services, ... and I was just so drained, like [another respondent] said, if we are all so drained, and we stop doing that, then we lose our connection. But I, physically, by the time I got home at the end of the day, I was just so tired. I didn't even feel like giving back to my community; I didn't feel like doing anything. And so I withdrew from church activities, to the point where I just really was not contributing anything. And it was pulling all that energy; I was exhausted from dealing with what I had to

> at work. And then whatever little bit was left went to my family, so there was nothing there to give.

The considerable impact of workplace racism is graphically described here, for even church activities become a challenge for this person. What energy there is left after struggles at work with racism is reserved for the family. These economically successful African Americans can be important role models in their local communities, but only if they have the energy to participate actively in churches and other community organizations. These accounts of withdrawal from, or lack of energy for, community activities are worth considering in the light of the many accomplishments of African Americans in organizing to improve both their own communities, in a variety of churches and other important local organizations, and also in working together in national organizations to improve the larger U.S. society. In spite of the great strain, pain, and energy loss stemming from individual and institutional racism, the majority of African Americans manage to strive, endure, and succeed in raising families and building communities. Being tired from the daily struggle against racism in the workplace, schools, and public accommodations has not stopped many African Americans from continuing to organize, a point we will accent in the conclusion that follows.

CONCLUSION: CREATING COMMUNITIES OF RESISTANCE

Dealing with the family and community impact of racism is a constant challenge for African Americans. The question of how to develop effective community strategies for development or change has been a recurring topic of concern and analysis for African American intellectuals and other leaders....

It is evident in the words of numerous respondents that interpersonal relationships, including family relationships, are sometimes jeopardized because one or both partners does not have much energy left at the end of a racially stressful day. Yet most seem to manage to overcome these huge challenges most of the time. Indeed, from their discussions of the energy-draining aspects of discrimination, one might wonder how African Americans have developed vibrant community organizations and successful resistance movements over nearly four centuries now. Most African Americans persevere through the barriers and manage to overcome the constant "rain" of antiblack racism enough to stay centered in their life struggles and, remarkably, thrive....

These organized efforts have not only liberated African Americans from legal segregation, but also the *country as a whole* from these enormous barriers to human progress. All Americans are the current beneficiaries of organized African American resistance to racial oppression. There is greater freedom in the U.S. today than there would have been without the civil rights movement of the 1960*s*. This is as true today as it has been in the past. Racism is a destructive force that ultimately affects all residents of this country. As the English poet John Donne said long ago, "Any man's death diminishes me, because I am involved in mankind; and therefore never send to know for whom the bell tolls; it tolls for thee."

REFERENCES

1. Daniel P. Moynihan, *The Negro Family: The Case for National Action* (Washington, D.C.: U.S. Department of Labor, 1965), p. 30.
2. This term was first coined by Oscar Lewis to refer to Mexicans, in Oscar Lewis, *The Children of Sanchez* (New York: Random House, 1961). He later applied the term to Puerto Ricans. Lewis saw the "culture of poverty," at least in part, as a response to oppressive circumstances. It was Moynihan and others who later applied the term to black Americans and used it in more of a victim-blaming manner.

DISCUSSION QUESTIONS

1. What does it mean to say that the common view of African American families has been one of "pathology?" How do Feagin and McKinney evaluate this assumption?
2. In what specific ways do African American families adapt to the racism they encounter?

39

Breaking the Last Taboo: Interracial Marriage in America

ZHENCHAO QIAN

Although interracial marriage is becoming more common, it is still quite rare. Qian's essay reviews the extent of and attitudes toward interracial dating and marriage and how these vary among different groups in the U.S. population.

Guess Who's Coming to Dinner, a movie about a white couple's reaction when their daughter falls in love with a black man, caused a public stir in 1967. That the African-American character was a successful doctor did little to lower the anxieties of white audiences. Now, almost four decades later, the public hardly reacts at all to interracial relationships. Both Hollywood movies and TV shows, including *Die Another Day, Made in America, ER, The West Wing*, and *Friends*, regularly portray interracial romance.

What has changed? In the same year that Sidney Poitier startled Spencer Tracy and Katherine Hepburn, the Supreme Court ruled, in *Loving v. Virginia*, that laws forbidding people of different races to marry were unconstitutional. The civil rights movement helped remove other blatant legal barriers to the integration of racial minorities and fostered the growth of minority middle classes. As racial minorities advanced, public opinion against interracial marriage declined, and rates of interracial marriage grew rapidly.

Between 1970 and 2000, black-white marriages grew more than fivefold from 65 to 363 thousand, and marriages between whites and members of other races grew almost fivefold from 233 thousand to 1.1 million. Proportionately, interracial marriages remain rare, but their rates increased from less than 1 percent of all marriages in 1970 to nearly 3 percent in 2000. This trend shows that the "social distance" between racial groups has narrowed significantly, although not nearly as much as the social distance between religious groups. Interfaith marriages have become common in recent generations. That marriages across racial boundaries remain much rarer than cross-religion marriages reflects the greater prominence of race in America. While the interracial marriage taboo seems to be gradually breaking down, at least for certain groups, intermarriage

SOURCE: From *Contexts*, 4 (4): 33–37, 2005. Copyright © American Sociological Association. Reprinted by permission of The University of California Press.

in the United States will not soon match the level of intermarriage that European immigrant groups have achieved over the past century.

PUBLIC ATTITUDES

Americans have become generally more accepting of other races in recent decades, probably as a result of receiving more education and meeting more people of other races. Americans increasingly work and go to school with people from many groups. As racial gaps in income narrow, more members of racial minorities can afford to live in neighborhoods that had previously been white. Neighbors have opportunities to reduce stereotypes and establish friendships. Tolerance also grows as generations pass; elderly people with racist attitudes die and are replaced by younger, more tolerant people. The general softening of racial antagonisms has also improved attitudes toward interracial marriage.

In 1958, a national survey asked Americans for the first time about their opinions of interracial marriage. Only 4 percent of whites approved of intermarriage with blacks. Almost 40 years later, in 1997, 67 percent of whites approved of such intermarriages. Blacks have been much more accepting; by 1997, 83 percent approved of intermarriage. Whites' support for interracial marriage—which may to some extent only reflect respondents' sense of what they should tell interviewers—lags far behind their support of interracial schools (96 percent), housing (86 percent), and jobs (97 percent). Many white Americans apparently remain uneasy about interracial intimacy generally, and most disapprove of interracial relationships in their own families. Still, such relationships are on the increase.

INTERRACIAL DATING

According to a recent survey reported by George Yancey, more than one-half of African-, Hispanic-, and Asian-American adults have dated someone from a different racial group, and even more of those who have lived in integrated neighborhoods or attended integrated schools have done so. Most dates, of course, are casual and do not lead to serious commitments, and this is especially true for interracial dating. Analyzing data from the National Longitudinal Study of Adolescent Health, Kara Joyner and Grace Kao find that 71 percent of white adolescents with white boyfriends or girlfriends have introduced them to their families, but only 57 percent of those with nonwhite friends have done so. Similarly, 63 percent of black adolescents with black boyfriends or girlfriends have introduced them to their families, but only 52 percent of those with nonblack friends have done so. Data from another national survey show similar patterns for young adults aged 18–29 (61 percent versus 51 percent introducing for whites, and 70 percent versus 47 percent for blacks).

While resistance to interracial relationships in principle has generally declined, opposition remains high among the families of those so involved. Interracial couples express concern about potential crises when their families become aware of such relationships. Their parents, especially white parents, worry about what those outside the family might think and fear that their reputations in the community will suffer. Maria Root notes that parents actively discourage interracial romance, often pointing to other peoples' prejudice—not their own—and expressing concern for their child's well being: "Marriage is hard enough; why make it more difficult?"

The dating and the parental reservations reveal a generation gap: Young men and women today are more open to interracial relationships than their parents are. This gap may be due simply to youthful experimentation; youngsters tend to push boundaries. As people age, they gradually learn to conform. Kara Joyner and Grace Kao find that interracial dating is most common among teenagers but becomes infrequent for people approaching 30. They attribute this shift to the increasing importance of family and friends—and their possible disapproval—as we age. When people are ready to be "serious," they tend to fall in love with people who are just like themselves.

INTERRACIAL COHABITATION AND MARRIAGE

Who pairs up with whom depends partly on the size of the different racial groups in the United States. The larger the group, the more likely members are to find marriageable partners of their own race. The U.S. Census Bureau classifies race into four major categories: whites, African Americans, Asian Americans and American Indians. Hispanics can belong to any of the four racial groups.... Whites form the largest group, about 70 percent of the population, and just 4 percent of married whites aged 20–34 in 2000 had nonwhite spouses. The interracial marriage rates are much higher for American-born racial minorities: 9 percent for African Americans, about 39 percent for Hispanics, 56 percent for American Indians, and 59 percent for Asian Americans (who account for less than 4 percent of the total population). Mathematically, one marriage between an Asian American and a white raises the intermarriage rate for Asian Americans much more than for whites, because whites are so much more numerous. Because of their numbers as well, although just 4 percent of whites are involved in interracial marriages, 92 percent of all interracial marriages include a white partner. About half of the remaining 8 percent are black-Hispanic couples. Racial minorities have more opportunities to meet whites in schools, workplaces, and neighborhoods than to meet members of other minority groups.

Some interracial couples contemplating marriage avoid family complications by just living together. In 2000, 4 percent of married white women had nonwhite husbands, but 9 percent of white women who were cohabiting had non-white partners (see figure 1). Similarly, 13 percent of married black men had nonblack spouses, but 24 percent of cohabiting black men lived with nonblack partners. Hispanics and Asian Americans showed the same tendency; only American Indians showed the opposite pattern. Black-white combinations are particularly notable. Black-white pairings accounted for 26 percent of all cohabiting couples but only 14 percent of all interracial marriages. They are more likely to cohabit than other minority-white couples, but they are also less likely to marry. The long history of the ban on interracial marriage in the United States, especially black-white marriage, apparently still affects black-white relationships today.

Given differences in population size, comparing rates of inter-marriage across groups can be difficult. Nevertheless, statistical models used by social scientists can account for group size, determine whether members of any group are marrying out more or less often than one would expect given their numbers, and then discover what else affects intermarriage. Results show that the lighter the skin color, the higher the rate of intermarriage with white Americans. Hispanics who label themselves as racially "white" are most likely to marry non-Hispanic whites. Asian Americans and American Indians are next in their levels of marriage with whites. Hispanics who do not consider themselves racially white have low rates of intermarriage with whites. African Americans are least likely of all racial minorities to marry whites. Darker skin in America is associated with discrimination, lower educational attainment, lower income, and segregation. Even among African Americans, those of lighter tone tend to experience less discrimination.

RACE AND EDUCATION

Most married couples have similar levels of education, which typically indicates that they are also somewhat similar in social position, background, and values. Most interracial couples also have relatively equal educational attainments.

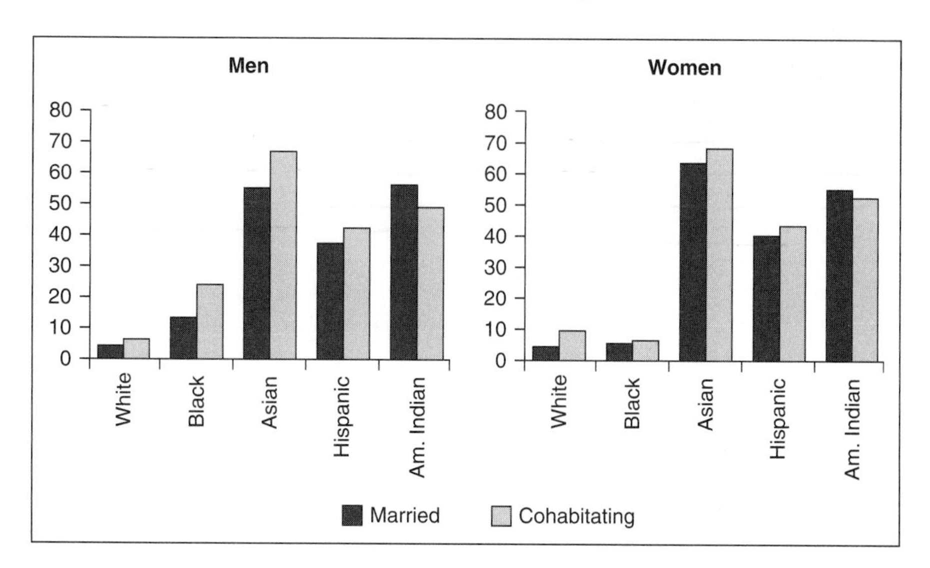

FIGURE 1 Interracial marriage remains rare among white and black Americans. This figure shows the percentage of Americans in couples married to (dark bars) or cohabiting with (light bars) someone of a different race.

SOURCE: 2000 Census.

However, when interracial couples do differ in their education, a hierarchy of color is apparent. The darker the skin color of racial minorities, the more likely they are to have married whites "below" them, that is, with less education than themselves. Six of ten African Americans who marry whites with different levels of education marry whites less educated than themselves. Hispanics also tend to marry whites less educated than themselves, but Asian Americans marry whites at about the same educational level.

Highly educated minority members often attend integrated colleges, and their workplaces and neighborhoods are integrated. Although they often develop a strong sense of their group identity in such environments, they also find substantial opportunities for interracial contact, friendship, romance, and marriage. College-educated men and women are more likely to marry interracially than those with less education. The fact that Asian Americans attend college at unusually high rates helps explain their high level of intermarriage with whites. The major exceptions to the interracial influence of higher education are African Americans.

Although middle-class African Americans increasingly live in integrated neighborhoods, they are still much more segregated than other minorities. Well-educated African Americans are less likely to live next to whites than are well-educated Hispanics and Asian Americans. One reason is that middle-class, black Americans are so numerous that they can form their own middle-class black neighborhoods, while middle-class Hispanic and Asian-American communities are smaller and often fractured by differences in national origin and language. In addition, studies show clearly that whites resist having black neighbors much more than they resist having Hispanic or Asian American neighbors....

Residential and school segregation on top of a long and relentless history of racial discrimination and inequality reduce African Americans' opportunities for interracial contact and marriage. The geographic distance between blacks and whites is in many ways rooted in the historical separation between the two groups. In contrast, the distance of Hispanics and Asian Americans from whites has more to do with their current economic circumstances; as those improve, they come nearer to whites geographically, socially, and matrimonially.

A MAN AND A WOMAN

Black-white couples show a definite pattern: 74 percent involve a black husband and a white wife. Asian American-white couples lean the other way; 58 percent involve an Asian-American wife. Sex balances are roughly even for couples that include a white and a Hispanic (53 percent involve a Hispanic husband) or a white and an American Indian (49 percent involve Indian husbands).

I mentioned before that most black-white couples have similar educations; nonetheless, white women who marry black men "marry up" more often than those who marry white men. This is especially striking because the pool of highly educated white men greatly outnumbers the pool of highly educated black men. More than half of black husbands of white women have at least some college education, but only two-fifths of black husbands of black women do. In that sense, white wives get more than their "share" of well-educated black husbands. This further reduces the chances that black women, especially highly educated black women, will marry, because they often face shortages of marriageable men. African-American women often resent this. Interviewed by Maria Root, one black man in such a relationship reported being accused of "selling out" and "dissing his black sisters."

Half a century ago, Robert Merton proposed a "status exchange" theory to explain the high proportion of marriages between black men and white women. He suggested that men with high economic or professional status who carry the stigma of being black in a racial caste society "trade" their social position for whiteness by marriage. On the other hand, some social scientists argue that racialized sexual images also encourage marriages between white women and black men. Throughout Europe and the West, people have long seen fair skin tone as a desirable feminine characteristic, and African Americans share those perceptions. For example, Mark Hill found that black interviewers participating in a national survey of African Americans rated black women interviewees with lighter skin as more attractive than those with darker skin. But they did not consider male interviewees with light skin any more attractive than darker-skinned men.

Asian Americans show a different pattern; in most of their marriages with whites, the husband is white. Although Asian-American men are typically more educated than white men, in the mixed couples, white husbands usually have more education than their Asian-American wives. As with white wives of black men, the wives have "married up" educationally. Some speculate that Asian-American women tend to marry white men because they perceive Asian-American men to be rigidly traditional on sex roles and white men as more nurturing and expressive. The emphasis in Asian cultures on the male line of descent may pressure Asian-American men to carry on the lineage by marrying "one of their own." But what attracts white men to Asian-American women? Some scholars suggest that it is the widespread image of Asian women as submissive and hyper-feminine (the "Madame Butterfly" icon).

THE FUTURE OF INTERRACIAL MARRIAGE

Rates of interracial marriage in the future will respond to some conflicting forces: the weakening of barriers between groups; increasing numbers of Hispanics and Asians in the nation; and possible rising ethnic consciousness. The continued progress of racial minorities in residential integration and economic achievement promotes contact between members of different races as equals. The color line, however, probably will not disappear. Marriage between African Americans and whites is likely to remain rare. Stubborn economic differences may be part of the reason for the persistence of this barrier, but cultural experiences also play a role.

In recent years, the middle-class African-American population has grown, yet the persistence of residential segregation reduces the opportunities for contact between blacks and whites. African Americans also maintain a strong racial identity compared to that of other minorities. In the 1990 census, for example, less than 25 percent of children born to a black-white couple were identified by their parents as white—a much lower percentage than for other biracial children. In the 2000 census, blacks identified themselves or their children as multiracial much less often than did other racial minorities. The stronger racial identities of African Americans, forged by persistent inequality, discrimination, and residential isolation, along with continued white resistance, will hold down the increase in marriages across the black-white divide.

Increases in the relatively high marriage rates of Hispanics and Asian Americans with whites may slow as new immigrants keep arriving from their homelands. Immigration expands the marriage pools for the native-born, who are more able to find spouses in their own racial or ethnic groups. These pools are expanded further by the way the wider society categorizes Hispanics and Asian Americans. They distinguish among themselves by national origin (Cuban versus Mexican or Thai versus Chinese), but whites tend to lump them into two large groups. Common experiences of being identified as the same, along with anti-Latino and anti-Asian prejudice and discrimination, help create a sense of pan-ethnic identity. This in turn inhibits marriage with whites, fosters solidarity within the larger group, and increases marriage rates between varieties of Hispanics and Asian Americans. Interethnic marriage is frequent among American-born Asians despite small group sizes and limited opportunities for contact. For example, in 1990, 18 percent of Chinese-Americans and 15 percent of Japanese-Americans aged 20–34 married spouses of other Asian ethnic groups (compared to 39 percent and 47 percent who married whites).

Many people view the increasing number of interracial marriages as a sign that racial taboos are crumbling and that the distances between racial groups in American society are shrinking. However, marriages across racial boundaries remain rarer than those that cross religious, educational, or age lines. The puzzle is whether interracial marriages will develop as marriages between people of different nationalities did among European immigrants and their descendants in the early 20th century. Diverse in many ways when they entered the country, these 20th-century European Americans, such as Italians, Poles, and Greeks, reached the economic level of earlier immigrants within a couple generations. Their success blurred ethnic boundaries and increased the rate of interethnic marriage. Many of their descendants now define themselves simply as white despite their diverse national origins. For most white Americans, ethnic identities have become largely symbolic.

Similar trends for interracial marriages are unlikely in the near future. The experiences of European Americans show the importance of equal economic achievement in dissolving barriers, so what happens economically to recent immigrants and African Americans will be important. Even then, the low levels of interracial marriage for middle-class African Americans suggest that this particular color line will persist as a barrier to marriage. And the continuing influx of Asian and Latino immigrants may reinforce those groups' barriers to intermarriage.

DISCUSSION QUESTIONS

1. How has acceptance of interracial dating and marriage changed over time? What obstacles remain?
2. What social factors are related to people's acceptance of interracial dating and marriage? Have you seen evidence of this in your own experience?

40

How Motherhood Changed My Life

KATHYRN EDIN AND MARIA KEFALAS

Based on their community study, Edin and Kefalas examine what motherhood means to young, poor women. Contrary to popular opinion, which stereotypes these young women as bad mothers whose early pregnancies ruin their lives, Edin and Kefalas find that the limits in these young women's lives come not from their choices about motherhood, but from the broader social conditions in which they live.

MILLIE AND CARLOS

Millie Acevedo is a diminutive, twenty-seven-year-old Puerto Rican mother of three who "came up" on Eighth and Indiana, one of the roughest corners in the West Kensington section of Philadelphia. She greets us at the door of her rowhouse with a well-scrubbed look—in a crisp white T-shirt with a face free of makeup and her hair pulled neatly back in a bun. A block and a half away, a bulldozer grinds noisily at the remains of another abandoned neighborhood factory. But Millie's block is relatively well-maintained and peaceful. The telephone poles up and down the street are plastered with advertisements for neighborhood events and give some indication that a community still exists here.

At fourteen, Millie believed she had found her future in Carlos, an older boy of nineteen whose best friend lived on her block. Carlos had a job and an apartment of his own. They had been together for a month when Millie moved in with him, and the couple began describing themselves as husband and wife, though they had no legally binding tie. Millie and Carlos were eager to have children but agreed to wait so that she could stay in school. Despite being on birth control and "taking my pills every day," she became a mother before her sixteenth birthday. Though the conception was not planned, the prospect of becoming parents delighted them both. The pair shared old-fashioned, Puerto Rican family values, and she willingly dropped out of school to care for their child full time. A year later, they conceived a second son, this

SOURCE: From Kathryn Edin and Maria Kefalas, *Promises I Can Keep* (Berkeley: University of California Press, 2005).

one planned, reasoning that as long as they had started a family, they might as well finish the job.

Millie and Carlos enjoyed a fairly stable relationship until she became pregnant a third time, a "total accident" in Millie's words, three months after the second child was born. When she told Carlos about the pregnancy, he "totally flipped." Though he'd been ecstatic about the first child and had been the one to push hard for a second, he immediately threatened to leave if she didn't end the third pregnancy. Millie relates the story this way, "He couldn't deal with [having another child], and he left. And he was with a couple of other girls out there. Then, after [I went through] the whole pregnancy by myself, he came back after the baby was born. He wanted to be with me again.... We tried to stay together for like a whole year after the baby was born."

But during this year, which Millie recalls as the worst in her life, Carlos "had so many jobs it wasn't even funny." His frequent conflicts with supervisors led to violent confrontations at home. "And when it got to that point, I was like, 'This is no good for my kids, this is no good for me, either he's gonna hurt me, I'm gonna hurt him, one of us is gonna be *dead*, one of us is gonna be in *jail*, and what's gonna happen with my kids *then?*' That's when I put an end to it," she says. "I got a restraining order on him, I got him out of the house, and that was the *end* of it. I never took him back." She told him, "'That's just it. I'm not taking abuse.'"

Millie is matter-of-fact when describing her failed relationship with Carlos, but visibly lights up when she talks about being a mother. As Millie imagines what life would be like had she not had children, she tells us her dream was to finish high school and enroll in college. Yet, like so many other mothers we met, Millie believes that her children have proved far more of a help than a hindrance. "My kids, they've matured me a lot. If I hadn't had them and had gone to college, I probably would have gotten introduced to the wrong crowd, and would have gotten lost because of the drugs and stuff." She believes having children was providence's way of saving her from this fate. "Maybe I needed my kids [to keep me safe]. They come first. I've always stayed off of drugs for them, and they helped me grow up.... I can't picture myself without them."

Millie's story shares many themes with other stories we heard. She believes that having children is a normal part of life, though she feels she and Carlos got started a year or two too early. Millie and Carlos's first and third pregnancies were both accidents, but poor women are often more favorably oriented toward having a child than not. Once pregnant, poor mothers pursue parenthood with few of the reservations that middle-class observers assume they must (and should) have about raising children when they are young, poor, and single.

Mothers raising children in the toughest sections of Philadelphia almost always hope and plan for their children's fathers to be part of their children's lives, just as Millie did. When a man and a woman cannot survive as a couple, though, it is an immense disappointment but not the end of the world. As Millie puts it, "[I've] got a good home for [my kids]. They have everything they need and I give them a lot of love and attention." When we ask about Carlos's ongoing role, she replies, "They don't *need* anything from him—you know what I'm saying—so I don't *ask* for nothing."

MOTHERHOOD AS A TURNING POINT

In an America that is profoundly unequal, the poor and rich alike are supposed to wait to bear children until they can complete their schooling, find stable employment, and marry a man who has done the same. Yet poor women realize they may never have children if they hold to this standard. Middle-class taxpayers see the children born to a young, poor, and unmarried mother as barriers to her future achievement, short circuiting her chances for what might have been a better life, while the mother herself sees children as the best of what life offers. Though some do express regret that an untimely birth robbed them of chances to improve their lot in life, most do not. Instead, they credit their children for virtually all that they see as positive in their lives. Even those who say they might have achieved more if they hadn't become parents when and how they did almost always believe the benefits of children far exceed the costs. As Celeste, a twenty-one-year-old white mother of a five-month-old, explains, "I'd have no direction [if I hadn't had a child]. I could sit here and say, 'Oh, I would have... gone to a four-year college,' [but] I probably wouldn't have." Like Celeste, many unmarried teens bear children that are conceived only after they've already experienced academic difficulties or dropped out of school.

Despite the ascent of feminism and the rapid entry of women into jobs formerly reserved for men, motherhood still offers a powerful source of meaning for American women. This is particularly true for low-income women living in the poorest sections of the Philadelphia area, who have little access to the academic degrees, high-status marriages, and rewarding professions that provide many middle- and upper-class women with gratifying social identities.

Poor youth are driven by a logic that is profoundly counterintuitive to their middle-class critics, who sometimes assume that poor women have children in a twisted competition with their peers to gain status, because they have an insufficient knowledge of—or access to—birth control, or so they can "milk" the welfare system. Yet our mothers almost never refer to these motivations. Rather, it is the perceived low costs of early childbearing and the high value that poor women place on children—and motherhood—that motivate their seemingly inexplicable inability to avoid pregnancy.

These poor young women are not unusually altruistic, though parenthood certainly requires self-sacrifice. What outsiders do not understand is that early childbearing does not actually have much effect on a low-skilled young woman's future prospects in the labor market. In fact, her life chances are so limited already that a child or two makes little difference..... What is even less understood, though, are the rewards that poor women garner from becoming mothers. These women rely on their children to bring validation, purpose, companionship, and order to their often chaotic lives—things they find hard to come by in other ways. The absolute centrality of children in the lives of low-income mothers is the reason that so many poor women place motherhood before marriage, even in the face of harsh economic and personal circumstances. For women like Millie, marriage is a longed-for luxury; children are a necessity.

A REASON TO GET UP IN THE MORNING

Children provide motivation and purpose in a life stalled by uncertainty and failure. As Adlyn, a pregnant nineteen-year-old Puerto Rican mother of a three-year-old and an eight-month-old, exclaims, "It's what gets [me] going.... It's like a *burst* of *energy*." Seventeen-year-old African American Kyra says her son, nearly two, gives her "something to look forward to. Like when I don't even have enough energy to get out of bed in the morning...I know I have to. When I turn over and look at him, it's like I'm trying to give him a better life, so I gotta get up and I gotta *do*."

Motherhood offers young women with limited options a valid role and a meaningful set of challenges. Zeyora, a white fifteen-year-old with a six-month-old, recalls, "I wanted a baby to take challenges into my *own hands*." Allison, twenty-eight and white, was a heroin addict who joined a methadone program when she learned she was pregnant. She says her life was going nowhere before her daughter, now nine months old, was born. "There was nothing to live for other than the next day getting high. [My life had] no *point*, there was no *joy*. I had lost all my friends—my friends were totally disgusted with me—I was about to lose my job, [and] I ended up dropping out of another college.... Now I feel like, 'I have a beautiful little girl!' I'm *excited* when I get up in the morning!"

Amanda, a twenty-year-old Puerto Rican mother of a three-year-old, says the birth of her son ignited her ambition and drive. She recalls, "When I had him was when I started thinking, *'Damn*, you know, I have to *change!'* When you have a kid, you really need—I think you should have an education so that [your kids] can look forward. You can look forward to telling him, 'Look, mommy works here.' He can go to school and tell his friends, 'Look my mom is *this!' not* 'Oh my mom works in a factory,' or 'We get food stamps every month.'" Destiny, an eighteen-year-old white mother of two girls, ages two and three, hopes her daughters "will grow up to *be* something and not depend on anybody. No man, no welfare, no nothing." She's doing her part by "making [the] kids smart and taking care of them, making them feel good." Thirty-year-old TJ, an African American mother of three, ages four, two, and four months, says that motherhood has completely reoriented her life. "I'm complete, and I've done what I am supposed to." She adds, "I don't see myself as being an individual anymore, really. Everything I do is mostly centered around my children, to make *their* lives better."...

Part of the reason that motherhood breeds such a strong sense of purpose is the high cost of failure. Mothers repeatedly offer horrific examples of neglectful and abusive mothers from well-publicized child abuse cases as haunting counterpoints to their own mothering. Jennifer, a twenty-three-year-old Puerto Rican mother of six children under age seven, told us, "At least I don't throw my children in the *trash* or drown them in the *bathtub*."...

"MY SON GIVES ME ALL THE LOVE I NEED."

Many Americans believe that though the poor don't have much in the way of economic resources, they compensate by forming unusually rich social and emotional ties. But in the neighborhoods we studied, nothing could have been further from the truth. Indeed, many mothers tell us they cannot name one person they would consider a friend, and the turmoil of adolescence often breeds a sense of alienation from family as well. Thus, mothers often speak poignantly about the strong sense of relational poverty they felt in the period before childbearing and believe they have forged those missing attachments through procreation, a self-made community of care. Brielle, a thirty-two-year-old African American mother of four children between the ages of three and eleven, says that few outsiders understand how central this motivation is for single mothers like herself. "A lot of people...say [young girls have babies] for money from welfare. It's not for that.... It's not even to keep the *guy*. It's just to have somebody...to take *care* of, or somebody to *love* or what ever." Nineteen-year-old Keisha, an African American mother of a one-year-old, paints the following picture of her bleak social landscape: "I don't have *nobody* that I can talk to. I don't have no friends, only got my baby. I can't even talk to my mom. I don't have nobody but my child."...

For many, not even a relationship with a man can fill the relational void. "When I didn't have kids," remembers Yolanda, a twenty-six-year-old Puerto Rican mother of two, ages three and four, "me and [my son's father] were together but something was missing.... It was like we *needed* something. And then there were babies, you know, to fill the void we were feeling." We ask her, "Is that the most important thing?" She replies, "Kids, I think so. In my life, yes." Beatrice, a twenty-year-old Puerto Rican mother of a three-month-old, simply states, "My son gives me all the love I need."...

The relationship between a mother and her child offers a haven from the often harsh world of adult relationships, especially those with men. Jen Burke remembers how her son was there for her after a particularly ugly fight with her boyfriend Rick. "After a fight with my boyfriend, I was crying, I was mad. And my son is sixteen months, and he came in and he was hugging me with his little arms, so I got happy!" "That's the one good thing about being a mother," she finishes. "Your baby is always there for you."...

In motherhood, young women who live in the city's hardscrabble core can find a powerful source of validation, for they believe that childrearing is something they can be good at, a meaningful and valued identity they can successfully realize. Pepper Ann, a forty-seven-year-old African American mother of two grown children and a twelve-year-old, recalls how as a child she "always wanted a baby." She says she thought "babies were the most precious thing on this earth," and she "wouldn't know what to do without [her daughter].... She is my little heart."

The birth of a baby also tends to mobilize others on the child's behalf. And while the baby takes center stage, the mother enjoys some of the limelight. Rosita, a twenty-three-year-old Puerto Rican mother of a two-year old, tells how becoming a mother made her feel "special." "Well, when I first became a mom, I got more attention, like everybody was closer to me. After I had her, I was coming out of the room, the hallway was packed with all my friends and family, and everybody was there, and I just felt real special, and everybody used to come every day to see the baby. Then when I went home, they was always around me and the baby all the time."

"BEING A MOM IS SOMETHING I KNEW I COULD DO."

But simply having a baby is not enough to earn social rewards; rather, a young mother must demonstrate that she has risen to the challenge of her new role. By presenting a clean, healthy, well-behaved child to the world, a young woman whose life may seem otherwise insignificant can prove her worth. Aliya, mentioned earlier, says that neighbors and kin see her differently now that she has a child and is managing to raise it on her own. "I guess they respect me more. I am taking care of my son myself."

There is no greater proof of a young woman's merit than the spontaneous praise of her mothering from a stranger on the street. The well-dressed child transforms the shabbily dressed mother. A child swathed in layers of warm clothing, even in a spring thaw, is testimony that an aimless teen is now a caring, competent, and responsible adult. The almost obsessive concern she has with her newborn's cleanliness, however, exposes the fragility of her new claim to respectability. According to Keisha, "It's *hard* when you don't have money to take care of your baby. People talk about you when—like if your child is dirty and stuff. So that's why I try to keep my child clean and buy...for Cheresa before I buy for myself. I don't have no clothes, she has everything."

Young women whose lives revolve around their children know that others are judging their failure or success at motherhood by these outside appearances. Santana, a thirty-four-year-old white mother of an eighteen-month-old girl, says, "When I go out with her, I think [about] the way she...will reflect on me. I feel that people look at her and then look at me.... Before, I went out alone, and I didn't think people were observing me... Now I feel like they do. Like if she starts crying on the bus, and they look at me. They're thinking, 'Can't you make her *quiet?*' And so how *she* is is about *me*."

This is why, though most mothers try to deemphasize material possessions, many still occasionally bypass the K-Mart on Broad Street for the designer discount chain across the street where Hilfiger and Polo abound, or even the Gap down on South Street. Having at least one "name brand" outfit for the baby is important to a young mother's quest for validation. "[It's] because of the status thing—to show people that you can afford to take care of your family," says Marilyn, a twenty-four-year-old white mother of two, ages four and five....

A well-cared-for child is the tangible evidence of a young mother's importance. *She* is the one raising the happy, healthy, carefully dressed child. *She* is the one who is teacher and guide. *She* is the one who is helping the child reach each developmental milestone: the first step, the first word, a dawning awareness of right and wrong. Her identity is secure as she basks in the glow of the child's accomplishments. Marilyn speaks passionately about the importance of a mother in her child's life. "When you have a baby, people don't realize you're raising not only the body, but the *mind*—the psychology, the mentality, the emotions, you're raising *all* that. You're actually teaching another person how to speak *English*, speak the *language*. I mean, to know that I brought this person into the world and they didn't know anything, and now I'm teaching and they're learning—that was a great feeling."

Mothers take enormous satisfaction from their child's accomplishments. Cheyenne, a twenty-five-year-old white mother of two, ages five and eight, starts out each school year by reserving a place on the wall of the front room for her eight-year-old daughter's school artwork and awards. "I like the rewards [from being a mother], the stuff they make for me to have. I'm so proud. Well, yeah, this is gonna be this year's wall [pointing to a blank space on the wall]. If I had the money, *every one* [of her pictures and awards] would be in a frame."...

These mothers see themselves in their offspring—often so much so that children become a reconstituted, yet more positive, image of themselves. The family resemblance between the two makes the child's accomplishments the mother's own. Carol brags about how her seven-year-old daughter Tabitha, the youngest, "is my prodigy child." Tabitha, she says, was "walking at nine months, talking at ten months. She's smart," Carol beams. "The kids fight over who gets to sit with her at school. I love it. First grade, and she was the most popular kid in the classroom! And she looks like me!" Kyra, talking about her one-year-old, says, "It's like I got this little *me* to take care of because he looks just *like* me."

Women near the bottom of the American class ladder hope their children will give them a vicarious second chance at the social mobility that has slipped out of their grasp. Even though a woman's own prospects might be limited, a new baby's life is a clean slate. As she "struggles and strives," to give them a better life, she heals the regret of having "messed up" her own. "I have so many aspirations for my daughter," proclaims Jerri, a thirty-five-year-old African American mother of two children just a year apart. She tells her older child, now eight, "'Look for something that you want to be, but be good at it and be useful to somebody.' A lot of times I've felt I was useless to people. I didn't matter."

"TO BE SOMETHING BETTER"

In these decaying, inner-city neighborhoods, motherhood is the primary vocation for young women, and those who strive to do it well are often transformed by the process. Nineteen-year-old Shonta, an African American with one child who became a mother at only fourteen, says she knows motherhood "has its ups and downs, [but] I never felt my daughter held me back from anything. If anything she taught me how to be responsible and mature."...

Middle-class observers often believe that the lives of poor youth could be salvaged if not for the birth of a child—but this is seldom the case. Our mothers' stories show that young women raised in poor neighborhoods can suffer far worse fates than having to drop out of school to care for a baby. The poor women we came to know often describe their lives prior to motherhood as spinning out of control. They recall an existence blemished by more than mere economic insecurity. For most, the "rippin' and runnin'" days before children were marked by depression, school failure, drug and alcohol use, and promiscuous sexual activity. Along with this self-made chaos were their sometimes troubled and abusive home lives and the danger, violence, and oppression of the neighborhood.

Over and over again, mothers tell us their children tamed or calmed their wild behavior, got them off of the street, and helped put their lives back together. Children can banish depression, calm a violent temper, or serve as do-it-yourself rehab from alcohol and drugs. Children—and the minute-by-minute demands they make on their mothers' time, energy, and emotions—bring order out of chaos. "I would still be wild and stuff, hanging in the streets, hanging on the corners and stuff," testifies fifteen-year-old Jessica, a Puerto Rican mother of a two-year-old. "[But] I didn't want to get in trouble and then DHS [Department of Human Services] come and take her away from me. I used to be *bad*, go around break windows and stuff. Now I don't do *nothing*. I be with *her* all day. I come to school, go home, be with her."...

WHAT MOTHERHOOD DOES FOR MOMS

Some readers may be deeply troubled by the idea that poor, unmarried women reap so many benefits from the children they bear. The idea of a woman viewing her offspring as a resource violates powerful social norms about how a mother should behave. Altruism, not need, ought to govern her relationship to her children. Yet, as we've shown, altruism is also common and strong among poor mothers who are raising children alone. In reality, the motivation to mother among women of all social classes is a mix of self-sacrifice and expected reward.

One of the many advantages of membership in the privileged classes is greater access to a range of satisfying social roles. All parents presumably derive some additional degree of validation, some greater sense of purpose, some new sense of connection, and at least some level of reorientation because of their children. Yet few middle-class women approach mothering with such a great sense of need, and few see their children as very nearly their sole source of fulfillment.

When we've told these mothers' stories to audiences around the country, many of our listeners are surprised to learn that, for young women on the economic edge, having a child can bring order to a life with no point or purpose. One woman, a new mother herself, conceded that having a child had improved her life, but found our claim that children bring order to the lives of poor mothers difficult to swallow. Her once-regimented, childless, middle-class lifestyle was more out of control since the birth of her child than she had ever dreamed possible. "How out of control must their lives *be?*" she asked, unable to conceive of how a child could *reduce* rather than *multiply* a mother's day-to-day level of chaos.

On another occasion, a listener told us she simply couldn't accept our contention that poor women express so little regret over having had children when and how they did. She demanded to know more about the "ambivalence" women must feel toward their children and toward motherhood. "There *has* to be some tension or sense of regret!" she exclaimed. But as the question-and-answer period ended and the crowd dispersed, a pregnant woman who had not spoken up earlier turned to her companion and asked, "Why is it so hard for people here to believe that the women would *want* their children?"

After spending six years talking about these issues with poor unwed mothers, the worldviews they hold no longer seem strange to us. But we must admit that at first, we were as astonished by their viewpoints as many of our listeners have been. Looking back over our own experiences on the street corners and stoops, the coffee shops and fast-food restaurants, and the front rooms and kitchens of these mothers, we can now identify the one question we asked that proved most revealing: "What would your life be like without children?" We assumed this question would prompt stories of regret over opportunities lost and ambitions foiled, and some did indeed say what we had expected to hear. But there were startlingly few "if only" tales of how "coming up pregnant" wrecked dreams, of education, career, marriage, or material success. Instead, mothers repeatedly offered refrains like these: "I'd be dead or in jail," "I'd be in the streets," "I wouldn't care about anything," "My child saved me," and "It's only because of my children that I'm where I am today." For all but a few, becoming a mother was a profound turning point that "saved" or "rescued" them from a life either leading nowhere or going very wrong. Rather than derailing their lives, they believed their children were what finally set them on the right track.

Aside from a few notable exceptions,…the mothers we met seemed, in our judgment, to be adequate parents. Our central point here is not that these mothers were exceptionally good, but that they use motherhood as a way to make meaning in a void. Their life stories are testimonials to motherhood's transforming influence, leading them to abandon their "drinking and drugging," to trade a wild life for one spent at home, to return to school, pursue employment, reconnect with family, and to find a new sense of hope and purpose. This does not mean we accept at face value that these mothers are right when they say they are better off having had their children when they did. Some could probably have benefited from waiting. What we are saying is that these mothers perceive these things to be true, and it is this *perception*, rather than reality, that guides their actions.…

Yet motherhood's influence may be fleeting for some. We're not claiming that every poor, young woman will stop "using," get her high school diploma, find a career, or never again have an abusive relationship with a man when she becomes a mother. What motherhood offers is the possibility—not the promise—of validation, purpose, connection, and order. More important, children allow mothers to transcend, at least psychologically and symbolically, the limitations of economic and social disadvantage. These women put motherhood before marriage not primarily out of welfare opportunism, a lack of discipline, or sheer resignation. Rather, the choice to mother in the context of personal difficulty is an affirmation of their strength, determination, and desire to offer care for another. In the end, establishing the primordial bonds of love and connection is the ultimate goal of their mothering.

DISCUSSION QUESTIONS

1. Thinking of someone you know who is a very young mother, how does her experience compare to the women in Edin and Kefalas's study? What does this teach you about the influence of race and class in how motherhood is experienced?
2. What do people in your surrounding environment say about teen pregnancy? Does Edin and Kefalas's research change your understanding of what people say?

41

Child Welfare as a Racial Justice Issue

DOROTHY ROBERTS

There are huge racial disparities in the placement of children into foster care, with the vast majority of such children coming from African American families. Roberts, a legal scholar, argues that this reflects the increased surveillance of Black families by the state and racist assumptions that Black families are somehow inadequate.

My main intellectual and activist project over the last five years has been to explain child welfare policy as an issue of racial justice. Child welfare was once viewed as a social issue, but by the 1970s the main mission of public child welfare departments had become protecting children against maltreatment inflicted by pathological parents. Child welfare decision making has an atomistic focus, that zooms in on the situation of individual children and their families. If child welfare is discussed as a matter of rights at all, it is usually framed as a contest between children's rights and parents' rights, falsely assuming that the interests of parents and children in the child welfare system are always in opposition to each other and that the system treats all parents and children equally badly.

Strangely, criticisms of the child welfare system are not placed among the burning social justice issues of our day. I say strangely because anyone who is familiar with the child welfare system in the nation's large cities knows that it is basically an apartheid institution. Spend a day at any urban dependency court and you will see a starkly segregated operation. If you came with no preconceptions about the purpose of the child welfare system, you would have to conclude that it is an institution designed to monitor, regulate, and punish poor minority families, especially Black families.

The number of Black children in state custody—those in foster care as well as those in juvenile detention, prisons, and other state institutions—is alarming.... Black children make up about two-fifths of the foster care population, although they represented less than one-fifth of the nation's children (Administration for Children and Families 2003). The color of child welfare is most apparent in big cities where there are sizeable Black and foster care populations. In Chicago, for example, almost all of the children in foster care are Black (Pardo 1999). The racial imbalance in New York City's foster care population is also mind-boggling: out of 42,000 children in the system at the end of 1997,

SOURCE: From Dorothy Roberts, "Child Welfare as a Racial Justice Issue," 2002.
Reprinted by permission of the author.

only about 1000 were white (Guggenheim 2000). Black children in New York were 10 times as likely as White children to be in state protective custody. Although the total numbers are smaller, this racial disproportionality extends to cities and states where Black children are less visible.

State agencies treat child maltreatment in Black homes in an especially aggressive fashion. They are far more likely to place Black children than other children who come to their attention in foster care instead offering their families less traumatic assistance. A national study of child protective services by the U.S. Department of Health and Human Services reported that "[m]inority children, and in particular African American children, are more likely to be in foster care placement than receive in-home services, even when they have the same problems and characteristics as white children" (U.S. Dept. HHS 1997). Most white children who enter the system are permitted to stay with their families, avoiding the emotional damage and physical risks of foster care placement, while most Black children are taken away from theirs. Foster care is the main "service" state agencies provide to Black children. And once removed from their homes, Black children remain in foster care longer, are moved more often, receive fewer services, and are less likely to be either returned home or adopted than any other children (Courtney & Wong 1996; Jones 1997).

In some cases, protecting children requires immediately removing them from their homes. But the public often overlooks the costs to children of separating them from their families. In 2001, Judge Jack Weinstein of the Eastern District of New York issued a blistering condemnation of New York City's Administration for Children's Services for automatically removing children from mothers who were victims of domestic violence (*Nicholson v. Scoppetta* 2001). Judge Weinstein's decision is especially noteworthy for its rare judicial recognition of the harm inflicted on children by unnecessarily taking them from their parents. "It hardly needs to be added that the exact language of the Thirteenth Amendment covers protection of the children's rights," Judge Weinstein wrote. "They are continually forcibly removed from their abused mothers without a court adjudication and placed in a forced state custody in either state or privately run institutions for long periods of time. They are disciplined by those not their parents. This is a form of slavery."

A new politics of child welfare threatens to intensify state supervision of Black children. In the last several years, federal and state policy has shifted away from preserving families toward "freeing" children in foster care for adoption by terminating parental rights. The Adoption and Safe Families Act, passed in 1997, imposes an expedited time frame for state agencies to file petitions to terminate parental rights and give states a financial bonus for increasing the number of adoptions of children in foster care—$4,000 per child, $6,000 if the child is classified as having special needs. The campaign to increase adoptions has hinged on the denigration of foster children's parents, the speedy destruction of their family bonds, and the rejection of family preservation as an important goal of child welfare practice. Adoption is increasingly presented not as an option for a minority of foster children who cannot be reunited with their parents, but as the preferred outcome for all children in foster care....

Welfare reform, by throwing many families deeper into poverty, heightens the risk that the most vulnerable children will be placed in foster care. A front page story in the *New York Times*, entitled "Side Effect of Welfare Law: the No-Parent Family," reported a study of census data in all 50 states that found that the rate of Black children in cities living without their parents has more than doubled as a result of welfare reform—an estimated 200,000 more Black children separated from their parents (Bernstein 2002).

In addition, tougher treatment of juvenile offenders, imposed most harshly on African American youth, is increasing the numbers incarcerated in juvenile detention facilities and adult prisons. These political trends are converging to address the deprivation of poor Black children by placing more of them in one form of state custody or another. Child welfare policy conforms to the current political climate, which embraces private solutions—and when those fail, punitive responses—to the seemingly intractable plight of America's isolated and impoverished inner cities. As welfare reform reduces the welfare rolls by promoting marriage and imposing sanctions for failing to find work, child welfare policy reduces the foster care rolls by terminating parental rights and promoting adoption of the "legal orphans" it creates.

The color of America's child welfare system undeniably shows that race matters to state interventions in families. So why have scholars and policymakers been slow to describe the disproportionate involvement of Black families in the system as a racial injustice? Let me suggest three related explanations.

First, there is profound confusion about the reasons for the system's racial disparity. The existing social science literature contains theories that attribute the racial disparity both to differences in the well being of children and to differences in the system's treatment of children. In other words, there is disagreement over whether the disparity stems from societal conditions outside the system, such as higher poverty rates among nonwhite children, or from racially biased practices *within* the system. Many experts believe that this distinction in causes—societal forces vs. child welfare practices—makes a crucial difference in how we should address the racial disparities (Courtney et al. 1998). If the cause of the system's racial imbalance is social and economic inequality, some say, we can't blame the system (Bartholet 1999).

This is related to a second reason for failing to see a racial justice issue—the concern for children's rights. If the child welfare system is simply reflecting inequities in children's living conditions, we would expect—we would even want—the state to intervene more often to protect Black children from the greater harm that they face. Indeed, wouldn't the government violate Black children's rights if it *failed* to intervene more often to protect them?...

A final stumbling block is the official understanding of racial discrimination. Under current civil rights jurisprudence, the racial disparity in the child welfare system may not constitute racial discrimination without a showing of racial motivation. The system is racist only if Black children are pulled out of their homes by bigoted caseworkers or as part of a deliberate government scheme to subjugate Black people. Any other explanation—such as higher rates of Black family poverty or unwed motherhood—negates the significance of race....

But these views of the child welfare system, children's rights, and racial discrimination fail to take into account the political dimension of child welfare policy. To begin with, which harms to children are detected, identified as abuse or neglect, and punished is determined by inequities based on race, class, and gender. The U.S. child welfare system is and always has been designed to deal with the problems of poor families (Pelton 1989). The child welfare system hides the systemic reasons for poor families' hardships by attributing them to parental deficits and pathologies that require therapeutic remedies rather than social change. The harms caused to children by uncaring, substance-abusing, mentally unstable, absentee parents in middle-class and affluent families usually go unheeded. Although these children from privileged homes might spend years in psychotherapy, it is unlikely they will spend any time in foster care. Most child maltreatment charges are for neglect and involve poor parents whose behavior was a consequence of economic desperation as much as lack of caring for their children.

The racial disparity in the child welfare system also reflects a political choice about how to address child neglect. It is no accident that child welfare philosophy became increasingly coercive as Black children made up a greater and greater share of the caseloads. In the past several decades, the number of children receiving child welfare services has declined dramatically, while the foster care population has skyrocketed (U.S. Dept. HHS 1997). As the child welfare system began to serve fewer white children and more Black children, state and federal governments spent more money on out-of-home care and less on in-home services. This mirrors perfectly the metamorphosis of welfare once the welfare rights movement succeeded in making AFDC available to Black families in the 1960s. As welfare became increasingly associated with Black mothers, it became increasingly burdened with behavior modification rules and work requirements until the federal entitlement was abolished altogether in 1996. Both systems responded to their growing Black clientele by reducing their services to families while intensifying their punitive functions.

The child welfare system's reliance on a disruptive, coercive, and punitive approach also inflicts a political harm. American constitutional jurisprudence defines the harm caused by unwarranted state interference in families in terms of individual rights. Wrongfully removing children from the custody of their parents violates parents' due process right to liberty. The earliest cases interpreting the due process clause to protect citizens against government interference in their substantive liberty involved parental rights. But these explanations of harm do not account for the particular injury inflicted by racially disparate state intervention. The over-representation of Black children in the child welfare system, especially foster care, represents massive state supervision and dissolution of families. This interference with families helps to maintain the disadvantaged status of Black people in the United States. The child welfare system not only inflicts general harms disproportionately on Black families. It also inflicts a particular harm—a racial harm—on Black people as a group....

State supervision of families is antithetical to the role families are supposed to play in a democracy, as critical components of civil society. Families are a principal form of "oppositional enclaves" that are essential to citizens' free participation in democratic institutions, to use Harvard political theorist Jane Mansbridge's term (Mansbridge 1996, 58). Family and community disintegration weakens Blacks' collective ability to overcome institutionalized discrimination and to work toward greater political and economic strength. Family integrity is crucial to group welfare and identity because of the role parents and other relatives play in transmitting survival skills, values, and self-esteem to the next generation. Placing large numbers of children in state custody interferes with the group's ability to form healthy and productive connections among its members. The system's racial disparity also reinforces the quintessential racist stereotype: that Black people are incapable of governing themselves and need state supervision.

The impact of state disruption and supervision of families is intensified when it is concentrated in inner-city neighborhoods. In 1998, one out of every ten children in Central Harlem had been taken from their parents and placed in foster care (Center for an Urban Future 1998, 6). In Chicago, almost all child protection cases are clustered in a few zip code areas, which are almost exclusively African American. The spatial concentration of child welfare supervision creates an environment in which state custody of children is a realistic expectation, if not the norm....

The racial disparity in the foster care population should cause us to reconsider the state's current response to child maltreatment. The price of present policies that rely on child removal rather than family support falls unjustly on Black families and communities. In part because of narrow conceptions of racial discrimination and children's rights, judges, politicians, and the public have a hard time seeing this as a racial injustice. I propose that we figure out better ways of measuring and explaining this type of systemic, community-wide racial harm.

REFERENCES

Administration for Children and Families, U.S. Department of Health & Human Services. 2000. *Child Maltreatment 1998: Reports from the States to the National Child Abuse and Neglect Data System.* Washington, D.C.: U.S. Government Printing Office.

——— 2003. *The AFCARS Report: Preliminary FY 2001 Estimates as of March 2003.* http://www.acf.hhs.gov/programs/cb/publications/afcars/report8.htm.

Bernstein, Nina. 2002, July 29. "Side Effect of Welfare Law: The No-Parent Family." *New York Times*, p. 1.

Bartholet, Elizabeth. 1999. *Nobody's Children.* Boston: Beacon Press.

Center for an Urban Future, 1998. "Race, Bias, and Power in Child Welfare." *Child Welfare Watch.* Spring/Summer: 1.

Courtney, Mark E., et al. 1998. "Race and Child Welfare Services: Past Research and Future Directions." *Child Welfare* 75:99.

Courtney, Mark E. and Wong, Vin-Ling Irene. 1996. "Comparing the Timing of Exits from Substitute Care." *Child & Youth Services Review* 18:307.

Guggenheim, Martin. 2000. "Somebody's Children: Sustaining the Family's Place in Child Welfare Policy." *Harvard Law Review* 113: 1716.

Jones, Loring P. 1997. "Social Class, Ethnicity, and Child Welfare." *Journal of Multicultural Social Work* 6:123.

Males, Mike and Dan Macallair. 2000. *The Color of Justice: An Analysis of Juvenile Adult Transfers in California.* San Francisco: Justice Policy Institute.

Mansbridge, Jane. 1996. "Using Power/Fighting Power: The Polity." In *Democracy and Difference: Contesting the Boundaries of the Political,* edited by Seyla Benhabib. Princeton: Princeton University Press.

Nicholson v. Scoppetta. 2001. 202 F.R.D. 377. United States District Court, Eastern District of New York.

Pardo, Natalie. 1999. "Losing Their Children." *Chicago Reporter* 28:1.

Pelton, LeRoy H. 1989. *For Reasons of Poverty: A Critical Analysis of the Public Child Welfare System in the United States.* New York: Praeger.

U.S. Department of Health and Human Services. 1997. *National Study of Protective, Preventive, and Reunification Services Delivered to Children and Their Families.* Washington, D.C.: U.S. Government Printing Office.

DISCUSSION QUESTIONS

1. What does Roberts mean by an apartheid institution? How does this apply to the child welfare systems in major cities?
2. How can children become legal orphans?
3. How does the impact of welfare policies on individual families influence the collective well-being of the Black community?

42

The Social Construction of the "Immoral" Black Mother

Social Policy, Community Policing, and Effects on Youth Violence

BETH E. RICHIE

Black women as mothers have been stereotyped as incompetent and/or domineering, but Richie's study of Black mothers whose children have witnessed or experienced violence shows that such stereotypes grossly misinterpret Black women's struggles with institutionalized racism. By studying these women on their own terms, Richie provides a much needed correction to public perceptions and policies about Black women and mothering.

Few areas of social life have been as contested in social policy debates as the concept of the family. Highly charged rhetoric about gender and generational relationships surrounds most recent proposals for reform. From nostalgic calls for conservative approaches by religious right-wing forces to seemingly progressive legislative initiatives advocating gay/lesbian marriages, debates about family life are played out on various ideological templates. Even in progressive contexts, such as the recent reconsiderations of adolescent pregnancy, the problem has been constructed as the need to "strengthen fragile families." Similarly, in the field of public health, we see an emphasis on the family as the cornerstone of emotional and social well-being, examined via resiliency factors that emerge from particular forms of household arrangements. In these and other examples, current social policy reform is increasingly attached to the organization and meaning of the role of the family in contemporary society, and overall the constructs have a distinctively conservative tendency.

Motherhood, as a subcategory of the family debates, is constituted through a similar vast range of intellectual, political, and popular rhetoric, and with similar conservative undertones. While characterized by mixed conceptual frames (ranging from the best practice of motherhood to the healthiest type of relationships

SOURCE: From *Revisioning Women, Health, and Healing: Feminist, Cultural, and Technoscience Perspectives*, ed. by Adele E. Clarke and Virginia L. Olesen (New York: Routledge, 1999), pp. 283–297. Copyright © 1999. Reprinted by permission of Routledge/Taylor & Francis Books, Inc.

between mothers and children), still at the center of the ideological debate are universalistic assumptions that revolve around "desirable" family forms, "appropriate" gender roles, and the maintenance of a separation between public and the private spheres. For example, policy makers continue to interrogate researchers about the effects on children when women work outside the home, and legislators argue about what single form family leave should take....

Instead of being featured in the debates, most black women in low-income communities fall far outside the normative, hegemonic parameters of such discussions. With noted exceptions (Dickerson 1995), most considerations of the mothering that poor black women do is introduced into the political, social, and empirical debates from a very different social location. At best, their mothering is studied as a culturally distinct add-on to the dominant inquiries. In its worst and far more common form, low-income black women's mothering is used as a not-so-coded metaphor for much of what is wrong with contemporary society (Hill 1997). Black women are portrayed as creating pathological forms of families as "single heads of households," as draining public resources, or as breeding too many children who pose physical, social and economic risks to others. Their mothering is viewed as something quite different from the mothering efforts of other groups—as a category of activities enacted in such dissimilar ways from the dominant model that they are constructed as confusing, atypical, and dysfunctional. Ultimately, I will argue here, this outside position renders black women's mothering immoral, if not criminal, in the perspective of those who formulate and enforce social policies.

My argument here is that, worse than simply ignoring the role that mothering assumes in poor black women's lives, the current analytical and ideological framework does great harm to these women, their children, and their communities. Rather than seeking to understand and then address the social needs of black women and their families within the contexts in which we actually live, current conceptualizations ignore the specificity of the micro processes of mothering and misinterpret key behaviors and actions of mothers. The social policies that ensue reinforce such conceptualizations pathologizing and stigmatizing effects. The overall result is increased marginalization, structural disenfranchisement, hypersurveillance and overregulation of poor black women's mothering in new and profound ways. The particular case that I will use to argue this point concerns the consequences that social policy on youth violence has on black women's mothering.

I frame this discussion with findings from a study of twenty-four adult women who are female caretakers of adolescent children in a low-income community in a major urban area where, like many other cities, youth violence is a devastating social and public health problem. The broader research project of which this is a part examines how youth violence is distinctively and decidedly gendered in nature, and how the interventions designed to address these problems ignore this important dimension. Hence the problem of youth violence, typically constructed as a problem that affects young men of color, is neither linked to the issue of gender violence nor understood to have any effects on girls and women when, in actuality, it certainly does....

METHODOLOGY

My study was designed to explore the impact of youth violence on mothers and mothering, using the life-history interview technique to elicit data on the ways that black women thought and felt about their experiences. The life-history method was selected because it is particularly useful in gathering information about stigmatized, uncomfortable, or difficult circumstances in the subjects' lives. Compared to other, more structured qualitative methods, conducting life-history interviews offers a more open and intense opportunity to learn about the subjects' backgrounds, opinions, and feelings, as well as the meanings they give to both the mundane and exceptional experiences in their lives. Mothering is obviously in this realm....

The people to be interviewed were drawn from populations of black women whose children are involved in or at serious risk of experiencing or witnessing violence in the private or public spheres of their lives. Twenty-four women agreed to participate, ranging in age from nineteen to sixty-nine years old. Included in the sample were (1) mothers or guardians of pregnant or parenting adolescents; (2) mothers who resided in public housing, subsidized housing or public shelters; and (3) primary caretakers of children detained in institutional settings for juveniles....

Four basic areas were covered by the interviews, beginning with an open-ended question: "Tell me the things about yourself that are important to you." Next the women were asked about factors they felt influenced their role as mothers: "What is it like raising children in your household and neighborhood?" Third, I sought to capture their experiences and perceptions of gender and youth violence and how it affected their lives: "In what ways does violence or the threat of violence affect you and your family or neighborhood?" Last, I asked how their life might be different in the future, what they wanted, dreamed, hoped for, and expected for themselves and their children....

THE FIVE THEMES

Theme 1: The Diminished Ability to Parent Due to Limited Economic and Social Supports within the Context of Urban Decay

The context within which the women lived was marked by their economic marginalization. They described the following characteristics of their world. First, observation and experience had led them to conclude that "doing the right thing for your children" would not necessarily work for them as members of a marginalized ethnic group. They described feeling that, as low-income women whose attempts to mother had repeatedly failed, they somehow fell outside of society's parameters of goodness or fairness. This sense translated into feelings of powerlessness, frustration, and discontent with their own mothering abilities. They were typically self-blaming even when, paradoxically, they articulated an insightful analysis of the social conditions that led to their marginalized position as women and as mothers.

A second dimension of this theme was the degree to which their household fabric has been limited by changes in their efficacy as adults in the social world. Their household composition changes frequently, usually in response to economic shifts, and this often limits important intergenerational contact. They and other adults lose jobs or are only marginally employed. Their families double up in inadequate housing. They simply do not have the resources to perform their parenting roles as well as they desire. Successful role models for both their children and themselves were limited, and extended family networks were quite tenuous.

While many of the women interviewed grew up in poor families, the effects of persistent multigenerational poverty are taking a toll on them. For while poverty may not be new, the level and the nature of hostile public sentiment, the prolonged feelings of despair, and the extent of the violence in their communities are new. They did not grow up watching and knowing of their friends being killed the way their children do, and this huge experiential gap has left most of them unprepared to help their children make sense of these tragic events or offer much support.

The women interviewed for this study also described how public socializing systems are failing them and their children. Schools are considered dangerous, rigid places where the mothers described feeling as alienated as the young people do. One woman said, "They look much like prisons, and I feel like they are holding my child captive there for some crime of going to school. I have no rights as a visitor, and definitely no input into what happens there."

A broader exploration of structural conditions reveals that community institutions and most public spaces are decaying. Businesses, movie theaters, libraries, and parks are closing, and services at hospitals and mental health facilities are being cut back.... The women reported that a decline in availability of public transportation (buses have changed their routes and cabs won't stop to transport them) has left them isolated within their communities....

The women understand these isolating and confining strategies as symbolic of the larger community's fear of them and their children. Many schools have metal detectors through which children must pass. Surprisingly, so do laundromats, video arcades, and the music stores young people frequent. Gated retail establishments favor merchandizing large bottles of beer and candy packaged like liquor over fresh produce. In these and other obvious ways, raising children is limited by perverse environmental conditions, lack of social support, symbolic fear, and persistent economic decay....

Theme 2: The Constant Fear of Losing Children to Public Agencies

This second theme can best be characterized by the words of one of the women, who said, "The state is actually raising our children, and as far as I can see they are not doing a very good job. Our job as mothers has therefore become to keep running from child protection, from truant and probation officers, from social workers and the like who are trying to take our kids from us. Family values, not! It's like the slave days...they want to take our kids." This was one of many moving testimonies to how women are struggling to escape the intervention of authorities and maintain their custodial rights.

The phenomenon of women being surveilled and monitored in their domestic activities as mothers has an important relationship to the problem of youth violence, which is obviously also a problem of policing. The impact of feeling monitored as a mother while your children are being policed is profound. One informant described it as a "land mine, where you are constantly chasing your kid through dangerous streets hoping you will catch him before the police do. In the meantime, though, you have to watch out for yourself too."

This impression of mothers being scrutinized while they themselves are at risk takes several forms. One form is related to the increasing public anxiety related to the safety of children whose mothers are being battered. On the one hand, this attention is important and long overdue. Yet in a more problematic sense, we see how concern for women has been placed in conflict with the needs of their children, thus positioning advocates for battered women at odds with child protective service workers in some communities.

A second manifestation of the policing of women's mothering is the rigid monitoring of women whose children have been identified as at risk of abuse because of a series of early juvenile offenses. Against the backdrop of the national trend to hold parents accountable, women whose children are in more trouble face increased jeopardy themselves. Paradoxically, the women in this study described feeling that when the "authorities are watching," their children feel even *more* inclined to act out, especially when custody issues are pending. The children then can manipulate their mothers, knowing that they are likely to get away with undercutting her parental authority. One woman said, "It's like the kids *know* what they are doing. And I find myself begging my kids to behave rather than rearing them in any strong way. I don't have any dignity left when the kids know that my ability to mother them has been called into question by outsiders."...

The relationships of mothers with other (non-child-specific) public institutions are also important. Most of the women interviewed considered law enforcement agencies dangerous, public assistance programs adversarial and human services typically unhelpful. These women don't feel there is much of a safety net that they can trust or depend on to support their families. Most described profound despair and were disheartened. Yet they persist in trying to raise their children with very limited resources and in dangerous isolation.

Theme 3: The Fear of Abuse and Injury

One of the consequences of women's...continued attempts to enhance the safety of their children is the considerable risk of violence the women themselves face. This finding had specific gendered dimensions. The women were at risk because they were women and mothers. The responses demonstrated keen awareness that their neighborhood or "the block" is dangerous for all community members. However, they accepted and espoused the rhetoric that considered boys and men at particular risk, and therefore they themselves took particular risks for their male children as an extension of their mothering role.

For example, the women described trying to intervene with other young people when their children were in trouble. The combined mistrust of outside agencies, the sense of community loyalty (which emerged as the fourth theme), and the subjective desire to enact some degree of agency in their family and community life compelled some of the women I interviewed to try to resolve conflicts on their children's behalf. This left them extremely vulnerable. In almost half of the reported cases the women were injured by young men when they tried to protect their children. These assaults usually involved a weapon....

Theme 4: Generalized, Culturally-Constructed Loyalty to Black Young Men

As one informant said, "The puddle is muddied by the position of black men in society, especially the 'endangered species' [meaning young black men]. But as a community we are as sick as our secrets." This powerful statement suggests that given the well-known effects that violence, poverty, racism, and lack of opportunity have had on black boys, it can be very difficult and problematic to raise the issue of the condition of black girls and the compromised positions of their mothers. More broadly, the frequently expressed sentiments of the women in this study suggest that the nature of gender relationships in the black community are complicated by cultural loyalties. The rhetoric sounds like this: "Men are vulnerable to societal abuse and women have had more opportunity than they." "Boys are the endangered species, and girls need to be more responsible." "It's black mothers who are raising these sons but no one pays attention to us."

These sentiments represent the opinion of a considerable segment of black communities in this country, and the extent to which this culturally constructed loyalty interacts with and is influenced by mothering warrants further investigation. In this study, it suggested a skewed set of community priorities bolstered by a simplistic public policy agenda that not only ignores the vulnerability of women and girls but also particularly punishes mothers for attempting to protect their daughters.

Theme 5: Involvement in Prevention Initiatives and Community Activism

The fifth theme concerns the problematic nature and outcomes of women's community activism to prevent youth violence. There is a long history of documentation and analysis of black women's activism that emerged, in part, from the unique position we've assumed vis-à-vis the labor force, constitutional rights, social justice initiatives, and reform movements. This literature has generally concluded that black women's community work has been an important source of empowerment and expression of agency.

The interviews revealed a different picture. The women's accounts of their actual experiences were full of powerlessness, a sense of failure, increased risk of injury and fear for their safety, and renewed pessimism regarding their ability to accomplish the role of mothering in ways they desired. In a troubling sense, what has historically been a source of liberation for black women has become, in the face of these contemporary problems, actually a way to further marginalize women and stigmatize their inability to protect and nurture their children. Now this is in the public sphere as well as at home. The combination of structural conditions and hostile relationships between outside agencies and community groups contaminates these initiatives and causes them to fail. Most regrettably, women are set up as scapegoats here as well.

This conclusion emerged from several accounts of women who had been convinced to report their children's criminal activities in exchange for some help or leniency. They quickly learned that with current enhanced prosecution practices, their children are facing very significant prison terms. Others described how their initial enthusiasm for working with the violence-prevention program associated with a law enforcement agency were tempered when they felt compelled to "set kids up." Many reported feeling alienated from their families and neighbors and afraid of retaliation because of their assumed cooperation with police. Three who testified about their role as community liaisons and mentors found themselves quoted in a legislative report supporting repressive welfare reforms. A simple case of tokenism? Perhaps. Certainly these stories indicate the clash between the women's subjective need to feel competent and recognized in their roles as mothers and the objective limitations of their power in the social worlds within which they live.... These findings describe more than the failure of programs to successfully engage low-income black women in violence prevention initiatives. Such programs set women up as local targets even more than they were in the first place.

CONCLUSION

In this article, I have attempted to explore how the micro process of black women's mothering is constrained by stigmatization, persistent social problems, and misguided social policy.... Uninformed social policy, which ignores such structural conditions, has profound and unchallenged effects on black women's efforts at the micro processes of mothering. Intervention programs are misguided, pathologically oriented, and dangerous for black women. In the case of youth violence, they have further stigmatized women and punished black mothers.

REFERENCES

Dickerson, B. (ed.). 1995. *African American Single Mothers: Understanding Their Lives and Families.* Thousand Oaks, CA: Sage Publications.

Hill, R. 1997. Social Welfare Policies and African American Families. Pp. 349–63 in Harriette McAdoo (ed.), *Black Families.* Thousand Oaks, CA: Sage Publications.

DISCUSSION QUESTIONS

1. According to the women Ritchie interviewed, how does living in a decaying urban community complicate mothering?
2. Why do many of the Black women Ritchie interviewed feel powerless as mothers?
3. How does the poverty these mothers experienced differ from the poverty that their children face?

43

Fences and Neighbors: Segregation in 21st-Century America

JOHN E. FARLEY AND GREGORY D. SQUIRES

Residential segregation is key to understanding a host of other dimensions of race relations, including access to schooling, as well as the formation of friendships. Farley and Squires review current trends in residential segregation and discuss the fair-housing movement.

"Do the kids in the neighborhood play hockey or basketball?"
—ANONYMOUS HOME INSURANCE AGENT, 2000

America became less racially segregated during the last three decades of the 20th century, according to the 2000 census. Yet, despite this progress, despite the Fair Housing Act, signed 35 years ago, and despite popular impressions to the contrary, racial minorities still routinely encounter discrimination in their efforts to rent, buy, finance, or insure a home. The U.S. Department of Housing and Urban Development (HUD) estimates that more than 2 million incidents of unlawful discrimination occur each year. Research indicates that blacks and Hispanics encounter discrimination in one out of every five contacts with a real estate or rental agent. African Americans, in particular, continue to live in segregated neighborhoods in exceptionally high numbers.

What is new is that fair-housing and community-development groups are successfully using antidiscrimination laws to mount a movement for fair and equal access to housing. Discrimination is less common than just ten years ago; minorities are moving into the suburbs, and overall levels of segregation have gone down. Yet resistance to fair housing and racial integration persists and occurs today in forms that are more subtle and harder to detect. Still, emerging coalitions using new tools are shattering many traditional barriers to equal opportunity in urban housing markets.

Segregation refers to the residential separation of racial and ethnic groups in different neighborhoods within metropolitan areas. When a metropolitan area is highly segregated, people tend to live in neighborhoods with others of their own group, away from different groups. The index of dissimilarity (*D*), a measure

SOURCE: From *Contexts* 4(1): 33–39, 2005. Reprinted by permission of University of California Press.

of segregation between any two groups, ranges from 0 for perfect integration to 100 for total segregation. For segregation between whites and blacks (imagining, for the sake of the example, that these were the only two groups), a *D* of 0 indicates that the racial composition of each neighborhood in that metropolitan area is the same as that of the entire area. If the metropolitan area were 70 percent white and 30 percent black, each neighborhood would reflect those percentages. A *D* of 100 would indicate that every neighborhood in the metropolitan area was either 100 percent white or 100 percent black. In real metropolitan areas, *D* always falls somewhere between those extremes. For example, the Chicago metropolitan area is 58 percent non-Hispanic white and 19 percent non-Hispanic black. Chicago's *D* was 80.8 in 2000. This means that 81 percent of the white or black population would have to move to another census tract in order to have a *D* of 0, or complete integration. On the other hand, in 2000 the Raleigh-Durham, N.C. metropolitan area, which is 67 percent non-Hispanic white and 23 percent non-Hispanic black, had a *D* of 46.2—a little more than half that of Chicago.

SEGREGATION: DECLINING BUT NOT DISAPPEARING

Although segregation has declined in recent years, it persists at high levels, and for some minority groups it has actually increased. Social scientists use a variety of measures to indicate how segregated two groups are from each other....

Although African Americans have long been and continue to be the most segregated group, they are notably more likely to live in integrated neighborhoods than they were a generation ago. For the past three decades, the average level of segregation between African Americans and whites has been falling, declining by about ten points on the *D* scale between 1970 and 1980 and another ten between 1980 and 2000. But these figures overstate the extent to which blacks have been integrated into white or racially mixed neighborhoods. Part of the statistical trend simply has to do with how the census counts "metropolitan areas." Between 1970 and 2000, many small—and typically integrated—areas "graduated" into the metropolitan category, which helped to bring down the national statistics on segregation. More significantly, segregation has declined most rapidly in the southern and western parts of the United States, but cities in these areas, especially the West, also tend to have fewer African Americans. At the same time, in large northern areas with many African-American residents, integration has progressed slowly. For example, metropolitan areas like New York, Chicago, Detroit, Milwaukee, Newark, and Gary all had segregation scores in the 80s as late as 2000. Where African Americans are concentrated most heavily, segregation scores have declined the least.... In places with the highest proportions of black population, segregation decreased least between 1980 and 2000. Desegregation has been slowest precisely in the places African Americans are most likely to live. There, racial isolation can be extreme. For example, in the Chicago, Detroit, and Cleveland metropolitan areas, most African Americans live in census tracts (roughly, neighborhoods) where more than 90 percent of the residents are black and fewer than 6 percent are white.

Other minority groups, notably Hispanics and Asian Americans, generally live in less segregated neighborhoods. Segregation scores for Hispanics have generally been in the low 50s over the past three decades, and for Asian Americans and Pacific Islanders, scores have been in the low 40s. Native Americans who live in urban areas also are not very segregated from whites (scores in the 30s), but two-thirds of the Native Americans who live in rural areas (about 40 percent of their total population) live on segregated reservations. Although no other minority group faces the extreme segregation in housing that African Americans do, other groups face segregation of varying levels and have not seen a significant downward trend.

CAUSES OF CONTINUING SEGREGATION

Popular explanations for segregation point to income differences and to people's preferences for living among their "own kind." These are, at best, limited explanations. Black-white segregation clearly cannot be explained by differences in income, education, or employment alone. Researchers have found that white and black households at all levels of income, education, and occupational status are nearly as segregated as are whites and blacks overall. However, this is not the case for other minority groups. Hispanics with higher incomes live in more integrated communities than Hispanics with lower incomes. Middle-class Asian Americans are more suburbanized and less segregated than middle-class African Americans. For example, as Chinese Americans became more upwardly mobile, they moved away from the Chinatowns where so many had once lived. But middle-class blacks, who have made similar gains in income and prestige, find it much more difficult to buy homes in integrated neighborhoods. For example, in 2000 in the New York metropolitan area, African Americans with incomes averaging above $60,000 lived in neighborhoods that were about 57 percent black and less than 15 percent non-Hispanic white—a difference of only about 6 percentage points from the average for low-income blacks.

Preferences, especially those of whites, provide some explanation for these patterns. Several surveys have asked whites, African Americans, and in some cases Hispanics and Asian Americans about their preferences concerning the racial mix of their neighborhoods. A common technique is to show survey respondents cards displaying sketches of houses that are colored-in to represent neighborhoods of varying degrees of integration. Interviewers then ask the respondents how willing they would be to live in the different sorts of neighborhoods. These surveys show, quite consistently, that the first choice of most African Americans is a neighborhood with about an equal mix of black and white households. The first choice of whites, on the other hand, is a neighborhood with a large white majority. Among all racial and ethnic groups, African Americans are the most disfavored "other" with regard to preferences for neighborhood racial and ethnic composition. Survey research also shows that whites are more hesitant to move into hypothetical neighborhoods with large African-American populations, even if those communities are described as having good schools, low crime rates, and other amenities. However, they are much less hesitant about moving into areas with significant Latino, Asian, or other minority populations.

Why whites prefer homogeneous neighborhoods is the subject of some debate. According to some research, many whites automatically assume that neighborhoods with many blacks have poor schools, much crime, and few stores; these whites are not necessarily responding to the presence of blacks per se. Black neighborhoods are simply assumed to be "bad neighborhoods" and are avoided as a result. Other research indicates that "poor schools" and "crime" are sometimes code words for racial prejudice and excuses that whites use to avoid African Americans.

These preferences promote segregation. Recent research in several cities, including Atlanta, Detroit, and Los Angeles, shows that whites who prefer predominantly white neighborhoods tend to live in such neighborhoods, clearly implying that if white preferences would change, integration would increase. Such attitudes also imply tolerance, if not encouragement, of discriminatory practices on the part of real estate agents, mortgage lenders, property insurers, and other providers of housing services.

HOUSING DISCRIMINATION: HOW COMMON IS IT TODAY?

When the insurance agent quoted at the beginning of this article was asked by one of his supervisors whether the kids in the neighborhood played hockey or basketball, he was not denying a home insurance policy to a particular black family because of race. However, he was trying to learn about the racial composition of the neighborhood in order to help market his policies. The mental map he was drawing is just as effective in discriminating as the maps commonly used in the past that literally had red lines marking neighborhoods—typically minority or poor—considered ineligible for home insurance or mortgage loans.

Researchers with HUD, the Urban Institute, and dozens of nonprofit fair housing organizations have long used "paired testing" to measure the pervasiveness of housing discrimination—and more recently in mortgage lending and home insurance. In a paired test, two people visit or contact a real estate, rental, home-finance, or insurance office. Testers provide agents with identical housing preferences and relevant financial data (income, savings, credit history). The only difference between the testers is their race or ethnicity. The testers make identical applications and report back on the responses they get.... Discrimination can take several forms: having to wait longer than whites for a meeting; being told about fewer units or otherwise being given less information; being steered to neighborhoods where residents are disproportionately of the applicant's race or ethnicity; facing higher deposit or down-payment requirements and other costs;

or simply being told that a unit, loan, or policy is not available, when it is available to the white tester.

In 1989 and 2000, HUD and the Urban Institute, a research organization, conducted nationwide paired testing of discrimination in housing. They found generally less discrimination against African Americans and Hispanics in 2000 than in 1989, except for Hispanic renters (see figure 1). Nevertheless, discrimination still occurred during 17 to 26 percent of the occasions when African Americans and Hispanics visited a rental office or real-estate agent. (The researchers found similar levels of discrimination against Asians and Native Americans in 2000; these groups were not studied in 1989.)

In 2000, subtler forms of discrimination, such as invidious comments by real estate agents, remained widespread. Even when whites and nonwhites were shown houses in the same areas, agents often steered white homeseekers to segregated neighborhoods with remarks such as "Black people do live around here, but it has not gotten bad yet;" "That area is full of Hispanics and blacks that don't know how to keep clean;" or "(This area) is very mixed. You probably wouldn't like it because of the income you and your husband make. I don't want to sound prejudiced."

Given the potential sanctions available under current law, including six- and seven-figure compensatory and punitive damage awards for victims, it seems

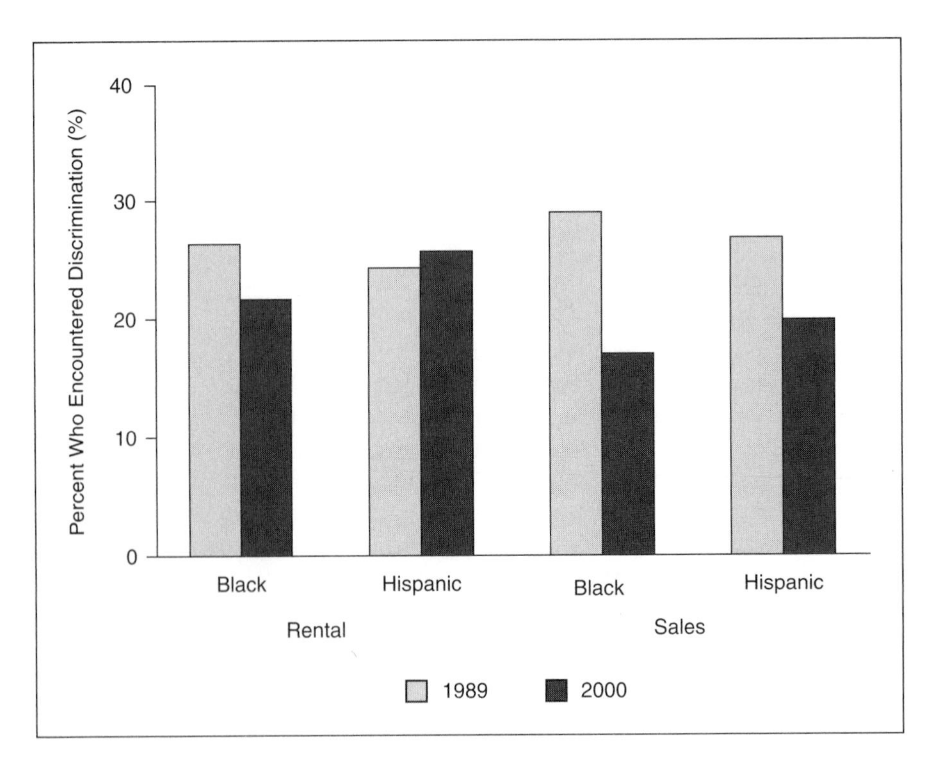

FIGURE 1 Percent of auditors who encountered discrimination, 1989 and 2000.

SOURCE: U.S. Department of Housing and Urban Development, 2000 Housing Discrimination Study.

surprising that an agent would choose to make such comments. However, research shows that most Americans are unfamiliar with fair housing rules, and even those who are familiar and believe they have experienced racial discrimination rarely take legal action because they do not believe anything would come of it. Most real estate professionals do comply with fair housing laws, but those who work in small neighborhoods and rely on word of mouth to get clients often fear losing business if they allow minorities into a neighborhood where local residents would not welcome them. In a 2004 study of a St. Louis suburb, a rental agent pointed out that there were no "dark" people in the neighborhood to a white tester. She said that she had had to lie to a black homeseeker and say that a unit was unavailable because she would have been "run out of" the suburb had she rented to a black family.

Discrimination does not end with the housing search. Case studies of mortgage lending and property insurance practices have also revealed discriminatory treatment against minorities. White borrowers are offered more choice in loan products, higher loan amounts, and more advice than minority borrowers. The Boston Federal Reserve Bank found that even among equally qualified borrowers in its region applications from African Americans were 60 percent more likely to be rejected than those submitted by whites. Other paired-testing studies from around the country conclude that whites are more likely to be offered home insurance policies, offered lower prices and more coverage, and given more assistance than African Americans or Hispanics.

THE CONTINUING COSTS OF SEGREGATION

Beyond constricting their freedom of choice, segregation deprives minority families of access to quality schools, jobs, health care, public services, and private amenities such as restaurants, theatres, and quality retail stores. Residential segregation also undercuts families' efforts to accumulate wealth through the appreciation of real estate values by restricting their ability both to purchase their own homes and to sell their homes to the largest and wealthiest group in the population, non-Hispanic whites. Just 46 percent of African Americans owned their own homes in 2000, compared to 72 percent of non-Hispanic whites. In addition, recent research found that the average value of single-family homes in predominantly white neighborhoods in the 100 largest metropolitan areas with significant minority populations was $196,000 compared to $184,000 in integrated communities and $104,000 in predominantly minority communities. As a result of the differing home values and appreciation, the typical white homeowner has $58,000 in home equity compared to $18,000 for the typical black homeowner. Segregation has broader effects on the quality of neighborhoods to which minorities can gain access. In 2000, the average white household with an income above $60,000 had neighbors in that same income bracket. But black and Hispanic households with incomes above $60,000 had neighbors with an average income of under $50,000. In effect, they lived in poorer neighborhoods, and this gap has widened since 1990.

Segregation restricts access to jobs and to quality schools by concentrating African Americans and Hispanics in central cities, when job growth and better schools are found in the suburbs. Amy Stuart Wells and Robert Crain found, for example, that black children living in St. Louis who attend schools in the suburbs are more likely to graduate and to go on to college than those attending city schools. Yet only busing makes it possible for these students to attend suburban schools, and America has largely turned away from this remedy to segregation. According to research by Gary Orfield at the Harvard Civil Rights Project, our nation's schools are as segregated today as they were 35 years ago. Most job growth also occurs in suburban areas, and difficulty in finding and commuting to those jobs contributes to high unemployment rates among African Americans and Latinos.

The risks of illness and injury from infectious diseases and environmental hazards are also greater in minority neighborhoods, while resources to deal with them are less available than in mostly white areas. For example, in Bethesda, Md., a wealthy and predominantly white suburb of Washington, D.C., there is one pediatrician for every 400 residents compared to one for every 3,700 residents in Washington's predominantly minority and poor southeast neighborhoods. As John Logan has argued, "The housing market and discrimination sort people into different neighborhoods, which in turn shape residents' lives—and deaths. Bluntly put, some neighborhoods are likely to kill you."...

Finally, segregation helps perpetuate prejudice, stereotypes, and racial tension. Several recent studies show that neighborhood-level contact between whites and African Americans reduces prejudice and increases acceptance of diversity. Yet with today's levels of housing segregation, few whites and blacks get the opportunity for such contact. More diverse communities generally exhibit greater tolerance and a richer lifestyle—culturally and economically—for all residents.

A GROWING MOVEMENT

In 1968, the U.S. Supreme Court ruled that racial discrimination in housing was illegal, characterizing it as "a relic of slavery." In the same year, Congress passed the Fair Housing Act, providing specific penalties for housing discrimination along with mechanisms for addressing individual complaints of discrimination. These legal developments laid the groundwork for a growing social movement against segregation that has brought limited but real gains. Members of the National Fair Housing Alliance, a consortium of 80 nonprofit fair housing organizations in 30 cities and the District of Columbia, have secured more than $190 million for victims of housing discrimination since 1990 by using the Federal Fair Housing Act and equivalent state and local laws. In addition, they have negotiated legal settlements that have transformed the marketing and underwriting activities of the nation's largest property insurance companies, including State Farm, Allstate, Nationwide, American Family, and Liberty. The key investigative technique members of the alliance have used to secure these victories is paired testing.

Community reinvestment groups have secured more than $1.7 trillion in new mortgage and small business loans for traditionally underserved low- and moderate-income neighborhoods and minority markets since the passage of the Community Reinvestment Act (CRA). The CRA was passed in order to prevent lenders from refusing to make loans, or making loans more difficult to get, in older urban communities, neighborhoods where racial minorities are often concentrated. Under the CRA, third parties (usually community-based organizations) can formally challenge lender applications or requests by lenders to make changes in their business operations. Regulators who are authorized to approve lender applications have, in some cases, required the lender to respond to the concerns raised by the challenging party prior to approving the request. In some cases, just the threat of making such challenges has provided leverage for community organizations in their efforts to negotiate reinvestment agreements with lenders. Community groups have used this process to generate billions of new dollars for lending in low-income and minority markets. Sometimes, in anticipation of such a challenge, lenders negotiate a reinvestment program in advance. For example, shortly after Bank One and JP Morgan Chase announced their intent to merge in 2004, the lenders entered into an agreement with the Chicago Reinvestment Alliance, a coalition of Chicago-area neighborhood organizations. The banks agreed to invest an average of $80 million in community development loans for each of the next six years. Research by the Joint Center for Housing Studies at Harvard University indicates that mortgage loans became far more accessible in low-income and minority neighborhoods during the 1990s and that the CRA directly contributed to this outcome.

Many housing researchers and fair-housing advocates have criticized fair-housing enforcement authorities for relying too heavily on individual complaints and lawsuits to attack what are deeper structural problems. Currently, most testing and enforcement occurs when individuals lodge a complaint against a business rather than as a strategic effort to target large companies that regularly practice discrimination. Reinvestment agreements recently negotiated by community groups and lenders illustrate one more systemic approach. More testing aimed at detecting what are referred to as "patterns and practices" of discrimination by large developers and rental management companies would also be helpful. Such an undertaking, however, would require more resources, which are currently unavailable. Despite the limits of current enforcement efforts, most observers credit these efforts with helping to reduce segregation and discrimination.

Resistance to fair housing and integration efforts persists. For example, lenders and their trade associations continually attempt to weaken the CRA and related fair-housing rules. Yet fair-housing and community-reinvestment groups like the National Fair Housing Alliance and the National Community Reinvestment Coalition have successfully blocked most such efforts in Congress and among bank regulators. As more groups refine their ability to employ legal tools like the CRA and to litigate complex cases under the jurisdiction of the Fair Housing Act, we can expect further progress. The struggle for fair housing is a difficult one, but with the available tools, the progress we have made since 1970 toward becoming a more integrated society should continue.

DISCUSSION QUESTIONS

1. Look at the neighborhood where you grew up. Would you describe it as racially segregated or racially integrated? What effect did this have on the kinds of relationships you formed with people as you grew up?
2. What if someone calculated the index of dissimilarity on your campus? In a general sense, what figure would result? Were you in charge of campus life, what might you do if you think this needs to change?

44

The Color Line in American Education

Race, Resources, and Student Achievement

LINDA DARLING-HAMMOND

Educational reforms that may sound appealing to some members of the public have different consequences in schools that are mostly attended by poor and minority students. Darling-Hammond reviews some of the issues being faced in education at a time when the No Child Left Behind Act is creating new challenges, especially for minority schools.

W. E. B. Du Bois argued strenuously over many decades for investment in the education of African American students. Moreover, he outlined the critical importance of investing in a kind of education that would provide much more than minimal skills—the kind of education that would enable students to think critically and take control of the course of their own learning. His prediction that the central issue of the twentieth century would be the problem of the color line follows us into the twenty-first, especially with regard to education. The color line divides us still. In recent years, one visible piece of evidence of this in the public policy arena has been the persistent attack on affirmation action in higher education and employment. A mere fifty years after *Brown v. Board of Education* many Americans who believe the vestiges of discrimination have disappeared take the position that affirmative action now provides an unfair advantage to minorities. From the perspective of others who daily experience the consequences of ongoing discrimination, affirmative action is needed to protect opportunities likely to evaporate if an affirmative obligation to act fairly does not exist. And for Americans of all backgrounds, the allocation of opportunity in a society that is becoming ever more dependent on knowledge and education is a source of great anxiety and concern, exacerbating debates about who is entitled to high quality educational opportunities at every level of schooling.

Interpretations of the gaps in educational achievement between White and non-Asian minority students as measured by standardized test scores are at the center of these debates. The presumption that guides much of the conversation is that equal educational opportunity now exists; therefore, continued low levels

SOURCE: From *Du Bois Review*, 2004. "The Color Line in American Education," 12: 213–246. Reprinted by permission of Cambridge University Press and the author.

of achievement on the part of minority students must be a function of genes, culture, or a lack of effort and will....

The assumptions that undergird this debate miss an important reality: educational outcomes for students of color are much more a function of their unequal access to key educational resources, including skilled teachers and quality curriculum, than they are a function of race. In fact, the United States educational system is one of the most unequal in the industrialized world, and students routinely receive dramatically different learning opportunities based on their social status. In contrast to European and Asian nations that fund schools centrally and equally, the wealthiest 10% of school districts in the United States spend nearly ten times more than the poorest 10%, and spending ratios of three to one are common within states (Educational Testing Services (ETS) 1991; Kozol, 1991). These disparities reinforce the wide inequalities in income among families, with the most resources being spent on children from the wealthiest communities and the fewest on the children of the poor, especially in high-minority communities.

Nonetheless, despite stark differences in funding, teacher quality, curriculum, support services, and class sizes, the prevailing view is that if students do not achieve it is their own fault. Until these inequalities are confronted and addressed, we will never get beyond the problem of the color line....

Disparities in resources are largely a function of how public education in the United States is funded. In most cases, education costs are supported by a system of general taxes—primarily local property taxes, along with state grants-in-aid. Because these funds are typically raised and spent locally, districts with higher property values have greater resources with which to fund their schools, even when poorer districts tax themselves at proportionally higher rates. In Texas, for instance, the 100 wealthiest districts taxed their local property at an average rate of 47 cents per $100.00 of assessed worth in 1989; at that level of effort, they were able to spend over $7000 per student. Meanwhile, the 100 poorest districts, taxing themselves at a rate of over seventy cents per $100.00, were able to raise only enough to spend some $3000 per student (Kozol, 1991, p. 225).

These disparities translate into real differences in the services provided in schools: higher spending districts have smaller classes, higher paid and more experienced teachers, and greater instructional resources, as well as better facilities, more up-to-date equipment, and a wider range of course offerings. Districts serving large proportions of poor children generally have the fewest resources. Thus, those students least likely to encounter a wide array of educational resources at home are also least likely to encounter them at school (ETS, 1991)....

Not only do funding systems and other policies create a situation in which urban districts receive fewer resources than their suburban neighbors, but schools with high concentrations of "minority" students receive fewer resources than other schools within these districts. And tracking systems exacerbate these inequalities by segregating many "minority" students within schools, allocating still fewer educational opportunities to them at the classroom level. As I describe below, these compounded inequalities explain much of the achievement gap that many have attributed to everything from genetic differences in intelligence to childrearing to a "culture of poverty."

Serious policy attention to these ongoing, systemic inequalities is critical for improving educational outcomes. If Americans do not recognize that students experience very different educational realities, policies will continue to be based on the presumption that it is students, not their schools or classroom circumstances, that are the sources of unequal educational attainment.

...Such profound inequalities in resource allocations are supported by the increasing re-segregation of schools over the decades of the 1980s and 1990s. In 1998–1999, 70% of the nation's Black students attended predominantly minority schools, up significantly from the low point of 63% in 1980. The proportion of students of color in intensely segregated schools also increased. More than a third of African American and Latino students (36.5% and 36.6% respectively) attended schools with a minority enrollment of 90–100%. Furthermore, for all groups except Whites, racially segregated schools are almost always schools with high concentrations of poverty (Orfield, 2001).

African American and Hispanic American students continue to be concentrated in central city public schools, many of which have become majority "minority" over the past decade while their funding has fallen further behind that of their suburbs. As of 1997, students of color comprised more than 55% of those served by school districts of more than 15,000 students (National Center for Education Statistics, 2000, p. 99). Due to federal and state disinvestment in city schools that has worsened since 1980, these central city schools are typically funded at levels substantially below those of neighboring suburban districts serving fewer students with special needs. The continuing segregation of neighborhoods and communities intersects with funding formulas and school administration practices that create substantial differences in the educational resources made available in different communities. Together, these conditions produce ongoing inequalities in educational opportunity by race and ethnicity.

UNEQUAL ACCESS TO QUALIFIED TEACHERS

In many cities, increasing numbers of unqualified teachers have been hired since the late 1980s, when teacher demand began to increase while resources were declining. In 1990, for example, the Los Angeles City School District was sued by students in predominantly minority schools because their schools were not only overcrowded and less well funded than other schools, but also disproportionately staffed by inexperienced and unprepared teachers hired on emergency credentials. Unequal assignment of teachers creates ongoing differentials in expenditures and access to educational resources, including curriculum offerings requiring specialized expertise and the knowledge well-prepared teachers rely on in offering high-quality instruction....

The disparities in access to well-qualified teachers are large and growing worse. In 2001, for example, students in California's most segregated minority schools were more than five times as likely to have uncertified teachers than those in predominantly White schools (Shields et al., 2001). Similar inequalities have been documented in lawsuits challenging school funding in Massachusetts, South Carolina, New York, and Texas, among other states. By every measure of qualifications—state certification, content background for teaching, pedagogical training, selectivity of college attended, test scores, or experience—less qualified teachers are found disproportionately in schools serving greater numbers of low-income or minority students (NCES, 1997; Lankford et al., 2002).

Oakes' (1990) nationwide study of the distribution of mathematics and science opportunities confirmed these pervasive patterns. Based on teacher experience, certification status, preparation in the discipline, degrees, self-confidence, and teacher and principal perceptions of competence, low-income and minority students had less contact with the best-qualified science and mathematics teachers. Students in high-minority schools had less than a 50% chance of being taught by a math or science teacher holding a degree and a license in the field they taught....

These disparities are most troubling given recent evidence about the influence of teacher quality on student achievement. A number of studies have found that teachers who lack preparation in either subject matter or teaching methods are significantly less effective in producing student learning gains than those who have a full program of teacher education and who are fully certified (see, e.g. Goldhaber & Brewer, 2000)....

Whether students have access to well-qualified teachers can be a critical determinant of whether they can be successful on the state tests often required for promotion from grade to grade, for placement into more academically challenging classes, and for graduation from high school....

Unequal access to well-qualified teachers, a major side effect of unequal expenditures, appears to be one of the most critical factors in the underachievement of African American and Latino/a students. In some districts, like Compton and Ravenswood, California, which serve exclusively students of color, the proportions of uncertified teachers have exceeded 50 percent, leaving most students without any access to teachers who know how to teach them to read, who have learned about up-to-date teaching methods or about how children grow, learn, and develop, and who have a repertoire of teaching skills to use if they are having difficulties (Futernick, 2001).

Furthermore, recruits who are not prepared for teaching are much more likely to leave teaching quickly (Henke et al., 2000; National Commission on Teaching and America's Future, 2003), many staying only a year or less. This adds additional problems of staff instability to the already difficult circumstances in which central city youth attend school. Where these hiring practices dominate, many children are taught by a parade of short-term substitute teachers, inexperienced teachers without support, and underqualified teachers who know neither their subject matter nor effective teaching methods well.

In addition, when faced with shortages, districts often assign teachers outside their fields of qualification, expand class sizes, or cancel course offerings. These strategies are used most frequently in schools serving large numbers of minority students (NCES, 1997; NCTAF, 1997). No matter what strategies are adopted, the quality of instruction suffers. This sets up the school failure that society predicts for low-income and minority children—a failure created by lack of effective measures to deal with the issues of teacher supply and quality.

ACCESS TO HIGH QUALITY CURRICULUM

In addition to being taught by less qualified teachers than their White counterparts, students of color face dramatic differences in courses, curriculum materials, and equipment. Unequal access to high-level courses and challenging curriculum explains another substantial component of the difference in achievement between minority students and White students. Analyses of data from the "High School and Beyond" surveys demonstrate dramatic differences among students of various racial and ethnic groups in course taking in such areas as mathematics, science, and foreign language (Pelavin & Kane, 1990). These data also demonstrate that, for students of all racial and ethnic groups, course taking is strongly related to achievement for students with similar course taking records, achievement test score differences by race or ethnicity narrow substantially (College Board 1985, p. 38; Jones, 1984; Jones et al., 1984; Moore & Smith, 1985)....

One source of inequality is the fact that high-minority schools are much less likely to offer advanced and college preparatory courses in mathematics and science than are schools that serve affluent and largely White populations of students (Matthews, 1984; Oakes, 1990). Schools serving predominantly minority and poor populations offer fewer advanced and more remedial courses in academic subjects, and they have smaller academic tracks and larger vocational programs (NCES, 1985; Oakes, 2003; Rock et al., 1985)....

When high-minority, low-income schools offer any advanced or college preparatory courses, they offer them to only a very tiny fraction of students. Thus, at the high school level, African Americans, Hispanics, and American Indians have traditionally been underrepresented in academic programs and overrepresented in general education or vocational education programs, where they receive fewer courses in areas such as English, mathematics, and science (Oakes, 1990). Even among the college bound, non-Asian minority students take fewer and less demanding mathematics, science, and foreign language courses (Pelavin & Kane, 1990).

The unavailability of teachers who could teach these upper level courses, or who can successfully teach heterogeneous groups of students, reinforces these inequalities in access to high quality curriculum. Tracking persists in the face of growing evidence that it does not substantially benefit high achievers and tends to put low achievers at a serious disadvantage (Oakes, 1985; 1986; Hoffer, 1992; Kulik and Kulik, 1982; Slavin, 1990), in part because good teaching is a scarce resource, and thus must be allocated. Scarce resources tend to get allocated to the students whose parents, advocates, or representatives have the most political clout. This typically results in the most highly qualified teachers teaching the most enriched curricula to the most advantaged students....

Tracking in U.S. schools starts much earlier and is much more extensive than in most other countries where sorting does not occur until high school. Starting in elementary schools with the designation of instructional groups and programs based on test scores and recommendations, it becomes highly formalized by junior high school. From "gifted and talented" programs at the elementary level through advanced courses in secondary schools, teachers who are generally the most skilled offer rich, challenging curricula to select groups of students, on the theory that only a few students can benefit from such curricula. Yet the distinguishing feature of such programs, particularly at the elementary level, is not their difficulty, but their quality. Students in these programs are given opportunities to integrate ideas across fields of study. They have opportunities to think, write, create, and develop projects. They are challenged to explore. Though virtually all students would benefit from being taught in this way, their opportunities remain acutely restricted. The result of this practice is that challenging curricula are rationed to a very small proportion of students, and far fewer U.S. students ever encounter the kinds of curriculum students in other countries typically experience (McKnight et al., 1987; Usiskin 1987; Useem 1990; Wheelock 1992)....

NEW STANDARDS AND OLD INEQUALITIES

While these inequalities in educational opportunity continue—and have actually grown worse in many states over the last two decades—the increasing importance of education to individual and societal well-being has spawned an education reform movement in the United States focused on the development of new standards for students. Virtually all states have created new standards for graduation, new curriculum frameworks to guide instruction, and new assessments to test students' knowledge. Many have put in place high-stakes testing systems that attach rewards and sanctions to students' scores on standardized tests. These include grade retention or promotion as well as graduation for students, merit pay awards or threats of dismissal for teachers and administrators, and extra funds or loss of registration, reconstitution, or loss of funds for schools. The recently enacted federal *No Child Left Behind Act* reinforces these systems, requiring all schools receiving funding to test students annually, and enforcing penalties for those that do not meet specific test score targets both for students as a whole and for subgroups defined by race/ethnicity, language, socioeconomic status, and disability.

The rhetoric of "standards-based" reforms is appealing. Students cannot succeed in meeting the demands of the new economy if they do not encounter much more challenging work in school, many argue, and schools cannot be stimulated to improve unless the real accomplishments—or deficits—of their students are raised to public attention. These arguments certainly have some merit. But standards and tests alone will not improve schools or create educational opportunities where they do not now exist. The implications of standards-based reform for students who have not received an adequate education are suggested by long-standing data from Texas where more than a decade of high-stakes testing has contributed to four-year graduation rates for African American and Hispanic students that hover near 50% (Haney 2000), as well as more recent data from Massachusetts, which began to implement high-stakes testing in the late 1990s. As Massachusetts's accountability system was phased in, there was a 300% increase in middle school dropouts between 1997–1998 and 1999–2000 while greater proportions of students began disappearing from schools in ninth and tenth grades, most of them African American and Latino/a....

These patterns raise the possibility that pressures to increase school average test scores may create incentives to hold students back or encourage them to leave school. Recent studies have found that systems that reward or sanction schools based on average student scores create incentives for pushing low-scorers into special education so that their scores will not count in school reports (Allington and McGill-Franzen, 1992; Figlio and Getzler, 2002), retaining students in grade so that their grade-level scores will look better (Jacob, 2002; Haney, 2000)—a practice that increases later dropout rates, excluding low-scoring students from admissions (Darling-Hammond, 1991; Smith et al., 1986), and encouraging such students to leave schools or drop out (Haney 2000; Orfield and Ashkinaze, 1991; Smith et al., 1986)....

Reform rhetoric notwithstanding, the key question for students, especially those of color, is whether investments in better teaching, curriculum, and schooling will follow the press for new standards, or whether standards built upon a foundation of continued inequality in education will simply certify student failure with greater certainty and reduce access to future education and employment. A related question, a half-century after *Brown v. Board of Education*, is what it will take to secure a constitutional right to equal educational opportunity for all children in the United States.

The advent of high-stakes testing reforms requiring students to achieve specific test score targets in order to advance in grade or graduate from school has occurred while educational experiences for "minority" students continue to be substantially separate and unequal. State efforts to set standards for all students for school progression and graduation while failing to offer equal opportunities to learn have stimulated a new spate of equity litigation in nearly twenty states across the country. These lawsuits—which may be said to constitute the next generation of efforts begun by *Brown v. Board of Education*—argue that if states require all students to meet the same educational standards, they must assume a responsibility to provide resources adequate to allow students a reasonable opportunity to achieve those standards, including well-qualified teachers, a curriculum that fully reflects the standards, and the materials, texts, supplies, and equipment needed to teach the curriculum. Though this would seem like an obvious proposition, in most states the suits have been defended by legal arguments that money "does not make a difference" in educational outcomes....

The federal government can play a leadership role in providing an adequate supply of well-qualified teachers just as it has in providing an adequate supply of well-qualified physicians for the nation. When shortages of physicians were a major problem more than twenty-five years ago, Congress passed the 1963

Health Professions Education Assistance Act to support and improve the caliber of medical training, to create and strengthen teaching hospitals, to provide scholarships and loans to medical students, and to create incentives for physicians to train in shortage specialties and to locate in underserved areas. Similarly, federal initiatives in education should seek to:

1. ***Recruit new teachers***, especially in shortage fields and in shortage locations, through scholarships and forgivable loans for high quality teacher education.
2. ***Strengthen and improve teachers' preparation*** through improvement incentive grants to schools of education and supports for certification reform.
3. ***Improve teacher retention and effectiveness*** by improving clinical training and support during the beginning teaching stage when 30% of teachers drop out. This would include funding internship programs for new teachers in which they receive structured coaching and mentoring, preferably in urban schools that are supported to provide state-of-the-art practice. (For a more complete discussion of the federal role in addressing teacher shortages, see Darling-Hammond & Sykes, 2003.)

If the interaction between teachers and students is the most important aspect of effective schooling, then reducing inequality in learning has to rely on policies that provide equal access to competent, well-supported teachers. The public education system ought to be able to guarantee that every child who is forced to go to school by public law is taught by someone who is prepared, knowledgeable, competent, and caring.

REFERENCES

Allington, R. and A. McGill-Franzen (1992). Unintended effects of educational reform in New York. *Educational Policy*, 6: 397–414.

College Board (1985). *Equality and Excellence: The Educational Status of Black Americans.* New York: College Entrance Examination Board.

Darling-Hammond, L. (1991). The implications of testing policy for quality and equality, *Phi Delta Kappan*, November 1991: 220–225.

Darling-Hammond, L. and G. Sykes, (2003). Wanted: A national teacher supply policy for education: The right way to meet the 'highly qualified teacher challenge.' *Educational Policy Analysis Archives*, 11(3). http://epaa.asu.edu/epaa/v11n33/.

Educational Testing Service (1991). *The State of Inequality*. Princeton, NJ: E.T.S.

Figlio, D. N. and L. S. Getzler (2002, April). *Accountability, Ability, and Disability: Gaming the System?* Cambridge, MA: National Bureau of Economic Research.

Futernick, K. (2001). *A district-by-district analysis of the distribution of teachers in California and an overview of the Teacher Qualification Index (TQI)*. Sacramento: California State University, Sacramento, April, 2000.

Gamoran, A. (1992). Access to excellence: Assignment to honors English classes in the transition from middle to high school. *Educational Evaluation and Policy Analysis*, 14: 185–204.

Goldhaber, D. D. and D. J. Brewer, (2000). Does teacher certification matter? High school certification status and student achievement. *Educational Evaluation and Policy Analysis*, 22: 129–145.

Haney, W. (2000). The myth of the Texas miracle in education. *Educational Policy Analysis Archives*, (41): http://epaa.asu.edu/epaa/v8n41/

Henke, R. R., X. Chen, S. Geis, and P. Knepper. (2000). *Progress Through the Teacher Pipeline: 1992–93 College Graduates and Elementary/Secondary School Teaching as of 1997*. NCES 2000–152. Washington, DC: National Center for Education Statistics.

Hoffer, T. B. (1992). Middle school ability grouping and student achievement in science and mathematics. *Educational Evaluation and Policy Analysis*, 14(3): 205–227.

Jacob, B. A. (2002). The impact of high-stakes testing on student achievement: Evidence from Chicago. Working Paper. Harvard University.

Jones, L. V. (1984). White-Black achievement differences: The narrowing gap. *American Psychologist*, 39: 1207–1213.

Jones, L. V., N. W. Burton, and E. C. Davenport (1984). Monitoring the achievement of Black students, *Journal for Research in Mathematics Education*, 15: 154–164.

Kozol, J. (1991). *Savage Inequalities*. New York: Crown.

Krueger, A. B. (2000). Economic Considerations and Class Size, Paper 447. Princeton University, Industrial Relations Section. www.irs.princeton.edu/pubs/working _papers.html.

Kulik, C. C. and J. A. Kulik (1982). Effects of ability grouping on secondary school students: A meta-analysis of evaluation findings. *American Education Research Journal*, 19: 415–428.

Lankford, H., S. Loeb, and J. Wyckoff (2002). Teacher sorting and the plight of urban schools: A descriptive analysis. *Education Evaluation and Policy Analysis*, 24: 37–62.

Matthews, W. (1984). Influences on the learning and participation of minorities in mathematics. *Journal for Research in Mathematics Education*, 15: 84–95.

McKnight, C. C., J. A. CrossWhite, J. A. Dossey, E. Kifer, S. O. Swafford, K. J. Travers, and T. J. Cooney, (1987). *The Underachieving Curriculum: Assessing U.S. School Mathematics from an International Perspective*. Champaign, IL: Stipes Publishing.

Moore, E. G. and A. W. Smith (1985). Mathematics aptitude: Effects of coursework, household language, and ethnic differences. *Urban Education*, 20: 273–294.

National Center for Education Statistics (NCES) (1997). *America's Teachers: Profile of a Profession, 1993–94*. Washington, DC: U.S. Department of Education.

National Center for Education Statistics (NCES) (2000). *Digest of Education Statistics, 1999*. Washington, DC: U.S. Department of Education.

National Commission on Teaching and America's Future (NCTAF) (1997). Unpublished tabulations from the 1993–94 Schools and Staffing Surveys.

Oakes, J. (1985). *Keeping Track*. New Haven, CT: Yale University Press.

Oakes, J. (1990). *Multiplying Inequalities: The Effects of Race, Social Class, and Tracking on Opportunities to Learn Mathematics and Science*. Santa Monica: The RAND Corporation.

Orfield, G. (2001). *Schools More Separate: Consequences of a Decade of Resegregation*, Cambridge, MA, The Civil Rights Project, Harvard University.

Orfield, G. and C. Ashkinaze (1991). *The Closing Door: Conservative Policy and Black Opportunity*. Chicago: University of Chicago Press, p. 139.

Pelavin, S. H. and M. Kane (1990). *Changing the Odds: Factors Increasing Access to College*. New York: College Entrance Examination Board.

Rock, D. A., T. L. Hilton, J. Pollack, R. B. Ekstrom, and M. E. Goertz (1985). *A Study of Excellence in High School Education: Educational Policies, School Quality, and Student Outcomes.* Washington, DC: National Center for Education Statistics.

Shields, P. M., D. C. Humphrey, M. E.Wechsler, L. M. Riel, J. Tiffany-Morales, K. Woodworth, V. M. Young, and T. Price (2001). *The Status of the Teaching Profession 2001.* Santa Cruz, CA: The Center for the Future of Teaching and Learning.

Slavin, R. E. (1990). Achievement effects of ability grouping in secondary schools: A best evidence synthesis. *Review of Educational Research,* 60: 471–500.

Smith, F., et al. (1986). *High School Admission and the Improvement of Schooling.* New York: New York City Board of Education.

Useem, E. L. (1990, Fall). You're good, but you're not good enough: Tracking students out of advanced mathematics. *American Educator,* 14: 24–27, 43–46.

Usiskin, Z. (1987). Why elementary algebra can, should, and must be an eighth grade course for average students. *Mathematics Teacher,* 80: 428–438.

Wheelock, A. (1992). *Crossing the Tracks.* New York: The New Press.

DISCUSSION QUESTIONS

1. Does Darling-Hammond's description of schools apply to the school you attended prior to college? Why or why not? What does this tell you about education, race, and inequality?
2. What does Darling-Hammond argue has been the impact of current educational reforms? Have you seen evidence of this in your own schooling?

45

Shame of the Nation

JONATHAN KOZOL

Data show that public schools are becoming increasingly segregated by race. Here, Jonathan Kozol discusses the devastating and long-term impact that segregation has on the lives of Black children, as well as the negative consequences for Black-White relations.

I spent, in all, about a decade working with black schoolchildren in Boston, ... first in the public schools themselves (I also taught for two years in an integrated public school in Newton, a suburban town that pioneered a voluntary integration program that continues to the present day), then in grassroots programs taking place in storefronts and black churches. By the later 1970s and then increasingly during the early 1980s, I began to visit schools in other sections of the nation. I was writing books during this time, and principals and teachers who were reading them would frequently invite me to their schools. Teachers would sometimes let me teach a lesson to their class and, in the upper elementary grades (fourth and fifth, the grades that I had taught), would sometimes organize a class discussion in which I would be invited to take part.

There was a more optimistic mood in many of the urban schools I visited during these years. Some had been desegregated by court order in the aftermath of the Supreme Court's ruling in *Brown v. Board of Education*; other schools had been desegregated voluntarily. Physical conditions in these newly integrated schools were generally more cheerful, the state of mind among the teachers and the children more high-spirited, the atmosphere less desultory, more enlivening, than in the school in which I'd started out in Boston. Other schools I visited, admittedly, were not so fortunate. Disrepair and overcrowding were familiar still in many districts; and most of the black children I was meeting lived in very poor communities where residential segregation was a permanent reality. As serious cuts in social services and federal assistance for low-income housing took effect during the years when Ronald Reagan was in office, physical conditions in these neighborhoods became appreciably worse.

By the end of the 1980s, the high hopes that I had briefly sensed a decade earlier were hard to find. Many of the schools I visited during this period seemed every bit as grim as those I'd seen in Boston in the 1960s, sometimes a good deal

SOURCE: From Jonathan Kozol, *The Shame of the Nation: The Restoration of Apartheid Schooling in America.* copyright © 2005 by Jonathan Kozol. Used by permission of Crown Publishers, a division of Random House, Inc.

worse. I visited a high school in East St. Louis, Illinois, where the lab stations in the science rooms had empty holes where pipes were once attached. A history teacher who befriended me told me of rooms that were so cold in winter that the students had to wear their coats to class while kids in other classes sweltered in a suffocating heat that could not be turned down. A foul odor filled much of the building because of an overflow of sewage that had forced the city to shut down the school the year before.

I visited, too, the bleak, unhappy schools of Paterson and Camden in New Jersey and similar schools in Washington, D.C., Chicago, San Antonio, and Cincinnati. Back in New England, I spent time with teachers, parents, and their teenage kids in Bridgeport, where the poverty levels, overcrowded public schools, and health conditions of the children, many of whom had been lead-poisoned in the city's public housing, had created a sense of quiet desperation in the all-black and Hispanic neighborhoods I visited. Wherever I could, I started in the schools. Where I could not, I started in the streets. Everywhere I went, I did my best to spend my evenings with schoolteachers.

At the start of the 1990s I began to visit schools in New York City, where I'd come to know a group of children living in a section of the South Bronx called Mott Haven, in which I would end up spending long-extended periods of time in the next 15 years. Yet even in this period in which I grew entangled in my friendships with these children and their parents and with priests and ministers and doctors in the neighborhood (HIV infection ripped its way across the South Bronx in those years, and pediatric and maternal AIDS were added to the routine sorrows many children had to bear), I continued to spend time in schools in other cities too.

And almost everywhere I went from this point on, no matter what the hopes that had been stirred in many cities only a short time before, no matter what the progress that had frequently been made in districts where court-ordered integration programs had been in effect or where a civic leadership had found the moral will to act without court orders in a principled attempt to integrate their neighborhoods and schools, a clear reality was now in place: Virtually all the children of black and Hispanic people in the cities that I visited, both large and small, were now attending schools in which their isolation was as absolute as it had been for children in the school in which I'd started out so many years before....

You go into these deeply segregated schools and do your best—in order to enjoy the kids you meet and to appreciate what's taking place within the classroom here and now—to disconnect the present from the past. Try as I do, however, as the years go by, I find that act of disconnection very, very hard. I walk into a class of 25 or 30 students and I look around me at the faces of the children, some of whom in New York City I have known since they were born, and look into their eyes, and often see them also searching into mine, and I cannot discern the slightest hint that any vestige of the legal victory embodied in *Brown v. Board of Education* or the moral mandate that a generation of unselfish activists and young idealists lived and sometimes died for has survived within these schools and neighborhoods. I simply never see white children.

"We owe a definite homage to the reality around us," Thomas Merton wrote, "and we are obliged, at certain times, to say what things are and to give them their right names." No matter how complex the reasons that have brought us to the point at which we stand, we have, it seems, been traveling a long way to a place of ultimate surrender that does not look very different from the place where some of us began.

There are those, of course, who see no reason to regret this pattern of reversion to an older order of accepted isolation of the children of minorities and even find it possible to ridicule the notion that apartheid schooling might have any damaging effects upon a child's intellectual development or any other aspect of a child's heart or mind. The supposition that "black students suffer an unspecified psychological harm from segregation...," as Justice Clarence Thomas wrote in an opinion on a case the high court heard in 1995, is not merely incorrect, relying upon "questionable social science research," in his words, but also represents a form of prejudice, reflecting "an assumption of black inferiority." This is not dissimilar to the idea, expressed sometimes by white conservatives as well, that arguments for racial integration of our schools insultingly imply that children of minorities will somehow "become smarter" if they're sitting with white children—an idea, which is indeed insulting, that no advocate for integrated education I have known has ever entertained. But arguments like these and the debates surrounding them, in any case, have only the most indirect connection with the thoughts that come into my mind when I am sitting with a group of children in a kindergarten class in the South Bronx.

What saddens me the most during these times is simply that these children have no knowledge of the other world in which I've lived most of my life and that the children in that other world have not the slightest notion as to who these children are and will not likely *ever* know them later on, not at least on anything like equal terms, unless a couple of these kids get into college. Even if they meet each other then, it may not be the same, because the sweetness of too many of these inner-city children will have been somewhat corroded by that time. Some of it may be replaced by hardness, some by caution, some by calculation rooted in unspoken fear. I have believed for 40 years, and still believe today, that we would be an infinitely better nation if they knew each other now.

...If we have agreed to live with this reality essentially unaltered for another of the several generations yet to come, I think we need at least to have the honesty to say so. I also think we need to recognize that our acceptance of a dual education system will have consequences that may be no less destructive than those we have seen in the past century.

I don't think you can discern these consequences solely by examination of statistics or the words of education analysts or highly placed officials in school systems. I think you need to go into the schools in which the isolation of our children is the most extreme, do so repeatedly but, where it's possible, informally and not obtrusively, and try to make sure that you are allowed the time to listen carefully to children. I have been criticized throughout the course of my career for placing too much faith in the reliability of children's narratives; but I have almost always found that children are a great deal more reliable in telling us what actually goes on in public school than many of the adult experts who develop policies that shape their destinies. Unlike these powerful grown-ups, children have no ideologies to reinforce, no superstructure of political opinion to promote, no civic equanimity or image to defend, no personal reputation to secure. They may err sometimes about the minuscule particulars but on the big things children rarely have much reason to mislead us. They are, in this respect, pure witnesses, and we will hear their testimony in these pages.

...One of the consequences of their isolation ... is that they have little knowledge of the ordinary reference points that are familiar to most children in the world.... In talking with adolescents, for example, who were doing relatively well in school and said they hoped to go to college, I have sometimes mentioned colleges such as Columbia, Manhattanville, Cornell, or New York University, and found that references like these were virtually unknown to them. The state university system of New York was generally beyond their recognition too. The name of a community college in the Bronx might be familiar to them—or, for the boys, perhaps a college that was known for its athletic teams.

Now and then, in an effort to expand their reference points, the pastor takes a group of children to an inter-racial gathering that may be sponsored by one of the more progressive churches in New York or to a similar gathering held in New England, for example. I have accompanied the St. Ann's children on a couple of these trips. The travel involved is usually fun, and simply getting outside the neighborhood in which they live is an adventure for most of the children in itself. But the younger children tend to hold back from attempting to make friends with the white children whom they meet, and many of the teenage kids behave with a defensive edginess, even a hint of mockery, not of the white kids themselves but of a situation that seems slightly artificial and contrived to them and is also, as they surely recognize, a one-time shot that will not change the lives they lead when they return to the South Bronx.

It might be very different if these kids had known white children early in their lives, not only on unusual occasions but in all the ordinary ways that children come to know each other when they go to school together and play games with one another and share secrets with each other and grow bonded to each other by those thousands of small pieces of perplexity and fantasy and sorrow and frivolity of which a child's daily life is actually made. I don't think that you change these things substantially by organizing staged events like "Inter-racial Days." Even the talks that certain of the children are selected to deliver on these rare occasions often have a rather wooden sound, like pieties that have been carefully rehearsed, no matter how sincere the children are. Not that it's not worth holding such events. They energize politically the adults who are present and sometimes, although frankly not too often, long-term friendships may be made. But token days are not the ebb and flow of life. They ease our feelings of regret about the way things have to be for the remainder of the year. They do not really change the way things are.

Many Americans I meet who live far from our major cities and who have no first-hand knowledge of realities in urban public schools seem to have a rather vague and general impression that the great extremes of racial isolation they recall as matters of grave national significance some 35 or 40 years ago have gradually, but steadily, diminished in more recent years. The truth, unhappily, is that the trend, for well over a decade now, has been precisely the reverse. Schools that were already deeply segregated 25 or 30 years ago, like most of the schools I visit in the Bronx, are no less segregated now, while thousands of other schools that had been integrated either voluntarily or by the force of law have since been rapidly resegregating both in northern districts and in broad expanses of the South.

"At the beginning of the twenty-first century," according to Professor Gary Orfield and his colleagues at the Civil Rights Project at Harvard University, "American public schools are now 12 years into the process of continuous resegregation. The desegregation of black students, which increased continuously from the 1950s to the late 1980s, has now receded to levels not seen in three decades.... During the 1990s, the proportion of black students in majority white schools has decreased ... to a level lower than in any year since 1968.... Almost three fourths of black and Latino students attend schools that are predominantly minority," and more than two million, including more than a quarter of black students in the Northeast and Midwest, "attend schools which we call apartheid schools" in which 99 to 100 percent of students are nonwhite. The four most segregated states for black students, according to the Civil Rights Project, are New York, Michigan, Illinois, and California. In California and New York, only one black student in seven goes to a predominantly white school.

During the past 25 years, the Harvard study notes, "there has been no significant leadership towards the goal of creating a successfully integrated society built on integrated schools and neighborhoods." The last constructive act by Congress was the 1972 enactment of a federal program to provide financial aid to districts undertaking efforts at desegregation, which, however, was "repealed by the Reagan administration in 1981." The Supreme Court "began limiting desegregation in key ways in 1974"—and actively dismantling existing integration programs in 1991....

Racial isolation and the concentrated poverty of children in a public school go hand in hand, moreover, as the Harvard project notes. Only 15 percent of the intensely segregated white schools in the nation have student populations in which more than half are poor enough to be receiving free meals or reduced price meals. "By contrast, a staggering 86 percent of intensely segregated black and Latino schools" have student enrollments in which more than half are poor by the same standards. A segregated inner-city school is "almost six times as likely" to be a school of concentrated poverty as is a school that has an overwhelmingly white population.

"So deep is our resistance to acknowledging what is taking place," Professor Orfield notes, that when a district that has been desegregated in preceding decades now abandons integrated education, "the actual word 'segregation' hardly ever comes up. Proposals for racially separate schools are usually promoted as new educational improvement plans or efforts to increase parental involvement.... In the new era of 'separate but equal,' segregation has somehow come to be viewed as a type of school reform"—"something progressive and new," he writes—rather than as what it is: an unconceded throwback to the status quo of 1954. But no matter by what new name segregated education may be known, whether it be "neighborhood schools, community schools, targeted schools, priority schools," or whatever other currently accepted term, "segregation is not new... and neither is the idea of making separate schools equal. It is one of the oldest and extensively tried ideas in U.S. educational history" and one, writes Orfield, that has "never had a systematic effect in a century of trials."

Perhaps most damaging to any effort to address this subject openly is the refusal of most of the major arbiters of culture in our northern cities to confront or even clearly name an obvious reality they would have castigated with a passionate determination in another section of the nation 50 years before and which, moreover, they still castigate today in retrospective writings that assign it to a comfortably distant and allegedly concluded era of the past. There is, indeed, a seemingly agreed-upon convention in much of the media today not even to use an accurate descriptor such as "racial segregation" in a narrative description of a segregated school. Linguistic sweeteners, semantic somersaults, and surrogate vocabularies are repeatedly employed. Schools in which as few as three or four percent of students may be white or Southeast Asian or of Middle Eastern origin, for instance—and where *every other child* in the building is black or Hispanic—are referred to, in a commonly misleading usage, as "diverse." Visitors to schools like these discover quickly the eviscerated meaning of the word, which is no longer a descriptor but a euphemism for a plainer word that has apparently become unspeakable.

School systems themselves repeatedly employ this euphemism in descriptions of the composition of their student populations. In a school I visited in fall 2004 in Kansas City, Missouri, for example, a document distributed to visitors reports that the school's curriculum "addresses the needs of children from diverse backgrounds." But as I went from class to class I did not encounter any children who were white or Asian—or Hispanic, for that matter—and when I later was provided with the demographics of the school, I learned that 99.6 percent of students there were African-American. In a similar document, the school board of another district, this one in New York State, referred to "the diversity" of its student population and "the rich variations of ethnic backgrounds...." But when I looked at the racial numbers that the district had reported to the state, I learned that there were 2,800 black and Hispanic children in the system, one Asian child, and three whites. Words, in these cases, cease to have real meaning; or, rather, they mean the opposite of what they say.

One of the most disheartening experiences for those who grew up in the years when Martin Luther King and Thurgood Marshall were alive is to visit public schools today that bear their names, or names of other honored leaders of the integration struggles that produced the temporary progress that took place in the three decades after *Brown*, and to find how many of these schools are bastions of contemporary segregation. It is even more disheartening when schools like these are not in segregated neighborhoods but in racially mixed areas in which the integration of a public school would seem to be most natural and where, indeed, it takes a conscious effort on the part of parents or of school officials in these districts to *avoid* the integration option that is often right at their front door.

In a Seattle neighborhood, for instance, where approximately half the families were Caucasian, 95 percent of students at the Thurgood Marshall Elementary School were black, Hispanic, Native American, or of Asian origin. An African-American teacher at the school told me of seeing clusters of white parents and their children on the corner of a street close to the school each morning waiting for a bus that took the children to a school in which she believed that the enrollment was predominantly white. She did not speak of the white families waiting for the bus to take their children to another public school with bitterness, but wistfully.

"At Thurgood Marshall," according to a big wall-poster in the lobby of the school, "the dream is alive." But school assignment practices and federal court decisions that have countermanded long-established policies that previously fostered integration in Seattle's schools make the realization of the dream identified with Justice Marshall all but unattainable today.

"Thurgood Marshall must be turning over in his grave," one of the teachers at the school had told the principal, as he reported this to me....

In the course of two visits to the school, I had a chance to talk with a number of teachers and to spend time in their classrooms. In one class, a teacher had posted a brief summation of the *Brown* decision on the wall; but it was in an inconspicuous corner of the room and, with that one exception, I could find no references to Marshall's struggle against racial segregation in the building.

When I asked a group of fifth grade boys who Thurgood Marshall was and what he did to have deserved to have a school named after him, most of the boys had no idea at all. One said that he used to run "a summer camp." Another said he was "a manager"—I had no chance to ask him what he meant by this, or how he'd gotten this impression. Of the three who knew that he had been a lawyer, only one, and only after several questions on my part, replied that he had "tried to change what was unfair"—and, after a moment's hesitation, "wanted to let black kids go to the same schools that white kids did." He said he was "pretty sure" that this school was not segregated because, in one of the other classrooms on the same floor, there were two white children.

There is a bit of painful humor that I've heard from black schoolteachers who grew up during the era of the integration movement and have subsequently seen its goals abandoned and its early victories reversed. "If you want to see a *really* segregated school in the United States today, start by looking for a school that's named for Martin Luther King or Rosa Parks."...

Many educators make the argument today that, given the demographics of large cities like New York and their suburban areas, our only realistic goal should be the nurturing of strong, empowered, and well-funded schools in segregated neighborhoods—an argument with which, in any given and specific local situation, it would seem impossible to disagree. Even if we have to doubt the likelihood that genuine empowerment and anything approaching full equality will ever be achieved on a broad scale, or long sustained, in the dynamics of the dual system as it stands, one also feels compelled to hope these reservations will be proven wrong and, therefore, to do everything we can to reinforce the efforts of the principals and teachers who devote their lives to working in these schools.

Black school officials in these situations have sometimes conveyed to me a bitter and clear-sighted recognition that they're being asked, essentially, to mediate and render functional an uncontested separation between children of their race and children of white people living sometimes in a distant section of their town and sometimes in almost their own immediate communities. Implicit in this mediation is a willingness to set aside the promises of *Brown* and, perhaps while never stating this or even thinking of it clearly in these terms, to settle for the promise made more than a century ago in *Plessy v. Ferguson*, the 1896 Supreme Court ruling in which "separate but equal" was accepted as a tolerable rationale for the perpetuation of a dual system in American society.

Equality itself—equality alone—is now, it seems, the article of faith to which increasing numbers of the principals of inner-city public schools subscribe. And some who are perhaps most realistic do not even ask for, or expect, complete equality, which seems beyond the realm of probability for many years to come, but look instead for only a sufficiency of means—"adequacy" is the legal term most often used today—by which to win those practical and finite victories that may appear to be within their reach. Higher standards, higher expectations, are insistently demanded of these urban principals, and of their teachers and the students in their schools, but far lower standards certainly in ethical respects appear to be expected of the dominant society that isolates these children in unequal institutions.

At an early-morning assembly at Seattle's Thurgood Marshall School, the entire student body stood and chanted, "I have confidence that I can learn!" exactly 30 times. Similar sessions of self-exhortation are familiar at innumerable inner-city schools: "Yes, I can! I know I can!" "If it is to be, it's up to me." In some schools, these chantings are accompanied by rhythmic clapping of the hands or snapping of the fingers or by stamping on the floor. It usually seems like an invigorating way to start the day. At the same time, politically conservative white people visiting these schools often seem to be almost too gratified to hear black and Hispanic children speaking in these terms. If it's up to "them," the message seems to be, it isn't up to "us," which appears to sweep the deck of many pressing and potentially disruptive and expensive obligations we may otherwise believe our nation needs to contemplate.

And, in plain honesty, when we invite these children to repeat in unison that "if it is to be, it's up to me," we are asking them to say something which, while they have no way of knowing this, is simply not the truth. It is, indeed, an odd thing, when one thinks of it, to ask a six- or seven-year-old child to believe. Does a school board or school system have no role in what this child is to be? Do taxpayers have no role in this? Do Congress and the courts and local legislators have no role in setting up the possibilities of what is "to be," or not to be, within these children's opportunities to learn? Why are the debates about state distribution of resources for our schools so heated, and the opposition to a fairer distribution on the part of wealthy districts so intense, if citizens do not believe that fiscal policies enacted by the government have a decisive role in the determination of the destinies of children?

One of the reasons for these incantations in the schools that serve black and Hispanic children is what is believed to be the children's loss of willingness "to try," their failure to believe they have the same abilities as do white children in more privileged communities. It is this attribution of a loss of faith in their potential and, as an adaptive consequence, a seeming "will to fail"—a psychological pathology—that justifies the hortatory slogans they are asked to chant and the multitude of posters, loaded with ambitious verbs such as "succeed," "attain," "achieve," that are found on classroom walls and sometimes even painted on the outside of a school.

...It is notable in this respect that, in all the many writings and proposals dedicated to the alteration of self-image among inner-city youth and the reversal of debilitating pressures from their peers, the suggestion is virtually never made that one of the most direct ways to reduce the damage done to children by peer pressure is to change the *make-up* of their peers by letting them go to schools where all their classmates are not black and brown and poor, and children and grandchildren of the poor, but where a healthy confidence that one can learn is rooted in the natural assumptions of Americans who haven't been laid waste by history.

When I was standing with the children at the Thurgood Marshall School and counting the number of times they chanted the word "confidence," I remember looking at the faces of the boys who stood the closest to me in the gym in which the morning chants were taking place and wondering what impact this was having on them inwardly. When you're among a group of children, you inevitably want to hope that rituals like these might really do some good, that they may make a difference that will last beyond the hour of exhilaration and hand-clapping. Still, these exercises are place-markers. They tell us we are in a world where hope must be constructed therapeutically because so much of it has been destroyed by the conditions of internment in which we have placed these children. It is harder to convince young people they "can learn" when they are cordoned off by a society that isn't sure they really can. That is, I am afraid, one of the most destructive and long-lasting messages a nation possibly could give its children.

DISCUSSION QUESTIONS

1. What does Kozol mean by describing U.S. schools as a system of apartheid?
2. What are the consequences of the inequality of race and education in the lives of young Black children?

46

The Significance of Race and Gender in School Success among Latinas and Latinos in College

HEIDI LASLEY BARAJAS AND JENNIFER L. PIERCE

Based on a study of Latina high school and college students, including women and men, Barajas and Pierce examine the different paths to success that Latinas and Latinos experience in schools. They find that the men more often succeed through their experience in sports and with significant mentors; Latinas, on the other hand, find support through their own social networks and relationships with other Latinas.

Assimilation in American society has long been a central concern of sociologists (Glazer and Moynihan 1963; Gordon 1964; Park 1950; Rumbaut and Portes 1990). In Robert Park's original and influential formulation, the process of assimilation or the acceptance of the dominant culture's norms and values comes about through an immigrant group's contact with a new culture. This concept is not only central to research on recent immigrants but to studies in the sociology of education where it is considered key to understanding the success or failure of students from racial ethnic minority and white working-class backgrounds. Students who succeed,...do so because they have assimilated to the dominant norms and values such as individualism, while those who fail do not. Thus, the path to student success is paved through the process of assimilation to an individualistic and meritocractic understanding of the social world.

Several assumptions inform this understanding of student success. First, success is predicated on assimilation. If students do not conform to the mainstream culture, they will fail. Such an assumption precludes other possible definitions of success, such as students who may be successful academically but are still strongly tied to a culture and an identity that is not white, Anglo-Saxon, Protestant, and individualistic....

This article provides an empirical and theoretical challenge to the logic of the conventional assimilationist argument by looking at the success of Latino students in college in a midwestern region of the United States. Currently, the high

SOURCE: From *Gender & Society* 15(6): 859–878. Copyright © 2001 Sociologists for Women in Society. Reprinted by permission of Sage Publications, Inc.

school drop-out rate of young Latinos nationwide is 46 percent (McMillen 1995). While the literature in the sociology of education suggests that students of color must adopt white middle-class behaviors to succeed, our research demonstrates that Latino students construct paths through the terrain of discrimination and prejudice they encounter in schools in much more complex and varied ways.... We specifically selected a group of *successful* Latino students, a group that has rarely been studied, because we were interested in addressing theoretical questions that this particular student population could help us answer.

We will demonstrate that their paths to success did not follow the typical assimilationist trajectory predicted by the literature. Furthermore, there are *gendered* patterns through which these students construct paths to success in college. Young Latinas in this study navigate successfully through and around negative stereotypes of Hispanics by maintaining positive definitions of themselves and by exphasizing their group membership as Latinas. Furthermore, their positive self-definition is reinforced through supportive relationships with other Latinas earlier in high school and now in college. On the other hand, young Latino men who also see themselves as part of a larger cultural group tend to have less positive racial and ethnic identities than the women. Typically, they are supported by mentors, such as white athletic coaches, and tend to see themselves as having "worked hard," thus they draw from the meritocractic ethos of sports and regard their success in more individualistic terms. While successful Latinas do not assimilate in the ways predicted by the literature, the young men in this study accept the individualistic and meritocractic ethos of the dominant culture, but not without a psychological price....

Data were collected by the first author during a two-year period from 1996 to 1998 through a mentor program called "The Bridge" at a large U.S. research university that we call Midwestern University. Latino college students volunteered to participate in the program and mentored Latinos in local area high schools. All 45 college student mentors and 27 high school student mentees who participated in the program were interviewed. Among the college students, 31 were young women and 14 were young men. Their ages ranged from 18 to 25. Among the high school students, 11 were women and 16 were men. Students who participated in the study came from various Hispanic backgrounds, primarily Mexican, Puerto Rican, and Honduran. The majority were from second- or third-generation immigrant and poor or working-class families....

LATINAS: SUCCEEDING THROUGH RELATIONSHIPS WITH OTHERS

When asked why they wanted to mentor to Latino kids, the young women in this study were prompted to speak candidly about their experiences in the larger social world and how these experiences informed their school experiences and their desire to become mentors. More than two-thirds of the Latinas said they enrolled in the mentor program because they had a strong desire to help someone like themselves. For example, Emilia, a 23-year-old university senior from Latin America describes herself as having lived two lives: one as a poor daughter of a single mom in Latin America and another as the privileged step-daughter of a white father in the United States. Emilia grew up in Latin America and came to the United States after her mother married an American working for the government. . . . She recounted the following story:

> My stepfather works for the government. When I was around high school age, he was transferred to an office [in another country]. He went first, and my mother, me, and my brother followed a short time later. We were at the airport in New York waiting to get on the plane. . . . Well, the man at the counter called for all family members of these government officials to begin boarding. My mom and my brother and I went to the door. But the man at the counter stopped us and told us this was boarding for special people and that we needed to wait. My mom tried to explain that we were family members, but he just wouldn't listen [she begins to cry]. I just remember him being so rude. He just assumed that because we are brown, because we weren't white, that we could not be family members of a government official....

Emilia's early experiences with discrimination prompted her desire to work with other Latinos so that they could learn that "brown people are successful" too. Moreover, like the other mentors, Emilia found the program to be a safe space for her. She enjoyed being part of a group where positive meanings were attached to brownness, she liked working with other Latinos, and she liked teaching others how to navigate the treacherous waters of a college that was unwelcoming to its students of color.

For Jennifer, a 25-year-old student of Mexican heritage and a senior in college, the mentor experience produced a heightened awareness about her own community....

> I have always lived in West Town. I have always lived around mostly Latinos and I never thought about it. I know what some people think about Mexicanos, but I never let it bother me. Then I started working in the elementary school with the teachers. I didn't know how much need there is out there. I mean, I never saw how little my community has—like resources, opportunities. And other people like them, the kids need to see people like them who are educated, who are going college. These kids are smart, they just don't have what other kids have. Going to college has really opened my eyes as to what other people have....

Although Jennifer is aware of her difference from white students, she describes herself as a Latina in very positive ways throughout her interview—"I know what some people think ... but I never let it bother me." Her conscious understanding of what being different meant appears to have changed when Jennifer worked in her own community. She was taught by her family and chose to think positively about her Mexicano background, and she did not use her difference as a way to explain her own difficulties in getting through school. However, after attending college, she became aware of the privileges people who were not from West Town enjoyed and became attuned to the lack of resources that were available in her own community....

The majority of mentors had similar reports about the importance of positive relationships with other Latinas in their lives. Marta, a 20-year-old Chicana and a college junior, thought the most important contribution of mentoring is the fact that it is relational, particularly because Latino backgrounds are so varied....

Marta believes that having a Latino mentor for Latino kids helps them to see themselves in positive ways, but this is only important to a point because each student is different. Her mentee is from Mexico, and she herself is from the southwestern region of the United States.... Despite her recognition of differences, Marta believes the most important part of mentoring high school Latinos is to help them understand why they are seen as different in school and to establish a real relationship with them....

Through her work with K-12 Latinos and her own experiences, she recognized that going to school at all levels is a family choice for Latinos, rather than an individual one. For Marta, social class and gender play an important part in how Latinos "think about themselves, and too, how other [white] people think about you." What frustrates her, however, is how little school personnel know about the dynamics of many Latino families, particularly poor families....

Several other Latina mentors expressed their disappointment with school authorities who do not understand the fact that Latino families make decisions about education for different reasons than white families do. Many said that school authorities consider going to school a taken-for-granted decision, failing to realize that for many poor and migrant Latino families, one child going to school may be a financial sacrifice for the entire family. High school and college attendance require money for clothes, school materials, lunch money, and transportation. Paying tuition or living expenses at college is rarely a possibility.

Like Jennifer and Marta, Gina emphasizes the importance of positive relationships to survive being considered different in school. A high-achieving college senior, Gina talked freely about being raised by her single mother on the West Coast and living with her extended family: her grandmother, her aunt, and her aunt's daughter....

She knows that being recognized as other opens the possibility of "thinking of myself negatively." However, she doesn't allow others to racialize her in negative ways. When asked how she handles the way she is seen as different by school authorities and mainstream peer culture, she says,

> I also feel like a misfit in [this Midwestern state]. I mentioned that it is obvious I am different because, well, I had someone ask, "What kind of food do you eat?" "Excuse me? The same kind of food you eat." I understand what they are getting at, but it is kind of insulting sometimes. People don't mean to be harmful, though sometimes they do and sometimes they just ask me questions because they are curious. I say it is not appropriate. I don't know. I think it has made me think about not having

> a day that I see my mom and grandma struggle. My mom and grandma are really strong women and so, being a woman, yeah, that affects me. We are doing fine....

Gina maintains that she chooses how to behave rather than allowing others to define her behavior to fit their assumptions. She does not allow herself "to feel oppressed by it." Futhermore, she emphasizes her "positive strong identity," something she hopes to convey to the mentee she works with.

These vignettes demonstrate how young Latinas maintain positive self-definitions and self-valuations in the face of racial discrimination, prejudice, and pejorative stereotypes. As Hill Collins (1990, 140–44) pointed out, when Black women have a safe space, they are able to create such definitions for one another. For these Latinas, safe spaces are created in relationships with friends, family, and community including association with other successful Latino students in spaces such as Latino organizations. These relationships with cultural translators become spaces in which Latinas learn positive meanings and valuations that counter the negative significations operating in schools. In addition, Latinas create new relationships as mentors and in the mentor program because they share what they have learned about being successful Latinas, and they add to their own positive self-understandings by acting as role models.

LATINOS: SUCCEEDING THROUGH ATHLETICS

Like Latinas, young Latino men talked about being made to feel different at school and refrained from talking directly about race or labeling school experiences as acts of prejudice, discrimination, or racism. They also discussed their desire to mentor and to help others like themselves. Unlike their Latina counterparts, however, these young men tended to talk about themselves in very singular ways, as individuals who worked very hard. Their focus was on ways they, as individuals, were able to change their attitudes about school and achieve school success because of support from a coach who was typically a white male....

Given that individualism and meritocracy are central American cultural ideals (Bellah et al. 1985), it is not surprising that Latinos held fast to these ideas. All students are socialized to accept the notion that the character and desire of the individual determines their destiny and that everyone will be rewarded for the hard work they perform. For young Latinos, these ideas were further encouraged through their participation in school athletics, and because most of these students were successful athletes, the notion that they were successful because they worked hard was strongly reinforced.

Ricky, an 18-year-old college freshman, was typical of the majority of young Latinos in this study. In high school, he experienced isolation from others like himself. He comments:

> I was not the type to have really good friends that I hang out with, that I call, things like that. I just had friends. People that I talk to. They were

> not really my type. I just don't like to get all personal, on a personal level with people because sometimes, I don't know, I just feel … I feel that sometimes you just find more differences and things you don't agree with that person....

The marginalization Ricky experienced in high school was common to almost all of the Latinos in this study. Few had friends who were Latino, and fewer still had close friends among other students of color or among white students in high school or in college....

When asked why he was so motivated to succeed in school, he said that his junior year of high school, he started working out of family necessity. That year he turned "away from school," but got back "on track" through sports. His senior year, he was recruited by a suburban high school to wrestle. Although he continued to work part-time, the coaches, acting as mentors, helped Ricky focus on both wrestling and school. At the same time, Ricky was greatly influenced by his new peers at school.

> I saw the success other people were having … how they kept going in 10th grade, 11th grade. And then I saw myself, and I was like wow, I dropped out of the race.... Most of my influence comes from the economic status that we are at, and like the way our lives are, and I just don't want to be like that when I grow up. I want to get out of school, get a job, buy a house, buy a car, you know, pay for all my things. Just live a normal life, and I know that a lot of the minority students are in the same situation....

Ricky attributes his success to his own initiative in taking advantage of the opportunities offered through sports. Furthermore, sports reinforces the idea that school success is based on merit and that these advantages are open to everyone equally. Consequently, he believes that any problems must lie with the individual or the individual's family background....

However, when talking about his mentor experience in the elementary school in the neighborhood where he grew up, Ricky contradicts himself:

> I think that a big part that [school] plays a role in shaping their [mentees'] character—because I was sitting there at school and I was looking around the walls, looking at pictures, and just the way the school was built. The resources that they had, classrooms, the desk, computers, it's like amazing. It's not fair. It's not equal. And it's all in the other school [where he had transferred]. It's amazing the difference, those kids have amazing resources compared to these kids. Over here, you basically have a teacher—and like the teacher has to purchase teaching aids herself. I felt bad just because there is such a difference there. And they are the ones that need most of the help....

The opportunity to return to a K-12 institution as a college student changed Ricky's perspective. As he reflects more on his opportunities, he discusses the ways they were made available to him. For instance, he thinks that one of the reasons he was able to go to a different and better school is because his mother drove him there every day. He was also able to participate in school activities because the white coach and athletic director made sure that he obtained financial waivers. "Everyone was making school and everything more convenient for me. They wanted me to succeed, also they made it easier. They helped me out and I took advantage [of the opportunity]."...

Brian, a 20-year-old college junior, and Reuben, a 22-year-old senior, both equate learning to succeed in school with their participation in sports. The difference between their experiences and Ricky's is that for Brian and Reuben, sports in their high school years was only one of many opportunities they had. As a swimmer, Reuben learned to compete and developed confidence about his abilities. Although he continued to swim in college, he did not hesitate to give it up when his swimming schedule interfered with his course work and extracurricular activities. He says,

> Swimming in high school, and even in college, was important. But what really made a difference is that my mom always taught me to try different things. Giving up swimming was a decision, but it wasn't like giving it up left me with nothing. I just moved on to the next thing—which is traveling and writing. My Mom, and my Dad too in a different way, encouraged me to try whatever I wanted. I guess what I am trying to say is that success is one thing, but having the experience is the important thing....

There are obvious social class differences in the lives of Ricky and Reuben, and their discussions about school and education reflect these differences. Reuben comes from a middle-class background with professional parents who both have an extensive education. Furthermore, Reuben's father is white and his mother is Latina.... Ricky, on the other hand, comes from a working-class, single-parent home where a high school diploma is considered a great accomplishment.

For Ricky, participation in sports was the opportunity to succeed. Had he not been exceptional at his sport, the opportunity would not have been there. On the other hand, participating in sports for Reuben was one among many choices for success. Had Reuben failed to excel in sports, he would not have been viewed as a failure by his parents, and this one failed opportunity would not have denied him success.

CONCLUSION

...Despite the negative stereotypes they faced, successful Latinas found ways to carve out safe spaces through their relationships with other Latinas and to maintain a positive sense of racial ethnic identity. Consequently, their success in school did not entail giving up their ethnic identity. On the other hand, as men, Latinos experienced certain opportunities and advantages through sports that most of the young women did not. Specifically, sports provided them with a valuable mentor such as a coach who encouraged them to do well in sports and academics. In addition, competition through sports supported and reinforced the notion that they alone were responsible for their success. At the same time, however, these young men often paid a psychological price for their conformity to these norms. The majority had strongly ambivalent feelings about their racial ethnic identities, and although they often associated with other Latinos on campus, they had less social support and shared understanding for being "different."

The gendered differences we have highlighted speak to the significance of race and gender as categories of analysis that operate together to produce divergent experiences for young Latinas and Latinos. While both Latino women and men faced racial prejudice, discrimination, and exclusion throughout their school years, young women were able to insulate themselves through supportive relationships with other Latinas in high school and in college, while young men were able to transcend some of these obstacles through participation in sports. Early in their schooling, Latinas sought out and found cultural translators who aided them in becoming bicultural, while Latinos found models from the dominant group who encouraged mainstream success but did not help them learn how to navigate between dominant and minority group cultures. These gendered strategies for success suggest that relationships and connection to others are more important to these young women and girls as Gilligan (1982) and others have argued. On the other hand, athletic ability is more highly valued and encouraged for boys in American culture than for girls regardless of race or ethnicity. Hence, participation in athletics becomes a vehicle for success for these racial ethnic minority boys, but not for girls.

Significantly, however, in contrast to studies that suggest that women's focus on relationships inhibits competitive achievement, our findings demonstrate how Latinas used relationship as a path to success.... Latinas experience a chilly climate in classrooms both as women and as members of a racial ethnic minority. However, rather than succumb to the pressures of this gendered and raced dynamic, they seek out protective relationships, support, and encouragement where they can achieve a positive sense of racial ethnic identity that they carry with them from high school to college. As members of a racial ethnic minority, young Latinos also encounter a chilly classroom, but as men they are encouraged to participate in sports, which becomes a springboard to success. However, once these young men enter college, the gendered advantages promised by sports diminish and race begins to take on more significance in their lives. Because they lacked cultural translators, they had not developed strong positive Latino identities in high school and found themselves at once confused and ambivalent about their racial identity, about other Latinos, and about the general fate of members from their own racial ethnic minority group. In this way, our analysis highlights how the privileges of masculinity promised through sport did not shield them from the psychological injuries and disadvantages shaped by race.

REFERENCES

Bellah, Robert, Richard Madsen, William Sullivan. Ann Swidler, and Steven Tipton. 1985. *Habits of the heart: Individualism and commitment in American life.* Berkeley and Los Angeles: University of California Press.

Chase, Susan. 1995. *Ambiguous empowerment: The work narratives of women school superintendents.* Amherst: University of Massachusetts Press.

Gilligan, Carol. 1982. *In a different voice: Psychological theory and women's development.* Cambridge, MA: Harvard University Press.

Glazer, Nathan, and Daniel P. Moynihan. 1963. *Beyond the melting pot: The Negroes, Puerto Ricans, Jews, Italians, and the Irish of New York City.* Cambridge: MIT Press.

Gordon, Milton. 1964. *Assimilation in American life: The role of race, religion and national origins.* New York: Oxford University Press.

Hill Collins, Patricia 1990. *Black feminist thought.* New York: Routledge.

McMillen, Mary. 1995. *National Center for Educational Statistics: Drop-out report.* Washington, DC: Government Printing Office.

Park, Robert.1950. *Race and culture.* Glencoe, IL: Free Press.

Rumbaut, Rubén, and Alexandro Portes. 1990. *Immigrant America.* Berkeley: University of California Press.

DISCUSSION QUESTIONS

1. How does assimilation theory explain educational success?
2. Why do the Latinas interviewed by Barajas and Pierce offer to mentor other students of color?
3. How do sports help young Latinos to achieve educational success?
4. Why are young Latinos more likely to embrace the ideology of assimilation than young Latinas?

47

Keeping Them in Their Place

The Social Control of Blacks Since the 1960s

EDUARDO BONILLA-SILVA

Bonilla-Silva argues that state policies toward African Americans are based on a model of social control. This control is manifested in the criminal justice system through differential arrest rates of African Americans, high rates of police brutality in African American communities, and the much greater imposition of the death penalty in cases involving African Americans.

All domination is, in the last instance, maintained through social control strategies. For example, during slavery whites used the whip, overseers, night patrols, and other highly repressive practices along with some paternalistic ones to keep blacks "in their place." After slavery was abolished, whites felt threatened by free blacks, hence very strict written and unwritten rules of racial contact (the Jim Crow laws) were developed to specify "the place" of blacks in the new environment of "freedom." And, as insurance, lynching and other terroristic forms of social control were used to guarantee white supremacy. In contrast, as Jim Crow practices subsided, the control of blacks is today chiefly attained through state agencies (e.g., the police, the criminal court system, and the FBI). Manning Marable describes the new system of control:

> The informal, vigilante-inspired techniques to suppress Blacks were no longer practical. Therefore, beginning with the Great Depression, and especially after 1945, white racists began to rely almost exclusively on the state apparatus to carry out the battle for white supremacy. Blacks charged with crimes would receive longer sentences than whites convicted of similar crimes. The police forces of municipal and metropolitan areas received a carte blanche in their daily acts of brutality against Blacks. The Federal and state government carefully monitored Blacks who advocated any kind of social change. Most important, capital punishment was used as a weapon against Blacks charged and convicted of major crimes. The criminal justice system, in short, became a modern

SOURCE: From Eduardo Bonilla-Silva, *White Supremacy and Racism in the Post-Civil Rights Era* (Boulder, CO: Lynne Reinner Publishers), pp. 103–111. Copyright © 2001 by Lynne Reinner Publishers, Inc. Reprinted by permission of the publisher.

instrument to perpetuate white hegemony. Extralegal lynchings were replaced by "legal lynchings" and capital punishment.[1]

In the following sections, I review data on social control to see how well they fit Marable's interpretation of post-civil rights dynamics.

THE STATE AS ENFORCER OF RACIAL ORDER

Data on arrest show that the contrast between black and white arrest rates since 1950 has been striking. The black rate increased throughout this period reaching almost 100 per 1,000 by 1978 compared to 35 per 1,000 for whites. In terms of how many blacks are incarcerated, we found a pattern similar to their arrest rates. Although blacks have always been overrepresented in the inmate population, this overrepresentation has skyrocketed since the late 1940s. In 1950, blacks were 29 percent of the prison population. Ten years later, their proportion reached 38 percent. By 1980, blacks made up 47 percent of the incarcerated population, six times that of whites. Today the incarceration rate of blacks has "stabilized" to constitute around 50 percent of the prison population.[2]

This dramatic increase in black incarceration has been attributed to legislative changes in the penal codes and the "get tough" attitude in law enforcement fueled by white fear of black crime. Furthermore, the fact that blacks are disproportionately convicted and receive longer sentences than whites for similar crimes contributes to their overrepresentation in the penal population.... This disparity in sentencing, in conjunction with the complex ways in which race works out in the criminal justice system, may explain why, although blacks made up 31 percent of those arrested in 1995, their incarceration rate was close to 50 percent. In comparison, whites constituted 67 percent of those arrested but had an incarceration rate of 50 percent.[3]

OFFICIAL STATE BRUTALITY AGAINST BLACKS

Police departments grew exponentially after the 1960s, particularly in large metropolitan areas with large concentrations of blacks.... Despite attempts in the 1970s and 1980s to reduce the friction between black communities and police departments by hiring more black police officers and, in some cases, even hiring black chiefs of police, "there has been little change in the attitudes of blacks toward the police, especially when the attitudes of black respondents are compared to those of white respondents."[4] A 1996 report by the Joint Center for Political Economic Studies confirmed this trend: 43 percent of blacks polled believed that police brutality and harassment of blacks were a serious problem where they lived. These numbers double when the black population polled resides in urban areas.

The level of police force used with blacks has always been excessive. However, since the police became the primary agent of social control of blacks, the level of violence against them has skyrocketed. For example, in 1975, 46 percent of all the people killed by the police in official action were black. That situation has not changed much since. Robert C. Smith reported recently that of the people killed by the police, over half are black; the police usually claim that when they killed blacks it was "accidental" because they thought that the victim was armed, although in fact the victims were unarmed in 75 percent of the cases; there was an increase in the 1980s in the use of deadly force by the police and the only ameliorating factor was the presence of a sensitive mayor in a city; and in the aftermath of the Rodney King verdict, 87 percent of civilian victims of police brutality reported in the newspapers of fifteen major U.S. cities were black, and 93 percent of the officers involved were white....

CAPITAL PUNISHMENT AS MODERN FORM OF LYNCHING

The raw statistics on capital punishment seem to indicate racial bias prima facie: "Of 3,984 people lawfully executed since 1930 [until 1980], 2,113 were black, over half of the total, almost five times the proportion of blacks in the population as a whole."[5] However, social scientific research on racial sentencing has produced mixed results....

There is a substantial body of research showing that blacks charged with murdering whites are more likely to be sentenced to death than with any other victim-offender dyad. Similarly, blacks charged with raping white women receive the death sentence at a much higher rate than whites charged with raping white women. The two tendencies were confirmed by Spohn in a 1994 article using data for Detroit in 1977 and 1978: "Blacks who sexually assaulted whites faced a greater risk of incarceration than either blacks or whites who sexually assaulted blacks or whites who sexually assaulted whites; similarly, blacks who murdered whites received longer sentences than did offenders in the other two categories."[6]

The most respected study on race and death penalty, carried out by David C. Baldus to support the claim of Warren McClesky, a black man convicted of murdering a white police officer in 1978, found that there was a huge disparity in the imposition of the death penalty in Georgia. The study found that in cases involving white victims and black defendants, the death penalty was imposed 22 percent of the time whereas with the reverse dyad, the death penalty was imposed in only 1 percent of the cases. Even after controlling for a number of variables, blacks were 4.3 times as likely as whites to receive a death sentence. In a 1990 review of 28 studies on death penalty sentencing, 23 of the studied showed that the fact that victims are white "influences the likelihood that the defendant will be charged with a capital crime or that death penalty will be imposed."[7]

It should not surprise anyone that in a racialized society, court decisions on cases involving the death penalty exhibit a race effect. Research on juries suggests that they tend to be older, more affluent, more educated, more conviction-prone, and more white than the average in the community....

Preliminary data from the Capital Jury Study[8]—ongoing interviews with more than 1,000 jurors who have served in death penalty trials in 14 states—reveal that deep-seated prejudice finds its way into the jury room. The following three statements by some of the jurors interviewed in this study chillingly illustrate this point:

> He [the defendant] was a big man who looked like a criminal.... He was big an' black an' kind of ugly. So, I guess when I saw him I thought this fits the part.
>
> You know, if they'd been white people. I would've had a different attitude. I'm sorry that I feel that way.
>
> Just a typical nigger. Sorry, that's the way I feel about it.[9]

HIGH PROPENSITY TO ARREST BLACKS

Blacks complain that police officers mistreat them, disrespect them, assume that they are criminals, violate their rights on a consistent basis, and are more violent when dealing with them than when dealing with whites. Blacks and other minorities are stopped and frisked by police in "alarmingly disproportionately numbers."[10] Why is it that minorities receive "special treatment" from the police? Studies on police attitudes and their socialization suggest that police officers live in a "cops' world" and develop a cop mentality. That cops' world is a highly racialized one: minorities are viewed as dangerous, prone to crime, violent, and disrespectful. Various studies have noted that the racist attitudes that police officers exhibit have an impact in their behavior toward minorities.... In terms of demographic bias, research suggests that because black communities are overpatrolled, officers patrolling these areas develop a stereotypical view of residents as more likely to commit criminal acts and are more likely to "see" criminal behavior than in white communities.

Thus it is not surprising that blacks are disproportionately arrested compared to whites. It is possible to gauge the level of overarrest endured by blacks by comparing the proportion of times that they are described by victims as the attackers with their arrest rates. Using this procedure, Farai Chideya contends,

> For virtually every type of crime, African-American criminals are arrested at rates above their commission of the acts. For example, victimization reports indicated that 33 percent of women who were raped said that their attacker was black; however, black rape suspects made up fully 43 percent of those arrested. The disproportionate arrest rate adds to the public perception that rape is a "black" crime.[11]

Using these numbers, the rate of overarrest for blacks in cases of rape is 30 percent. As shocking as this seems to be, the rate for cases wherein the victim is white is even higher....

POST–CIVIL RIGHTS SOCIAL CONTROL AND THE NEW RACISM

Police brutality, overarrest, racial profiling, and many of the other social control mechanisms used to keep blacks "in their (new) place" in the contemporary United States are not overwhelmingly covert.... These practices are invisible to vast numbers of U.S. citizens. They are rendered invisible in four ways. First, because the enforcement of the racial order from the 1960s onward has been institutionalized, individual whites can express a detachment from the racialized way in which social control agencies operate in the United States. Second, because these agencies are legally charged with maintaining order in society, their actions are deemed neutral and necessary. Thus, it is no surprise that, whereas blacks mistrust the police in surveys, whites consistently support them. Third, journalists and academicians investigating crime are central agents in the reproduction of distorted views on crime. Few report the larger facts of crime in the United States (e.g., most crime is committed by whites: so-called white-collar crime costs us ten times as much as street crime; youth crime, which accounts for most crime, is directly connected to the "structure of opportunity" youngsters face).... Instead, thanks to their efforts, "The public's perception is that crime is violent, Black, and male, [trends that] have converged to create the criminal blackman."[12] Finally, incidents that seem to indicate racial bias in the criminal justice system are depicted by white-dominated media as isolated incidents. For example, cases that presumably expose the racial character of social control agencies (e.g., the police beating of Rodney King,...the acquittal or lenient sentences received by officers accused of police brutality, etc.) are viewed as "isolated" incidents and are separated from the larger social context in which they transpire. Therefore, these mechanisms fit my claim about the new racism because they are largely undetected and ignored.

REFERENCES

1. Manning Marable. *How Capitalism Underdeveloped Black America* (Boston: South End Press, 1983), pp. 120–121.
2. Gerald Jaynes and Robin M. Williams, *A Common Destiny* (Washington, DC: National Academies Press, 1989), 457–459. Trend data suggest that the arrest rate for blacks and whites has stabilized. Sixty-seven to 70 percent of those arrested are whites and 29 to 31 percent are black.
3. K. Russell, *The Color of Crime* (New York: New York University Press), 114.
4. Mark S. Rosentraub and Karen Harlow, "Police Policies and the Black Community: Attitude Toward the Police," pp. 107–121 in *Contemporary Public Policy Perspectives and Black Americans*, edited by Mitchell F. Rice and Woodrow Jones, Jr. (Westport, CT and London: Greenwood Press, 1984), p. 119.
5. Samuel R Gross and Robert Mauro, *Death and Discrimination: Racial Disparities in Capital Sentencing* (Boston: Northeastern University Press, 1989).
6. Cassia Spohn, "Crime and the Social Control of Blacks: Offender/Victim Race and the Sentencing of Violent Offenders," pp. 249–268 in *Inequality, Crime, and Social*

Control, edited by George S. Bridges and Martha A. Myers (Boulder, San Francisco, and Oxford: Westview Press. 1994), 264.

7. Derrick Bell, *Race, Racism, and American Law* (Boston, Toronto, and London: Little, Brown and Company, 1992), pp. 332–333.
8. William J. Bowers, Maria Sandys, and Benjamin D. Steiner, "Foreclosed Impartiality in Capital Sentencing: Jurors' Predispositions, Guilt Trial Experience, and Premature Decision Making." *Cornell Law Review* 83 (1998): 1476–1556.
9. Amnesty International, *Killing with Prejudice: Race and the Death Penalty in the USA, 1999*. Available online at http://www.amnesty-usa.org/rightsforall/dp/race (May 6, 2001).
10. Bell, *Race, Racism, and American Law*, 340.
11. Farai Chideya, *Don't Believe the Hype* (New York: Penguin Books 1995), 194.
12. K. Russell. *The Color of Crime*, 114.

DISCUSSION QUESTIONS

1. What does Bonilla-Silva mean when he says that racial minorities receive special treatment from the police and the courts?
2. Why do many scholars see the death penalty as modern-day lynching?
3. Why do Black people and White people have different views of the police?

48

Hidden Politics

Discursive and Institutional Policing of Rap Music

TRICIA ROSE

Rose uses the example of rap and hip-hop music to discuss how minority communities are over-policed. Her essay thus provides a good illustration of the process of social control discussed in the previous article by Bonilla-Silva.

The way rap and rap-related violence are discussed in the popular media is fundamentally linked to the larger social discourse on the spatial control of black people. Formal policies that explicitly circumscribe housing, school, and job options for black people have been outlawed; however, informal, yet trenchant forms of institutional discrimination still exist in full force. Underwriting these de facto forms of social containment is the understanding that black people are a threat to social order. Inside of this, black urban teenagers are the most profound symbolic referent for internal threats to social order. Not surprisingly, then, young African Americans are in fundamentally antagonistic relationships to the institutions that most prominently frame and constrain their lives. The public school system, the police, and the popular media perceive and construct young African Americans as a dangerous internal element in urban America; an element that if allowed to roam freely, will threaten the social order; an element that must be policed. Since rap music is understood as the predominant symbolic voice of black urban males, it heightens this sense of threat and reinforces dominant white middle-class objections to urban black youths who do not aspire to (but are haunted by) white middle-class standards.

My experiences and observations while attending several large-venue rap concerts in major urban centers serve as disturbingly obvious cases of how black urban youth are stigmatized, vilified, and approached with hostility and suspicion by authority figures. I offer a description of my confrontation and related observations not simply to prove that such racially and class-motivated hostility exists but, instead, to use it as a case from which to tease out how the public space policing of black youth and rap music feeds into and interacts with other media, municipal, and corporate policies that determine who can publicly gather and how.

SOURCE: From Tricia Rose, *Black Noise: Rap Music and Black Culture in Contemporary America* (Middletown, CT: Wesleyan University Press). Copyright © 1994 by Tricia Rose. Reprinted by permission of Wesleyan University Press.

Thousands of young black people milled around waiting to get into the large arena. The big rap summer tour was in town, and it was a prime night to see and be seen. The "pre-show show" was in full effect. Folks were dressed in the latest fly-gear: bicycle shorts, high-top sneakers, chunk jewelry, baggie pants, and polka-dotted tops. Hair style was a fashion show in itself: high-top fade designs, dreads, corkscrews, and braids with gold and purple sparkles. Crews of young women were checking out the brothers; posses of brothers were scooping out the sisters, each comparing styles among themselves. Some wide-eyed preteenyboppers were soaking in the teenage energy, thrilled to be out with the older kids.

As the lines for entering the arena began to form, dozens of mostly white private security guards hired by the arena management (many of whom are off-duty cops making extra money), dressed in red polyester V-neck sweaters and gray work pants, began corralling the crowd through security checkpoints. The free-floating spirit began to sour, and in its place began to crystallize a sense of hostility mixed with humiliation. Men and women were lined up separately in preparation for the weapon search. Each of the concertgoers would go through a body patdown, pocketbook, knapsack, and soul search. Co-ed groups dispersed, people moved toward their respective search lines. The search process was conducted in such a way that each person being searched was separated from the rest of the line. Those searched could not function as a group, and subtle interactions between the guard and person being searched could not be easily observed. As the concertgoers approached the guards, I noticed a distinct change in posture and attitude. From a distance, it seemed that the men were being treated with more hostility than the women in line. In the men's area, there was an almost palpable sense of hostility on behalf of the guards as well as the male patrons. Laughing and joking among men and women, which had been loud and buoyant up until this point, turned into virtual silence.

As I approached the female security guards, my own anxiety increased. What if they found something I was not allowed to bring inside? What was prohibited, anyway? I stopped and thought: All I have in my small purse is my wallet, eyeglasses, keys, and a notepad—nothing "dangerous." The security woman patted me down, scanned my body with an electronic scanner while she anxiously kept an eye on the other black women in line to make sure that no one slipped past her. She opened my purse and fumbled through it pulling out a nail file. She stared at me provocatively, as if to say "Why did you bring this in here?" I didn't answer her right away and hoped that she would drop it back into my purse and let me go through. She continued to stare at me, sizing me up to see if I was "there to cause trouble." By now, my attitude had turned foul; my childlike enthusiasm to see my favorite rappers had all but fizzled out. I didn't know the file was in my purse, but the guard's accusatory posture rendered such excuses moot. I finally replied tensely, "It's a nail file, what's the problem?" She handed it back to me, satisfied, I suppose, that I was not intending to use it as a weapon, and I went in to the arena. As I passed her, I thought to myself, "This arena is a public place, and I am entitled to come here and bring a nail file if I want to." But these words rang empty in my head; the language of entitlement couldn't erase my sense of alienation. I felt harassed and unwanted. This arena wasn't mine, it was hostile, alien territory. The unspoken message hung in the air: "You're not wanted here, let's get this over with and send you all back to where you came from."

I recount this incident for two reasons. First, a hostile tenor, if not actual verbal abuse, is a regular part of rap fan contact with arena security and police. This is not an isolated or rare example, incidents similar to it continue to take place at many rap concerts. Rap concertgoers were barely tolerated and regarded with heightened suspicion. Second, arena security forces, a critical facet in the political economy of rap and its related sociologically based crime discourse, contribute to the high level of anxiety and antagonism that confront young African Americans. Their military posture is a surface manifestation of a complex network of ideological and economic processes that "justify" the policing of rap music, black youths, and black people in general. Although my immediate sense of indignation in response to public humiliation may be related to a sense of entitlement that comes from my status as a cultural critic, thus separating me from many of the concertgoers, my status as a young African American woman is a critical factor in the way I was treated in this instance, as well as many others.

Rap artists articulate a range of reactions to the scope of institutional policing faced by many young African Americans. However, the lyrics that address the police directly—what Ice Cube has called "revenge fantasies"—have caused the most extreme and unconstitutional reaction from law enforcement officials in metropolitan concert arena venues....

Rap music is by no means the only form of expression under attack. Popular white forms of expression, especially heavy metal, have recently been the target of increased sanctions and assaults by politically and economically powerful organizations, such as the Parent's Music Resource Center, The American Family Association, and Focus on the Family. These organizations are not fringe groups, they are supported by major corporations, national-level politicians, school associations, and local police and municipal officials.

However, there are critical differences between the attacks made against black youth expression and white youth expression. The terms of the assault on rap music, for example, are part of a long-standing sociologically based discourse that considers black influences a cultural threat to American society. Consequently, rappers, their fans, and black youths in general are constructed as co-conspirators in the spread of black cultural influence. For the anti-rock organizations, heavy metal is a "threat to the fiber of American society," but the fans (e.g., "our children") are *victims* of its influence. Unlike heavy metal's victims, rap fans are the youngest representatives of a black presence whose cultural difference is perceived as an internal threat to America's cultural development. *They* victimize us. These differences in the ideological nature of the sanctions against rap and heavy metal are of critical importance, because they illuminate the ways in which racial discourses deeply inform public transcripts and social control efforts. This racial discourse is so profound that when Ice-T's speed metal band (*not rap group*) Body Count was forced to remove "Cop Killer" from its debut album because of attacks from politicians, these attacks consistently referred to it as a rap song (even though it in no way can be mistaken for rap) to build a negative head of steam in the public. As Ice-T describes it, "There is absolutely no way to listen to the song 'Cop Killer' and call it a rap record. It's so far from rap. But, politically, they know by saying the word rap they can get a lot of people who think, 'Rap-black-rap-black-ghetto,' and don't like it. You say the word *rock*, people say, 'Oh, but I like Jefferson Airplane, I like Fleetwood Mac—that's rock.' They don't want to use the word rock & roll to describe this song."[1]...

The social construction of "violence," that is, when and how particular acts are defined as violent, is part of a larger process of labeling social phenomena. Rap-related violence is one facet of the contemporary "urban crisis" that consists of a "rampant drug culture" and "wilding gangs" of black and Hispanic youths. When the *Daily News* headline reads, "L.I. Rap-Slayers Sought" or a *Newsweek* story is dubbed "The Rap Attitude," these labels are important, because they assign a particular meaning to an event and locate that event in a larger context. Labels are critical to the process of interpretation, because they provide a context and frame for social behavior. As Stuart Hall et al. point out in *Policing the Crisis*, once a label is assigned, "the use of the label is likely to mobilize this whole referential context, with all its associated meaning and connotations."[2] The question then, is not "is there really violence at rap concerts," but how are these crimes contextualized, labeled?... Whose interests do these interpretive strategies serve? What are the repercussions?

Venue owners have the final word on booking decisions, but they are not the only group of institutional gatekeepers. The other major powerbroker, the insurance industry, can refuse to insure an act approved by venue management. In order for any tour to gain access to a venue, the band or group hires a booking agent who negotiates the act's fee. The booking agent hires a concert promoter who "purchases" the band and then presents the band to both the insurance company and the venue managers. If an insurance company will not insure the act, because they decide it represents an unprofitable risk, then the venue owner will not book the act. Furthermore, the insurance company and the venue owner reserve the right to charge whatever insurance or permit fees they deem reasonable on a case-by-case basis. So, for example, Three Rivers Stadium in Pittsburgh, Pennsylvania, tripled its normal $20,000 permit fee for the Grateful Dead. The insurance companies who still insure rap concerts have raised their minimum coverage from about $500,000 to between $4 and $5 million worth of coverage per show.[3] Several major arenas make it almost impossible to book a rap show, and others have refused outright to book rap acts at all.

These responses to rap music bear a striking resemblance to the New York City cabaret laws instituted in the 1920s is response to jazz music. A wide range of licensing and zoning laws, many of which remained in effect until the late 1980s, restricted the places where jazz could be played and how it could be played. These laws were attached to moral anxieties regarding black cultural effects and were in part intended to protect white patrons from jazz's "immoral influences." They defined and contained the kind of jazz that could be played by restricting the use of certain instruments (especially drums and horns) and established elaborate licensing policies that favored more established and mainstream jazz club owners and prevented a number of prominent musicians with minor criminal records from obtaining cabaret cards.

During an interview with "Richard" from a major talent agency that books many prominent rap acts, I asked him if booking agents had responded to venue bans on rap music by leveling charges of racial discrimination against venue owners. His answer clearly illustrates the significance of the institutional power at stake:

> These facilities are privately owned, they can do anything they want. You say to them: "You won't let us in because you're discriminating against black kids." They say to you, "Fuck you, who cares. Do whatever you got to do, but you're not coming in here. You, I don't need you, I don't want you. Don't come, don't bother me. I will book hockey, ice shows, basketball, country music and graduations. I will do all kinds of things 360 days out of the year. But I don't need you. I don't need fighting, shootings and stabbings." Why do they care? They have their image to maintain.[4]

Richard's imaginary conversation with a venue owner is a pointed description of the scope of power these owners have over access to large public urban spaces and the racially exclusionary silent policy that governs booking policies....

Because rap has an especially strong urban metropolitan following, freezing it out of these major metropolitan arenas has a dramatic impact on rappers' ability to reach their fan base in live performance. Public Enemy, Queen Latifah, and other rap groups use live performance settings to address current social issues, media miscoverage, and other problems that especially concern black America. For example, during a December 1988 concert in Providence, R.I., Chuck D from Public Enemy explained that the Boston arena refused to book the show and read from a *Boston Herald* article that depicted rap fans as a problematic element and that gave its approval of the banning of the show. To make up for this rejection, Chuck D called out to the "Roxbury crowd in the house," to make them feel at home in Providence. Each time Chuck mentioned Roxbury, sections of the arena erupted in especially exuberant shouts and screams. Because black youths are constructed as a permanent threat to social order, large public gatherings will always be viewed as dangerous events. The larger arenas possess greater potential for mass access and unsanctioned behavior. And black youths, who are highly conscious of their alienated and marginalized lives, will continue to be hostile toward those institutions and environments that reaffirm this aspect of their reality.

The presence of a predominantly black audience in a 15,000 capacity arena, communicating with major black cultural icons whose music, lyrics, and attitude illuminate and affirm black fears and grievances, provokes a fear of the consolidation of black rage. Venue owner and insurance company anxiety over broken chairs, insurance claims, or fatalities are not important in and of themselves, they are important because they symbolize a loss of control that might involve challenges to the current social configuration. They suggest the possibility that black rage can be directed at the people and institutions that support the containment and oppression of black people. As West Coast rapper Ice Cube points out in "The Nigga Ya Love to Hate," "Just think if niggas decided to retaliate?"[5]

The coded familiarity of the rhythms and hooks that rap samples from other black music, especially funk and soul music, carries with it the power of black collective memory. These sounds are cultural markers, and responses to them are not involuntary at all but in fact densely and actively intertextual; they immediately conjure collective black experience, past and present....

Rap music is fundamentally linked to larger social constructions of black culture as an internal threat to dominant American culture and social order. Rap's capacity as a form of testimony, as an articulation of a young black urban critical voice of social protest, has profound potential as a basis for a language of liberation. Contestation over the meaning and significance of rap music and its ability to occupy public space and retain expressive freedom constitutes a central aspect of contemporary black cultural politics.

During the centuries-long period of Western slavery, there were elaborate rules and laws designed to control slave populations. Constraining the mobility of slaves, especially at night and in groups, was of special concern; slave masters reasoned that revolts could be organized by blacks who moved too freely and without surveillance. Slave masters were rightfully confident that blacks had good reason to escape, revolt, and retaliate. Contemporary laws and practices curtailing and constraining black mobility in urban America function in much the same way and for similar reasons. Large groups of African Americans, especially teenagers, represent a threat to the social order of oppression. Albeit more sophisticated and more difficult to trace, contemporary policing of African Americans resonates with the legacy of slavery.

Rap's poetic voice is deeply political in content and spirit, but rap's hidden struggle, the struggle over access to public space, community resources, and the interpretation of black expression, constitutes rap's hidden politics.

REFERENCES

1. Light, Alan. "Ice-T." *Rolling Stone*, 20 August 1992, pp. 32, 60.
2. Stuart Hall et al., *Policing the Crisis* (London: Macmillan, 1977), p. 19.
3. Interview with "Richard," a talent agency representative from a major agency that represents dozens of major rap groups, October 1990.
4. Rose interview with "Richard." I have decided not to reveal the identity of this talent agency representative, because it serves no particular purpose here and may have a detrimental effect on his employment.
5. Ice Cube, "The Nigga Ya Love to Hate," *AmeriKKKa's Most Wanted (Priority Records,* 1990).

DISCUSSION QUESTIONS

1. Why does Rose think that rap music events are heavily policed?
2. According to Rose, how does rap music challenge the dominant culture? Are there ways that it also reflects the dominant culture?
3. When was the last time you went to a venue to listen to live music? How were you treated? What do your observations suggest about the nature of music and treatment of the audience?

49

The Uneven Scales of Capital Justice

CHRISTINA SWARNS

Despite constitutional rights to equal protection under the law, the facts show that race, especially when combined with class, has a significant impact on the likelihood of receiving capital punishment (that is, the death penalty).

In 1972, the U.S. Supreme Court declared the death penalty unconstitutional. The Court found that because the capital-punishment laws gave sentencers virtually unbridled discretion in deciding whether or not to impose a death sentence, "The death sentence [was] disproportionately carried out on the poor, the Negro, and the members of unpopular groups."

In 1976, the Court reviewed the revised death-penalty statutes—which are in place today—and concluded that they sufficiently restricted sentencer discretion such that race and class would no longer play a pivotal role in the life-or-death calculus. In the 28 years since the reinstatement of the death penalty, however, it has become apparent that the Court was wrong. Race and class remain critical factors in the decision of who lives and who dies.

Both race and poverty corrupt the administration of the death penalty. Race severely disadvantages the black jurors, black defendants, and black victims within the capital-punishment system. Black defendants are more likely to be executed than white defendants. Those who commit crimes against black victims are punished less severely than those who commit crimes against white victims. And black potential jurors are often denied the opportunity to serve on death-penalty juries. As far as the death penalty is concerned, therefore, blackness is a proxy for worthlessness.

Poverty is a similar—and often additional—handicap. Because the lawyers provided to indigent defendants charged with capital crimes are so uniformly undertrained and undercompensated, the 90 percent of capitally charged defendants who lack the resources to retain a private attorney are virtually guaranteed a death sentence. Together, therefore, race and class function as an elephant on death's side of the sentencing scale.

When and how does race infect the death-penalty system? The fundamental lesson of the Supreme Court's 1972 decision to strike down the death penalty is that discretion, if left unchecked, will be exercised in such a manner that arbitrary and irrelevant factors like race will enter into the sentencing decision. That

SOURCE: From *The American Prospect*, June 18, 2004. Reprinted by permission.

conclusion remains true today. The points at which discretion is exercised are the gateways through which racial bias continues to enter into the sentencing calculation.

Who has the most unfettered discretion? Chief prosecutors, who are overwhelmingly white, make some of the most critical decisions vis-à-vis the death penalty. Because their decisions go unchecked, prosecutors have arguably the greatest unilateral influence over the administration of the death penalty.

Do prosecutors exercise their discretion along racial lines? Unquestionably yes. Prosecutors bring more defendants of color into the death-penalty system than they do white defendants. For example, a 2000 study by the U.S. Department of Justice reveals that between 1995 and 2000, 72 percent of the cases that the attorney general approved for death-penalty prosecution involved defendants of color. During that time, statistics show that there were relatively equal numbers of black and white homicide perpetrators.

Prosecutors also give more white defendants than black defendants the chance to avoid a death sentence. Specifically, prosecutors enter into plea bargains—deals that allow capitally charged defendants to receive a lesser sentence in exchange for an admission of guilt—with white defendants far more often than they do with defendants of color. Indeed, the Justice Department study found that white defendants were almost twice as likely as black defendants to enter into such plea agreements.

Further, prosecutors assess cases differently depending upon the race of the victim. Thus, the Department of Justice found that between 1995 and 2000, U.S. attorneys were almost twice as likely to seek the death penalty for black defendants accused of killing nonblack victims than for black defendants accused of killing black victims.

And, finally, prosecutors regularly exclude black potential jurors from service in capital cases. For example, a 2003 study of jury selection in Philadelphia capital cases, conducted by the Pennsylvania Supreme Court Commission on Race and Gender Bias in the Justice System, revealed that prosecutors used peremptory challenges—the power to exclude potential jurors for any reason aside from race or gender—to remove 51 percent of black potential jurors while excluding only 26 percent of nonblack potential jurors. Such bias has a long history: From 1963 to 1976, one Texas prosecutor's office instructed its lawyers to exclude all people of color from service on juries by distributing a memo containing the following language: "Do not take Jews, Negroes, Dagos, Mexicans or a member of any minority race on a jury, no matter how rich or how well educated." This extraordinary exercise of discretion harms black capital defendants because statistics reveal that juries containing few or no blacks are more likely to sentence black defendants to death.

Such blatant prosecutorial discretion has significantly contributed to the creation of a system that is visibly permeated with racial bias. Black defendants are sentenced to death and executed at disproportionate rates. For example, in Philadelphia, African American defendants are approximately four times more likely to be sentenced to death than similarly situated white defendants. And nationwide, crimes against white victims are punished more severely than crimes against black victims. Thus, although 46.7 percent of all homicide victims are black, only 13.9 percent of the victims of executed defendants are black. In some jurisdictions, all of the defendants on death row have white victims; in other jurisdictions, having a white victim exponentially increases a criminal defendant's likelihood of being sentenced to death. It is beyond dispute, therefore, that race remains a central factor in the administration of the death penalty.

Socioeconomic status also plays an inappropriate yet extremely influential role in the determination of who receives the death penalty. The vast majority of the people who are sentenced to death and executed in the United States come from a background of poverty. Indeed, as noted by the Supreme Court in 1972, "One searches our chronicles in vain for the execution of any member of the affluent strata of this society. The Leopolds and Loebs are given prison terms, not sentenced to death."

The primary reason for this economic disparity is that the poor are systematically denied access to well-trained and adequately funded lawyers. "Capital defense is now a highly specialized field requiring practitioners to successfully negotiate minefield upon minefield of exacting and arcane death-penalty law," according to the Pennsylvania commission. "Any misstep along the way can literally mean death for the client." It is therefore critical that lawyers appointed to represent poor defendants facing death possess the requisite compensation, training, and skill to mount a meaningful challenge to the government's case.

Unfortunately, few if any of the defendants on death row are provided with lawyers possessing the requisite skills and resources. Instead, poorly trained and underfunded court-appointed lawyers who provide abysmal legal assistance typically represent those death-sentenced prisoners. Tales of the pathetic lawyering provided by appointed counsel to their capitally charged clients are legion. Perhaps the most famous example is that of Calvin Burdine, whose court-appointed lawyer slept through significant portions of his trial. Another example is the case of Vinson Washington, whose court-appointed lawyer suggested to the defense psychiatrist that Vinson "epitomized the banality of evil." Death-sentenced defendants are so frequently provided with poor representation that, in 2001, Supreme Court Justice Ruth Bader Ginsberg commented that she had never seen a death-penalty defendant come before the Supreme Court in search of an eve-of-execution stay "in which the defendant was well-represented at trial."

One reason that appointed counsel perform so poorly is that they are grossly undercompensated. In some cases, capital-defense attorneys have been paid as little as $5 an hour. Not surprisingly, these paltry rates of compensation have yielded an equally paltry quality of representation. As was succinctly noted by the 5th U.S. Circuit Court of Appeals in its review of the quality of representation provided by a court-appointed lawyer to a capitally charged defendant in Texas: "The state paid defense counsel $11.84 per hour. Unfortunately, the justice system got only what it paid for."

Lawyers appointed to handle capital trials also often lack the expertise necessary to appropriately defend capitally charged defendants. Many states fail to provide appointed counsel with the training necessary to handle these complex cases, and many fail to impose minimum qualifications for lawyers handling capital cases. As a result, capital defendants have been represented by lawyers with absolutely no experience in criminal, much less capital, law. Although the American Bar Association has promulgated standards for the representation of indigent defendants charged with capital offenses, and although those guidelines have been endorsed by the Supreme Court, no death-penalty jurisdiction has implemented a system that meets these requirements. Thus, lawyers without meaningful training or expertise in the area of capital punishment continue to represent defendants facing death.

Because race and class continue to play a powerful role in the administration of the death penalty, it is clear that the current system is as broken today as it was in 1972. As the Supreme Court explained at the time, "A law that stated that anyone making more than $50,000 would be exempt from the death penalty would plainly fall, as would a law that in terms said that blacks, those who never went beyond the fifth grade in school, those who made less than $3,000 a year, or those who were unpopular or unstable should be the only people executed. A law which in the overall view reaches that result in practice has no more sanctity than a law which in terms provides the same."

Because the current death-penalty law, while neutral on its face, is applied in such a manner that people of color and the poor are disproportionately condemned to die, the law is legally and morally invalid.

DISCUSSION QUESTIONS

1. Why did the Supreme Court declare the death penalty unconstitutional in 1972?
2. What does discretion mean and why is it important in the criminal justice system? How does it influence who ends up on a jury?
3. Why does Swarns look at both race and social class in exploring the experiences of individuals in the criminal justice system?

50

The Mark of a Criminal Record

DEVAH PAGER

What are the consequences of the very high rate of incarceration of young Black men? In what is known as an audit study—a study in this case where pairs of Black and White men are identically matched in their social characteristics and compared in employment outcomes following incarceration. Pager shows the barrier that a criminal record creates for subsequent employment, showing the influence of race in the likelihood of former prisoners finding jobs.

While stratification researchers typically focus on schools, labor markets, and the family as primary institutions affecting inequality, a new institution has emerged as central to the sorting and stratifying of young and disadvantaged men: the criminal justice system. With over 2 million individuals currently incarcerated, and over half a million prisoners released each year, the large and growing numbers of men being processed through the criminal justice system raises important questions about the consequences of this massive institutional intervention.

This article focuses on the consequences of incarceration for the employment outcomes of black and white men. While previous survey research has demonstrated a strong *association* between incarceration and employment, there remains little understanding of the mechanisms by which these outcomes are produced. In the present study, I adopt an experimental audit approach to formally test the degree to which a criminal record affects subsequent employment opportunities. By using matched pairs of individuals to apply for real entry-level jobs, it becomes possible to directly measure the extent to which a criminal record—in the absence of other disqualifying characteristics—serves as a barrier to employment among equally qualified applicants. Further, by varying the race of the tester pairs, we can assess the ways in which the effects of race and criminal record interact to produce new forms of labor market inequalities.

TRENDS IN INCARCERATION

Over the past three decades, the number of prison inmates in the United States has increased by more than 600%, leaving it the country with the highest incarceration rate in the world (Bureau of Justice Statistics 2002a; Barclay, Tavares,

SOURCE: From 2003. *American Journal of Sociology* 108 (March): 937–975. Reprinted by permission of University of Chicago Press.

and Siddique 2001). During this time, incarceration has changed from a punishment reserved primarily for the most heinous offenders to one extended to a much greater range of crimes and a much larger segment of the population. Recent trends in crime policy have led to the imposition of harsher sentences for a wider range of offenses, thus casting an ever-widening net of penal intervention.

While the recent "tough on crime" policies may be effective in getting criminals off the streets, little provision has been made for when they get back out. Of the nearly 2 million individuals currently incarcerated, roughly 95% will be released, with more than half a million being released each year (Slevin 2000). According to one estimate, there are currently over 12 million ex-felons in the United States, representing roughly 8% of the working-age population (Uggen, Thompson, and Manza 2000). Of those recently released, nearly two-thirds will be charged with new crimes and over 40% will return to prison within three years (Bureau of Justice Statistics 2000). Certainly some of these outcomes are the result of desolate opportunities or deeply ingrained dispositions, grown out of broken families, poor neighborhoods, and little social control (Sampson and Laub 1993; Wilson 1997). But net of these contributing factors, there is evidence that experience with the criminal justice system in itself has adverse consequences for subsequent opportunities. In particular, incarceration is associated with limited future employment opportunities and earnings potential (Freeman 1987; Western 2002), which themselves are among the strongest predictors of recidivism (Shover 1996; Sampson and Laub 1993; Uggen 2000).

The expansion of the prison population has been particularly consequential for blacks. The incarceration rate for young black men in the year 2000 was nearly 10%, compared to just over 1% for white men in the same age group (Bureau of Justice Statistics 2001). Young black men today have a 28% likelihood of incarceration during their lifetime (Bureau of Justice Statistics 1997), a figure that rises above 50% among young black high school dropouts (Pettit and Western 2001). These vast numbers of inmates translate into a large and increasing population of black ex-offenders returning to communities and searching for work. The barriers these men face in reaching economic self-sufficiency are compounded by the stigma of minority status and criminal record. The consequences of such trends for widening racial disparities are potentially profound (see Western and Pettit 1999; Freeman and Holzer 1986)....

RESEARCH QUESTIONS

There are three primary questions I seek to address with the present study. First, in discussing the main effect of a criminal record, we need to ask whether and to what extent employers use information about criminal histories to make hiring decisions. Implicit in the criticism of survey research in this area is the assumption that the signal of a criminal record is not a determining factor. Rather, employers use information about the interactional styles of applicants, or other observed characteristics—which may be correlated with criminal records—and this explains the differential outcomes we observe. In this view, a criminal record does not represent a meaningful signal to employers on its own. This study formally tests the degree to which employers use information about criminal histories in the absence of corroborating evidence. It is essential that we conclusively document this effect before making larger claims about the aggregate consequences of incarceration.

Second, this study investigates the extent to which race continues to serve as a major barrier to employment. While race has undoubtedly played a central role in shaping the employment opportunities of African-Americans over the past century, recent arguments have questioned the continuing significance of race, arguing instead that other factors—such as spatial location, soft skills, social capital, or cognitive ability—can explain most or all of the contemporary racial differentials we observe (Wilson 1987; Moss and Tilly 1996; Loury 1977; Neal and Johnson 1996). This study provides a comparison of the experiences of equally qualified black and white applicants, allowing us to assess the extent to which direct racial discrimination persists in employment interactions.

The third objective of this study is to assess whether the effect of a criminal record differs for black and white applicants. Most research investigating the differential impact of incarceration on blacks has focused on the differential *rates* of incarceration and how those rates translate into widening racial disparities. In addition to disparities in the rate of incarceration, however, it is also important to consider possible racial differences in the *effects* of incarceration. Almost none of the existing literature to date has explored this issue, and the theoretical arguments remain divided as to what we might expect.

On one hand, there is reason to believe that the signal of a criminal record should be less consequential for blacks. Research on racial stereotypes tells us that Americans hold strong and persistent negative stereotypes about blacks, with one of the most readily invoked contemporary stereotypes relating to perceptions of violent and criminal dispositions (Smith 1991; Sniderman and Piazza 1993; Devine and Elliott 1995). If it is the case that employers view all blacks as potential criminals, they are likely to differentiate less among those with official criminal records and those without. Actual confirmation of criminal involvement then will provide only redundant information, while evidence against it will be discounted. In this case, the outcomes for all blacks should be worse, with less differentiation between those with criminal records and those without.

On the other hand, the effect of a criminal record may be worse for blacks if employers, already wary of black applicants, are more hesitant when it comes to taking risks on blacks with proven criminal tendencies. The literature on racial stereotypes also tells us that stereotypes are most likely to be activated and reinforced when a target matches on more than one dimension of the stereotype (Quillian and Pager 2002; Darley and Gross 1983; Fiske and Neuberg 1990). While employers may have learned to keep their racial attributions in check through years of heightened sensitivity around employment discrimination, when combined with knowledge of a criminal history, negative attributions are likely to intensify.

A third possibility, of course, is that a criminal record affects black and white applicants equally. The results of this audit study will help to adjudicate between these competing predictions.

THE AUDIT METHODOLOGY

...The basic design of an employment audit involves sending matched pairs of individuals (called testers) to apply for real job openings in order to see whether employers respond differently to applicants on the basis of selected characteristics.

The appeal of the audit methodology lies in its ability to combine experimental methods with real-life contexts. This combination allows for greater generalizability than a lab experiment and a better grasp of the causal mechanisms than what we can normally obtain from observational data. The audit methodology is particularly valuable for those with an interest in discrimination. Typically, researchers are forced to infer discrimination indirectly, often attributing the residual from a statistical model—which is essentially all that is not directly explained—to discrimination. This convention is rather unsatisfying to researchers who seek empirical documentation for important social processes. The audit methodology therefore provides a valuable tool for this research....

STUDY DESIGN

The basic design of this study involves the use of four male auditors (also called testers), two blacks and two whites. The testers were paired by race; that is, unlike in the original Urban Institute audit studies, the two black testers formed one team, and the two white testers formed the second team. The testers were 23-year-old college students from Milwaukee who were matched on the basis of physical appearance and general style of self-presentation. Objective characteristics that were not already identical between pairs—such as educational attainment and work experience—were made similar for the purpose of the applications. Within each team, one auditor was randomly assigned a "criminal record" for the first week; the pair then rotated which member presented himself as the ex-offender for each successive week of employment searches, such that each tester served in the criminal record condition for an equal number of cases. By varying which member of the pair presented himself as having a criminal record, unobserved differences within the pairs of applicants were effectively controlled. No significant differences were found for the outcomes of individual testers or by month of testing.

Job openings for entry-level positions (defined as jobs requiring no previous experience and no education greater than high school) were identified from the Sunday classified advertisement section of the *Milwaukee Journal Sentinel*. In addition, a supplemental sample was drawn from *Jobnet*, a state-sponsored web site for employment listings, which was developed in connection with the W-2 Welfare-to-Work initiatives.

The audit pairs were randomly assigned 15 job openings each week. The white pair and the black pair were assigned separate sets of jobs, with the same-race testers applying to the same jobs. One member of the pair applied first, with the second applying one day later (randomly varying whether the ex-offender was first or second). A total of 350 employers were audited during the course of this study: 150 by the white pair and 200 by the black pair. Additional tests were performed by the black pair because black testers received fewer callbacks on average, and there were thus fewer data points with which to draw comparisons. A larger sample size enables me to calculate more precise estimates of the effects under investigation.

Immediately following the completion of each job application, testers filled out a six-page response form that coded relevant information from the test. Important variables included type of occupation, metropolitan status, wage, size of establishment, and race and sex of employer. Additionally, testers wrote narratives describing the overall interaction and any comments made by employers (or included on applications) specifically related to race or criminal records.

One key feature of this audit study is that it focuses only on the first stage of the employment process. Testers visited employers, filled out applications, and proceeded as far as they could during the course of one visit. If testers were asked to interview on the spot, they did so, but they did not return to the employer for a second visit. The primary dependent variable, then, is the proportion of applications that elicited callbacks from employers. Individual voicemail boxes were set up for each tester to record employer responses. If a tester was offered the job on the spot, this was also coded as a positive response....

THE EFFECT OF A CRIMINAL RECORD FOR WHITES

I begin with an analysis of the effect of a criminal record among whites. White noncriminals can serve as our baseline in the following comparisons, representing the presumptively nonstigmatized group relative to blacks and those with criminal records. Given that all testers presented roughly identical credentials, the differences experienced among groups of testers can be attributed fully to the effects of race or criminal status.

Figure 1 shows the percentage of applications submitted by white testers that elicited callbacks from employers, by criminal status. As illustrated below, there is a large and significant effect of a criminal record, with 34% of whites without criminal records receiving callbacks, relative to only 17% of whites with criminal records. A criminal record thereby reduces the likelihood of a callback by 50%....

The results here demonstrate that criminal records close doors in employment situations. Many employers seem to use the information as a screening mechanism, without attempting to probe deeper into the possible context or complexities of the situation. As we can see here, in 50% of cases, employers were unwilling to consider equally qualified applicants on the basis of their criminal record.

Of course, this trend is not true among all employers, in all situations. There were, in fact, some employers who seemed to prefer workers who had been recently released from prison. One owner told a white tester in the criminal record condition that he "like[d] hiring people who ha[d] just come out of prison because they tend to be more motivated, and are more likely to be hard workers [not wanting to return to prison]."...

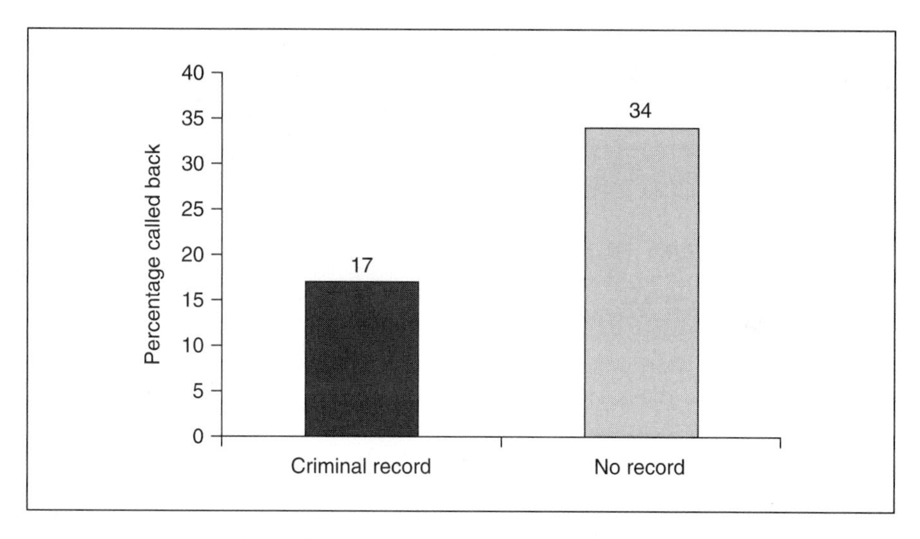

FIGURE 1 The effect of a criminal record on white job applicants.

THE EFFECT OF RACE

Figure 2 presents the percentage of callbacks received for both categories of black testers relative to those for whites. The effect of race in these findings is strikingly large. Among blacks without criminal records, only 14% received callbacks, relative to 34% of white noncriminals.... In fact, even whites *with* criminal records received more favorable treatment (17%) than blacks *without*

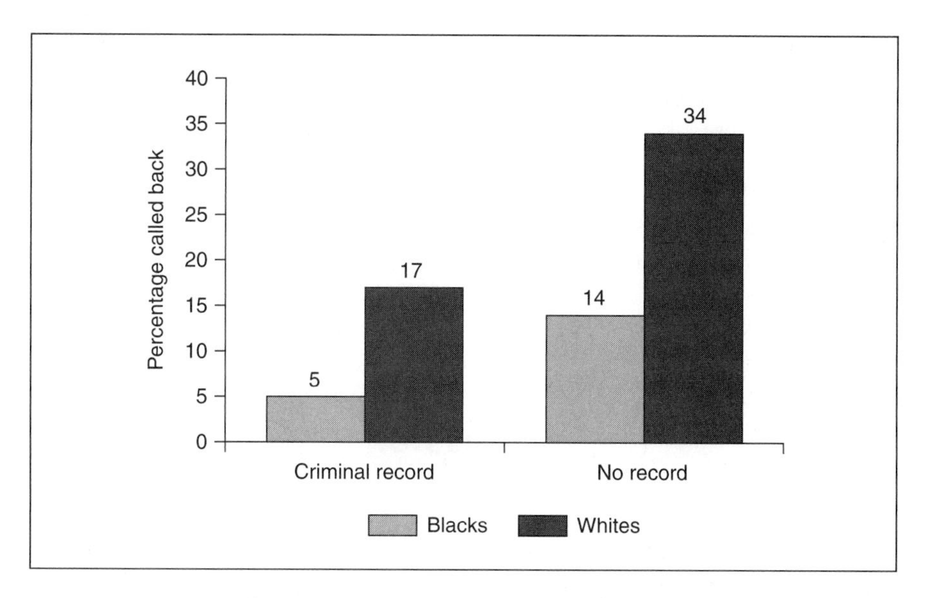

FIGURE 2 The effect of a criminal record for black and white job applicants.

criminal records (14%). The rank ordering of groups in this graph is painfully revealing of employer preferences: race continues to play a dominant role in shaping employment opportunities, equal to or greater than the impact of a criminal record....

RACIAL DIFFERENCES IN THE EFFECTS OF A CRIMINAL RECORD

The final question this study sought to answer was the degree to which the effect of a criminal record differs depending on the race of the applicant. Based on the results presented in Figure 2, the effect of a criminal record appears more pronounced for blacks than it is for whites. While this interaction term is not statistically significant, the magnitude of the difference is nontrivial. While the ratio of callbacks for nonoffenders relative to ex-offenders for whites is 2:1, this same ratio for blacks is nearly 3:1. The effect of a criminal record is thus 40% larger for blacks than for whites.

This evidence is suggestive of the way in which associations between race and crime affect interpersonal evaluations. Employers, already reluctant to hire blacks, appear even more wary of blacks with proven criminal involvement. Despite the face that these testers were bright articulate college students with effective styles of self-presentation, the cursory review of entry-level applicants leaves little room for these qualities to be noticed. Instead, the employment barriers of minority status and criminal record are compounded, intensifying the stigma toward this group.

The salience of employers' sensitivity toward criminal involvement among blacks was highlighted in several interactions documented by testers. On three separate occasions, for example, black testers were asked in person (before submitting their applications) whether they had a prior criminal history. None of the white testers were asked about their criminal histories up front.

The strong association between race and crime in the minds of employers provides some indication that the "true effect" of a criminal record for blacks may be even larger than what is measured here. If, for example, the outcomes for black testers *without* criminal records were deflated in part because employers feared that they may nevertheless have criminal tendencies, then the contrast between blacks with and without criminal records would be suppressed....

DISCUSSION

There is serious disagreement among academics, policy makers, and practitioners over the extent to which contact with the criminal justice system in itself leads to harmful consequences for employment. The present study takes a strong stand in this debate by offering direct evidence of the causal relationship between a criminal record and employment outcomes.... The finding that ex-offenders are only one-half to one-third as likely as nonoffenders to be considered by employers suggests that a criminal record indeed presents a major barrier to employment. With over 2 million people currently behind bars and over 12 million people with prior felony convictions, the consequences for labor market inequalities are potentially profound.

Second, the persistent effect of race on employment opportunities is painfully clear in these results. Blacks are less than half as likely to receive consideration by employers, relative to their white counterparts, and black nonoffenders fall behind even whites with prior felony convictions. The powerful effects of race thus continue to direct employment decisions in ways that contribute to persisting racial inequality. In light of these findings, current public opinion seems largely misinformed. According to a recent survey of residents in Los Angeles, Boston, Detroit, and Atlanta, researchers found that just over a quarter of whites believe there to be "a lot" of discrimination against blacks, compared to nearly two-thirds of black respondents (Kluegel and Bobo 2001). Over the past decade, affirmative action has come under attack across the country based on the argument that direct racial discrimination is no longer a major barrier to opportunity. According to this study, however, employers, at least in Milwaukee, continue to use race as a major factor in their hiring decisions. When we combine the effects of race and criminal record, the problem grows more intense. Not only are blacks much more likely to be incarcerated than whites; based on the findings presented here, they may also be more strongly affected by the impact of a criminal record. Previous estimates of the aggregate consequences of incarceration may therefore underestimate the impact on racial disparities.

Finally, in terms of policy implications, this research has troubling conclusions. In our frenzy of locking people up, our "crime control" policies may in fact exacerbate the very conditions that lead to crime in the first place. Research consistently shows that finding quality steady employment is one of the strongest predictors of desistance from crime (Shover 1996; Sampson and Laub 1993; Uggen 2000). The fact that a criminal record severely limits employment opportunities—particularly among blacks—suggests that these individuals are left with few viable alternatives.

As more and more young men enter the labor force from prison, it becomes increasingly important to consider the impact of incarceration on the job prospects of those coming out. No longer a peripheral institution, the criminal justice system has become a dominant presence in the lives of young disadvantaged men, playing a key role in the sorting and stratifying of labor market opportunities. This article represents an initial attempt to specify one of the important mechanisms by which incarceration leads to poor employment outcomes. Future research is needed to expand this emphasis to other mechanisms (e.g., the transformative effects of prison on human and social capital), as well as to include other social domains affected by incarceration (e.g., housing, family formation, political participation, etc.); in this way, we can move toward a more complete understanding of the collateral consequences of incarceration for social inequality.

At this point in history, it is impossible to tell whether the massive presence of incarceration in today's stratification system represents a unique anomaly of the late 20th century or part of a larger movement toward a system of stratification based on the official certification of individual character and competence. Whether this process of negative credentialing will continue to form the basis of emerging social cleavages remains to be seen....

REFERENCES

Barclay, Gordon, Cynthia Tavares, and Arsalaan Siddique. 2001. "International Comparisons of Criminal Justice Statistics, 1999." U.K. Home Office for Statistical Research. London.

Bureau of Justice Statistics. 1997. *Lifetime Likelihood of Going to State or Federal Prison*, by Thomas P. Bonczar and Allen J. Beck. Special report, March. Washington, D.C.

——. 2001. *Prisoners in 2000*, by Allen J. Beck and Paige M. Harrison. August. Bulletin. Washington, D.C.: NCJ 188207.

——. 2002a. *Sourcebook of Criminal Justice Statistics*. Last accessed March 1, 2003. Available http://www.albany.edu/sourcebook/

Darley, J. M., and P. H. Gross. 1983. "A Hypothesis-Confirming Bias in Labeling Effects." *Journal of Personality and Social Psychology* 44:20–33.

Fiske, Susan, and Steven Neuberg. 1990. "A Continuum of Impression Formation, from Category-Based to Individuating Processes." Pp. 1–63 in *Advances in Experimental Social Psychology*, vol. 23. Edited by Mark Zanna. New York: Academic Press.

Freeman, Richard B. 1987. "The Relation of Criminal Activity to Black Youth Employment." *Review of Black Political Economy* 16 (1–2): 99–107.

Freeman, Richard B., and Harry J. Holzer, eds. 1986. *The Black Youth Employment Crisis*. Chicago: University of Chicago Press for National Bureau of Economic Research.

Kluegel, James, and Lawrence Bobo. 2001. "Perceived Group Discrimination and Policy Attitudes: The Sources and Consequences of the Race and Gender Gaps." Pp. 163–216 in *Urban Inequality: Evidence from Four Cities*, edited by Alice O'Connor, Chris Tilly, and Lawrence D. Bobo. New York: Russell Sage Foundation.

Loury, Glenn C. 1977. "A Dynamic Theory of Racial Income Differences." Pp. 153–86 in *Women, Minorities, and Employment Discrimination*, edited by P. A. Wallace and A. M. La Mond. Lexington, Mass.: Heath.

Moss, Philip, and Chris Tilly. 1996. "'Soft Sills' and Race: An Investigation of Black Men's Employment Problems." *Work and Occupations* 23 (3): 256–76.

Neal, Derek, and William Johnson. 1996. "The Role of Premarket Factors in Black-White Wage Differences." *Journal of Political Economy* 104 (5): 869–95.

Pettit, Becky, and Bruce Western. 2001. "Inequality in Lifetime Risks of Imprisonment." Paper presented at the annual meetings of the American Sociological Association. Anaheim, August.

Quillian, Lincoln, and Devah Pager. 2002. "Black Neighbors, Higher Crime? The Role of Racial Stereotypes in Evaluations of Neighborhood Crime." *American Journal of Sociology* 107 (3): 717–67.

Sampson, Robert J., and John H. Laub. 1993. *Crime in the Making: Pathways and Turning Points through Life*. Cambridge, Mass.: Harvard University Press.

Shover, Neil. 1996. *Great Pretenders: Pursuits and Careers of Presistent Thieves.* Boulder, Colo.: Westview.

Slevin, Peter. 2000. "Life after Prison: Lack of Services Has High Price." *Washington Post,* April 24.

Smith, Tom W. 1991. *What Americans Say about Jews.* New York: American Jewish Committee.

Sniderman, Paul M., and Thomas Piazza. 1993. *The Scar of Race.* Cambridge, Mass.: Harvard University Press.

Uggen, Christopher, Melissa Thompson, and Jeff Manza. 2000. "Crime, Class, and Reintegration: The Socioeconomic, Familial, and Civic Lives of Offenders." Paper presented at the American Society of Criminology meetings, San Francisco, November 18.

Western, Bruce. 2002. "The Impact of Incarceration on Wage Mobility and Inequality." *American Sociological Review* 67 (4): 526–46.

Western, Bruce, and Becky Pettit. 1999. "Black-White Earnings Inquality, Employment Rates, and Incarceration." Working Paper no. 150. New York: Russell Sage Foundation.

Wilson, William Julius. 1987. *The Truly Disadvantaged: The Inner City, the Underclass, and Public Policy.* Chicago: University of Chicago Press.

——. 1997. *When Work Disappears: The World of the New Urban Poor.* New York: Vintage Books.

DISCUSSION QUESTIONS

1. What effect does Pager find of the impact of having a criminal record on one's employment status? What is the effect of race on this relationship?
2. Audit studies reveal the impact of race in a variety of settings. In what other context might you do an audit study to conduct research on the impact of race? See if you can design such a study, being careful to identify all of the things you would have to consider in setting it up.

51

Detaining Minority Citizens, Then and Now

ROGER DANIELS

By examining the detention of Japanese Americans during World War II, Daniels shows links between state policies of detention then and state policies that target Arab Americans now.

Sixty years ago, on February 19, 1942, in the aftermath of Pearl Harbor, Franklin Delano Roosevelt signed Executive Order 9066. Although it mentioned no ethnic group by name, it was the instrument by which some 120,000 Japanese Americans, more than two-thirds of them American citizens, were incarcerated, without indictment or trial, in ten desolate concentration camps in the interior of the United States.

In the months since the destruction of the World Trade Center and damage to the Pentagon, we are told that more than 1,000 aliens, largely of Middle Eastern nationalities, have been locked up. Many commentators have compared the two cases—some seeing a disturbingly similar pattern in the reaction against a feared nonwhite population, others praising what they see as the relative moderation of today's government.

Historical analogies are always tricky, particularly when one of the things being compared is a current event. Contemporary history is, after all, a contradiction in terms. Nevertheless, the ways in which our memory of what was done to Japanese Americans has evolved over six decades can shed some light on our contemporary situation.

In early 1942, there was almost no negative reaction to Roosevelt's order. The American Civil Liberties Union refused to protest, and Congress passed without a single negative vote a law imposing criminal penalties on anyone refusing to comply with the order....

Later during World War II, however, government attitudes to Japanese Americans began to soften: For example, by 1943 those who were American citizens were allowed to volunteer for military service, and, a year later, the draft was reimposed on them—even on those still behind barbed wire. And while the U.S. Supreme Court handed down, with three dissents, the now-discredited

SOURCE: From *Chronicle of Higher Education*, Feb. 15, 2002, p. 10. Reprinted by permission of the author.

Korematsu v. United States decision in December 1944—in effect sanctioning the government's right to put American citizens in concentration camps without indictment or trial—at the same time, indeed on the same day, it affirmed in the companion Ex parte Mitsuye Endo decision that citizens of undoubted loyalty should be free to come and go as they pleased.

After the war, President Harry S. Truman staged a welcome-home ceremony for members of the fabled Japanese American 442nd Regimental Combat Team and told the Nisei soldiers that "you have fought prejudice and won." Two years later, Truman successfully urged Congress to pass the Japanese American Evacuation Claims Act of 1948, which appropriated $38 million for all property of Japanese Americans that had been damaged or lost during their exile. That was nowhere near the true value, but no further government redress was made for nearly 40 years.

In the meantime, historians all but forgot what *Eugene V. Rostow*, in 1945, had called "our worst wartime mistake." In what was probably the most-liberal, general American-history text published in the 1950s, Richard Hofstadter and two collaborators, who devoted almost a full page to American intolerance in 1917–18, could say only: "Since almost no one doubted the necessity for the war, there was much less intolerance than there had been in World War I, although large numbers of Japanese Americans were put into internment camps under circumstances that many Americans were later to judge unfair or worse."

In the same decade, a security-minded Congress passed the Emergency Detention Act of 1950, authorizing, in the event of any future, "Internal Security Emergency," the apprehension and confinement of persons who "probably will engage in, or probably will conspire with others to engage in, acts of espionage or of sabotage...." Passed over Truman's veto and never invoked, the act nevertheless stayed on the books for 17 years. Then, in the civil-rights-conscious and protest-prone 1960s, a campaign for repeal was mounted, culminating in a law sponsored by two Japanese American legislators from Hawaii in 1971. Five years later, as a kind of bicentennial confession of political sin, President Gerald R. Ford used the 34th anniversary of Executive Order 9066 to issue a proclamation marking a "sad day in American history."

However, at the start of the 1970s a few Japanese American activists were already beginning to urge the Japanese American Citizens League to support a campaign for some kind of more tangible public atonement. The struggle was long and difficult. First, dubious Japanese Americans, who feared a backlash, had to be persuaded; then a presidential Commission on the Wartime Relocation and Internment of Civilians had to be created to study the matter. In 1983, that body called for a formal apology and a tax-free payment of $20,000 to each surviving internee.

And that, finally, set off a national examination of the wartime atrocity. After five years of debate, and formal opposition from the Reagan administration, the Civil Liberties Act of 1988 passed Congress and received Ronald Reagan's signature—although no money was appropriated, and no payments were made or individual apologies issued, until late 1990. The last of more than 80,000 payments was made early in the Clinton administration. The total cost was some $1.6 billion.

One of the recurrent themes—heard at first intermittently, more strongly since Ford's 1976 proclamation—has been: This kind of thing must never happen again. Anti-Asian activities have not been completely absent in recent decades, but few would disagree with the notion that there is less prejudice against Asians and other nonwhites than there was six decades ago. Yet, although nothing even resembling the mass roundup of Japanese Americans in 1942 has taken place, there has, nevertheless, been a persistent willingness to blame groups of "others" for the actions of individuals.

That is why, despite its phased-in atonements for the events of 1942, the U.S. government could continue to behave as if the civil rights of some persons—mostly aliens—were less to be regarded than those of others. During the crisis triggered by the takeover of the American Embassy in Tehran, for example, the Carter White House asked the Immigration and Naturalization Service to give it the names and addresses of all Iranian students in the United States—something the inept and underfinanced bureaucracy could not do. Then the administration asked colleges and universities to provide such lists, and most obediently complied. (Harvard was a notable exception).

During the Reagan administration, the inhumane treatment of illegal Haitian immigrants, incarcerated in Miami, was eventually enjoined by a federal judiciary unconstrained by a wartime emergency; the administration of the first President George Bush moved some of those immigrants to the American military base at Guantanamo Bay, thus largely avoiding the jurisdiction of federal courts, a policy continued by the Clinton and present Bush administrations. In a similar action singling out a group of people, just prior to the start of Desert Storm federal agents interrogated Muslim leaders in the United States—both citizens and aliens. (The questioning largely ceased after complaints; the government lamely explained that the interviews were to provide protection for those questioned.)

Today, we once again read about the detention of resident aliens for questioning; about plans to bypass normal legal procedures and create military tribunals to try "any individual who is not a United States citizen;" about federal requests to colleges and universities for the names of all foreign students. But when compared with what was done to Japanese. Americans during World War II, government actions before and after September 11 do not seem to amount to very much. Indeed, many media commentators have objected that even to mention them in connection with the massive violations of civil liberties by the Roosevelt administration is inappropriate.

That is an evasion: the kind of evasion that has allowed us to offer apologies for the actions we have taken against those whom we perceive to be outsiders, and then do the same thing to a different group. Time and again, scholars (if not the government) have eventually acknowledged that we, as a nation, have violated the spirit of our Constitution. Time and again, we have gone on to violate it again.

If that history isn't enough to give us pause about where our current actions might lead, consider this: Roosevelt's order was issued in wartime; the subsequent government actions have occurred in societies in which both racial discrimination and xenophobia have been greatly reduced, and they have often been accompanied (as Roosevelt's was) by reassuring statements from the highest levels of government about religious, racial, and ethnic egalitarianism. Even in the best of times, we have been able to justify abridging some peoples' rights.

Moreover, despite those laudable statements about nondiscrimination from the top of the chain of command, practice at the lower levels of government has often been blatantly biased. Nowhere has that been more evident than in the immigration service. And in no instance that I know has any lower government official been punished or even reprimanded for improper actions. As I was writing last week, we were still hearing of the roundup of suspects by law-enforcement officials and the assertions by INS officials of their plans to concentrate on "Middle Eastern" students and others who overstay their visas.

Moreover, private groups like airlines have forced citizens and aliens who look like the "enemy" to leave flights for which they had tickets—sometimes even winning praise for doing so. "I was relieved at the story of the plane passengers a few weeks ago who refused to board if some Mideastern-looking guys were allowed to board," Peggy Noonan, a contributing editor, wrote in *The Wall Street Journal.* "I think we're going to require a lot of patience from a lot of innocent people.... And you know, I don't think that's asking too much." Even more disturbing than such blathering is the fact that the cabinet officer responsible for aviation, Secretary of Transportation Norman Mineta—himself a child victim of wartime incarceration, who often recounts going off in his Cub Scout uniform to a 1942 government roundup—has done nothing about such blatant discrimination, at least not in public.

Optimists assure us that a mass incarceration of American citizens in concentration camps will not recur. But reflection on our past suggests we ought not to be so sanguine. To be sure, it was not just the disaster at Pearl Harbor, but the subsequent sequence of Japanese triumphs that triggered Executive Order 9066. But shouldn't we then ask, If successful terrorist attacks hadn't abated after September 11, would the current government reaction have been so moderate?

Many Japanese Americans, the only group of our own citizens ever incarcerated because of who they were as opposed to what they did, argue that the optimists are wrong and that what has happened before can surely, given the right triggering circumstances, happen again. As a student of Japanese American history, I can only agree with them.

DISCUSSION QUESTIONS

1. What actions did the government take with regard to Japanese immigrants and Japanese citizens after the bombing of Pearl Harbor in 1941?
2. What restrictions did Arab Americans face in the aftermath of the events on September 11, 2001?
3. What current issues do Arab Americans face as citizens of the United States?

Part VIII
Student Exercises

1. Think about your first experience of finding employment. Did you start in the informal economy, like babysitting or doing yardwork, or the formal economy, like the Taco Bell or the mall? At what age did you start to work? How much money did you earn and what did you do with it? What role did your community or your family network play in your securing of employment? Having thought about this, share your experiences with the whole class, or in groups of eight or ten, so that there will be some diversity of experiences heard. How do you think your own work experience is related to your race, gender, social class, and residential location? What lessons do you learn from this exercise about how race, gender, social class, residential location, and other social factors are related to the young person's job search?
2. In 2007, the state of New Jersey voted to eliminate the death penalty; instead they will sentence people who are found guilty of certain crimes to life without parole. The death penalty is very controversial in the United States—with supporters and critics. What are the practices with regard to the death penalty in your state? What issues about race and class bias are embedded in debates about the death penalty?

PART IX

INTRODUCTION

Mobilizing for Change: Looking Forward and Learning from the Past

Racial hierarchies, like all social hierarchies, are unstable and fluid. They change over time, but how do they change? They can change because the political economy needs different groups of workers with particular skills, which can result in dominant groups changing their views of others. For example, African Americans were barred from industrial work in one era but then were recruited for those same jobs when immigration was curtailed. Access to industrial work not only changed their opportunities but brought them into the cities in great numbers and gave them the resources to create new representations of themselves, as in the flowering of the Harlem Renaissance (Marks 1989).

More often, however, change occurs because of actions taken by oppressed people. Dominant groups use their economic and political power to develop ideologies that make the current racial order appear natural. Disadvantaged groups in turn mobilize not only to challenge those ideologies but even to change laws and create a more inclusive and just society. Such actions show us that, although systems of inequality may be established by those with the most power, they are not just quietly accepted. Rather, they are contested by oppressed groups and their allies. For example, notions of racial and cultural inferiority justified slavery, the mistreatment of American Indians, and the denial of rights to Chinese immigrants. Members of these groups had different interpretations of their situations. For example, many slaves rejected the system, escaping to form Maroon colonies in the South, and some even engaged in armed rebellions. American Indians knew that European settlers had guns while they had only bows and arrows and knew that the laws of the land gave advantages to White people, who had economic and political power. Still, they continued to contest and challenge these views. Denied the opportunity to become citizens, the Chinese used the state and federal court systems to fight the erosion of their rights as human beings (Takaki 1993). Among all oppressed racial and ethnic groups, there is a long history of active resistance, even as the elite of the nation used the power of government to control those who would not conform.

Racism is a tool for exclusion, but it also can be used to build community. People who are denied participation in the mainstream society build their own communities; segregation itself can end up fostering community building. Such communities are places from which people question the controlling images and negative stereotypes that the dominant group employs to justify inequality. Within their own communities, young people of color can learn affirming messages about themselves, as well as structural explanations for their lack of resources and opportunities. Such messages contradict explanations from the ruling system of power that often define those with fewer resources as less able. Alternative explanations motivate oppressed people and those who support oppressed people's goals to demand that all people have opportunities, not just the advantaged group. Although some members of oppressed groups internalize the negative messages from the dominant group, most develop a counter (or an oppositional) perspective.

Earlier, in Part II, Michael Omi and Howard Winant ("On Racial Formation") presented the concept of *racial projects*, defined as organized efforts to distribute social and economic resources along racial lines. However, racial projects can also be organized to redistribute resources with equality in mind as disadvantaged people challenge interpretations of themselves as undeserving. The history of resistance reveals that racial projects waged by disadvantaged groups are often undertaken to build and sustain community, so that they can survive and raise the next generation. With these actions, people challenge dominant, controlling images with new ones that better represent themselves (Omi and Winant 1994; Takaki 1993). Sharing new images with the majority groups and others in the society challenges the ideological justification for their unequal treatment.

Throughout this volume, we have seen that African Americans, Native Americans, Chinese Americans, Japanese Americans, Latinos, and others have been displeased with their status in U.S. society. Earlier European immigrant groups and their descendants were also unhappy with their status as exploited workers denied full participation in the society. Many survived in ethnic enclaves, but some found options for mobility during World War II, even while people of color remained economically and socially marginalized. In the twentieth century, people of color used different strategies to initiate social change. People waged legal struggles to abolish laws legalizing racial segregation and to create new ones that protected them from discrimination. For example, the National Association for the Advancement of Colored People (NAACP) brought together a legal team that worked to equalize salaries for teachers in segregated school systems, to guarantee Black workers' rights, and to challenge the system of Jim Crow segregation. Other racial groups also worked through the courts. As laws changed, so did expectations, but stakeholders in the system do not change immediately. As Martin Luther King, Jr. wrote in his famous "Letter from a Birmingham Jail," no one gives up privilege willingly (Williams 1987).

New arrangements are not simply a result of changes in the law. They take determined people actively pressing for their rights. Part IX looks at how people mobilize to bring those rights into being, thereby pushing the nation to face up to its mission to be a society in which all people are given, not just promised, rights. It is through social movements—groups that organize and act to bring about social change—that our nation has been reshaped. Is the nation now ready to move forward toward more equality?

Writing as the twenty-first century approached, Aldon Morris ("The Genius of the Civil Rights Movement") has reviewed the history of resistance that launched a major grassroots movement, as ordinary African American men and women engaged in nonviolent social protests, often facing violent resistance. They boycotted buses and stores, held sit-ins at lunch counters and movie theaters, and marched for voting rights. Their very actions challenged mainstream images of inferiority and their presumed satisfaction with second-class citizenship. There was power in resistance. It forced a nation to look at the consequences of imposed segregation and to take action against it. In addition to building support for civil rights legislation, the Civil Rights Movement was an inspiration for other groups, both within the United States and abroad.

Change does not always come through the efforts of mass-based movements. Many people work for change every day within the context of their work in nonprofit organizations, private companies, schools, and other organizations. What they can do, though, is often hampered by the workings of bureaucratic systems, as Harvey Molotch describes in his article ("Death on the Roof: Race and Bureaucratic Failure"), which discusses the federal failures in the aftermath of Hurricane Katrina. The failure of the government to respond quickly and effectively is well known and resulted in the deaths of many poor, African American people trapped by the rising floodwaters when the levees broke. Molotch shows that this disaster was not just the failure of individuals, but the failure of entire social systems, crippled by bureaucratic obstacles. While Katrina provides one horrifying example of such organizational failure, many toil every day to make social systems more responsive to people's needs that stem from racism, poverty, and other social ills.

David Naguib Pellow and Robert J. Brulle ("Poisoning the Planet: The Struggle for Environmental Justice") show how social change can occur when people mobilize to address a problem in their local community. Their article describes how people have organized to eliminate some of the toxic waste dumping that plagues their neighborhoods. Now known as the *environmental justice movement*, a coalition of local groups decided to take action against the vulnerability of their communities to toxic waste. The article shows how today's efforts to establish racial justice must be based on the particular problems in our changing racial landscape.

How do we work to end racial injustice and build a society that is a community of equals? We can begin by acknowledging our racial legacy and the impact of race on both institutions and individuals. Even as the media articulates a colorblind framework, we have learned from this volume that race matters and our racial identities are significant. However, powerful institutions and people still benefit from racial stratification and use their resources to distort that reality. Only when you see the power arrangements and systemic racism in society, can you begin to dismantle it. As in the past, changing the landscape requires a great deal of effort. This book gives you knowledge to inform your understanding of race in society. How do you work for change?

Jacqueline Johnson, Sharon Rush, and Joe Feagin ("Reducing Inequalities") discuss the need for institutional change and antiracism work on the part of many to build a new society. They argue that action is necessary on many fronts. The public educational system can play a role in redressing patterns of segregation that keep us from knowing each other. Media constructions of people of color are distortions that are kept alive when people lack direct contact as equals. The commitment to building an antiracist society requires cognitive and emotional work for us all. In the essay by Chesler, Peet, and Sevig in Part III ("Blinded by Whiteness"), we learned about racial identity and how learning about privilege can help people make new decisions. If we change both institutions and individual behavior, we can work to build an antiracist society. People of color can reject ideologies of inferiority. Whites can gain new insights into their own situation and that of others when they form friendships across racial-ethnic lines. Such relationships can help them recognize their own privileges and the cost of those privileges for others. Johnson, Rush, and Feagin's vision is a powerful one that can be shared with others who are interested in actively shifting the landscape towards greater racial equality.

REFERENCES

Marks, Carole. 1989. *Farewell—We're Good and Gone: The Great Black Migration*. Bloomington: Indiana University Press.

Omi, Michael, and Howard Winant. 1994. *Racial Formation in the United States*. New York: Routledge.

Takaki, Ronald. 1993. *A Different Mirror: A History of Multicultural America*. Boston: Little, Brown and Company.

Williams, Juan. 1987. *Eyes on the Prize: America's Civil Rights Years, 1954–1965*. New York: Penguin Books.

52

The Genius of the Civil Rights Movement

Can It Happen Again?

ALDON MORRIS

The Civil Rights Movement, Morris argues, succeeded in dismantling the explicit and overtly visible practices of Jim Crow racial segregation. Because this was such a visible target, it was possible to mount this historic mass movement for social change. Morris questions what mobilization for racial justice will look like given the racial realities and politics of the twenty-first century.

It is important for African Americans, as well as all Americans, to take a look backward and forward as we approach the turn of a new century, indeed a new millennium. When a panoramic view of the entire history of African Americans is taken into account, it becomes crystal clear that African American social protest has been crucial to Black liberation. In fact, African American protest has been critical to the freedom struggles of people of color around the globe and to progressive people throughout the world.

The purpose of this essay is: 1) to revisit the profound changes that the modern Black freedom struggle has achieved in terms of American race relations; 2) to assess how this movement has affected the rise of other liberation movements both nationally and internationally; 3) to focus on how this movement has transformed how scholars think about social movements; 4) to discuss the lessons that can be learned from this groundbreaking movement pertaining to future African American struggles for freedom in the next century.

It is hard to imagine how pervasive Black inequality would be today in America if it had not been constantly challenged by Black protests throughout each century since the beginning of slavery. The historical record is clear that slave resistance and slave rebellions and protest in the context of the Abolitionist movement were crucial to the overthrow of the powerful slave regime.

The establishment of the Jim Crow regime was one of the great tragedies of the late nineteenth and early twentieth centuries. The overthrow of slavery represented one of those rare historical moments where a nation had the opportunity to embrace a democratic future or to do business as usual by reinstalling undemocratic practices. In terms of African Americans, the White North and South chose to embark along undemocratic lines.

SOURCE: From Aldon Morris, "The Genius of the Civil Rights Movement: Can It Happen Again?" Reprinted by permission of the author.

For Black people, the emergence of the Jim Crow regime was one of the greatest betrayals that could be visited upon a people who had hungered for freedom so long: what made it even worse for them is that the betrayal emerged from the bosom of a nation declaring to all the world that it was the beacon of democracy.

The triumph of Jim Crow ensured that African Americans would live in a modern form of slavery that would endure well into the second half of the twentieth century. The nature and consequences of the Jim Crow system are well known. It was successful in politically disenfranchising the Black population and in creating economic relationships that ensured Black economic subordination. Work on wealth by sociologists Melvin Oliver and Thomas Shapiro (1995), as well as Dalton Conley (1999), are making clear that wealth inequality is the most drastic form of inequality between Blacks and Whites. It was the slave and Jim Crow regimes that prevented Blacks from acquiring wealth that could have been passed down to succeeding generations. Finally, the Jim Crow regime consisted of a comprehensive set of laws that stamped a badge of inferiority on Black people and denied them basic citizenship rights.

The Jim Crow regime was backed by the iron fist of southern state power, the United States Supreme Court, and white terrorist organizations. Jim Crow was also held in place by white racist attitudes. As Larry Bobo has pointed out, "The available survey data suggests that anti-Black attitudes associated with Jim Crow were once widely accepted... [such attitudes were] expressly premised on the notion that Blacks were innately, intellectually, culturally, and temperamentally inferior to Whites" (Bobo, 1997: 35). Thus, as the twentieth century opened, African Americans were confronted with a powerful social order designed to keep them subordinate. As long as the Jim Crow order remained intact, the Black masses could breathe neither freely nor safely. Thus, nothing less than the overthrow of a social order was the daunting task that faced African Americans during the early decades of the twentieth century.

The voluminous research on the modern civil rights movement has reached a consensus: That movement was the central force that toppled the Jim Crow regime. To be sure, there were other factors that assisted in the overthrow including the advent of the television age, the competition for Northern Black votes between the two major parties, and the independence movement in Africa which sought to overthrow European domination. Yet it was the Civil Rights Movement itself that targeted the Jim Crow regime and generated the great mass mobilizations that would bring it down.

What was the genius of the Civil Rights Movement that made it so effective in fighting a powerful and vicious opposition? The genius of the Civil Rights Movement was that its leaders and participants recognized that change could occur if they were able to generate massive crises within the Jim Crow order—crises of such magnitude that the authorities of oppression must yield to the demands of the movement to restore social order. Max Weber defined power as the ability to realize one's will despite resistance. Mass disruption generated power. That was the strategy of nonviolent direct action. By utilizing tactics of disruption, implemented by thousands of disciplined demonstrators who had been mobilized through their churches, schools, and voluntary associations, the Civil Rights Movement was able to generate the necessary power to overcome the Jim Crow regime. The famous crises created in places like Birmingham and Selma, Alabama, coupled with the important less visible crises that mushroomed throughout the nation, caused social breakdown in Southern business and commerce, created unpredictability in all spheres of social life, and strained the resources and credibility of Southern state governments while forcing white terrorist groups to act on a visible stage where the whole world could watch. At the national level, the demonstrations and repressive measures used against them generated foreign policy nightmares because they were covered by foreign media in Europe, the Soviet Union, and Africa. Therefore what gave the mass-based sit-ins, boycotts, marches, and jailing their power was their ability to generate disorder.

As a result, within ten years—1955 to 1965—the Civil Rights movement had toppled the Jim Crow order. The 1964 Civil Rights Bill and the 1965 Voting Rights Act brought the regime of formal Jim Crow to a close.

The Civil Rights Movement unleashed an important social product. It taught that a mass-based grassroots social movement that is sufficiently organized, sustained, and disruptive is capable of generating fundamental social change. In other words, it showed that human agency could flow from a relatively powerless and despised group that was thought to be backward, incapable of producing great leaders.

Other oppressed groups in America and around the world took notice. They reasoned that if American Blacks could generate such agency they should be able to do likewise. Thus the Civil Rights movement exposed the agency available to oppressed groups. By agency I refer to the empowering beliefs and action of individuals and groups that enable them to make a difference in their own lives and in the social structures in which they are embedded.

Because such agency was made visible by the Civil Rights movement, disadvantaged groups in America sought to discover and interject their agency into their own movements for social change. Indeed, movements as diverse as the student movement, the women's movement, the farm worker's movement, the Native American movement, the gay and lesbian movement, the Environmental movement, and the disability rights movement all drew important lessons and inspiration from the Civil Rights Movement. From that movement other groups discovered how to organize, how to build social movement organizations, how to mobilize large numbers of people, how to devise appropriate tactics and strategies, how to infuse their movement activities with cultural creativity, how to confront and defeat authorities, and how to unleash the kind of agency that generates social change.

For similar reasons, the Black freedom struggle was able to effect freedom struggles internationally. For example, nonviolent direct action has inspired oppressed groups as diverse as Black South Africans, Arabs of the Middle East, and pro-democracy demonstrators in China to engage in collective actions. The sit-in tactic made famous by the Civil Rights Movement, has been used in liberation movements throughout the third world, in Europe, and in many other foreign countries. The Civil Rights Movement's national anthem "We Shall Overcome" has been interjected into hundreds of liberation movements both nationally and internationally. Because the Civil Rights Movement has been so important to international struggles, activists from around the world have invited civil rights participants to travel abroad. Thus early in Poland's Solidarity movement Bayard Rustin was summoned to Poland by that movement. As he taught the lessons of the Civil Rights Movement, he explained that "I am struck by the complete attentiveness of the predominantly young audience, which sits patiently, awaiting the translations of my words," (Rustin, undated).

Therefore, as we seek to understand the importance of the Black Freedom Struggle, we must conclude the following: the Black Freedom Struggle has provided a model and impetus for social movements that have exploded on the American and international landscapes. This impact has been especially pronounced in the second half of the twentieth century.

What is less obvious is the tremendous impact that the Black Freedom Struggle has had on the scholarly study of social movements. Indeed, the Black freedom struggle has helped trigger a shift in the study of social movements and collective action. The Black movement has provided scholars with profound empirical and theoretical puzzles because it has been so rich organizationally and tactically and because it has generated unprecedented levels of mobilization. Moreover, this movement has been characterized by a complex leadership base, diverse gender roles, and it has revealed the tremendous amount of human agency that usually lies dormant within oppressed groups. The empirical realities of the Civil Rights Movement did not square with the theories used by scholars to explain social movements prior to the 1960s.

Previous theories did not focus on the organized nature of social movements, the social movement organizations that mobilize them, the tactical and strategic choices that make them effective, nor the rationally planned action of leaders and participants who guide them. In the final analysis, theories of social movements lacked a theory that incorporated human agency at the core of their conceptual apparatuses. Those theories conceptualized social movements as spontaneous, largely unstructured, and discontinuous with institutional and organizational behavior. Movement participants were viewed as reacting to various forms of strain and doing so in a non-rational manner. In these frameworks, human agency was conceptualized as reactive, created by uprooted individuals seeking to reestablish a modicum of personal and social stability. In short, social movement theories prior to the Civil Rights Movement operated with a vague, weak vision of agency to explain phenomena that are driven by human action.

The predictions and analytical focus of social movement theories prior to the 1970s stood in sharp contrast to the kind of theories that would be needed to capture the basic dynamics that drove the Civil Rights movement. It became apparent to social movement scholars that if they were to understand the Civil Rights Movement and the multiple movements it spun, the existing theoretical landscape would have to undergo a radical process of reconceptualization.

As a result, the field of social movements has been reconceptualized and this retheorization will effect research well into the new millennium. To be credible in the current period any theory of social movements must grapple conceptually with the role of rational planning and strategic action, the role of movement leadership, and the nature of the mobilization process. How movements are gendered, how movement dynamics are bathed in cultural creativity, and how the interactions between movements and their opposition determine movement outcomes are important questions. At the center of this entire matrix of factors must be an analysis of the central role that human agency plays in social movements and in the generation of social change.

Thanks, in large part, to the Black freedom struggle, theories of social movements that grapple with real dynamics in concrete social movements are being elaborated. Intellectual work in the next century will determine how successful scholars will be in unravelling the new empirical and theoretical puzzles thrust forth by the Black freedom movement. Although it was not their goal, Black demonstrators of the Civil Rights movement changed an academic discipline.

A remaining question is: Will Black protest continue to be vigorous in the twenty-first century, capable of pushing forward the Black freedom agenda? It is not obvious that Black protest will be as sustainable and as paramount as it has been in previous centuries. To address this issue we need to examine the factors important to past protests and examine how they are situated in the current context.

Social movements are more effective when they can identify a clear-cut enemy. Who or what is the clear-cut enemy of African Americans of the twenty-first century? Is it racism, and if so, who embodies it? Is it capitalism, and if so, how is this enemy to be loosened from its abstract perch and concretized? In fact, we do not currently have a robust concept that grasps the modern form of domination that Blacks currently face. Because the modern enemy has become opaque, slippery, illusive, and covert, the launching of Black protest has become more difficult because of conceptual fuzziness.

Second, during the closing decades of the twentieth century the Black class structure has become more highly differentiated and it is no longer firmly anchored in the Black community. There is some danger, therefore, that the cross-fertilization between different strata within the Black class structure so important to previous protest movements may have become eroded to the extent that it is no longer fully capable of launching and sustaining future Black protest movements.

Third, will the Black community of the twenty-first century possess the institutional strength required for sustaining Black protest? Black colleges have been weakened because of the racial integration of previously all white institutions of higher learning and because many Black colleges are being forced to integrate. The degree of institutional strength of the church has eroded because some of them have migrated to the suburbs in an attempt to attract affluent Blacks. In other instances, the Black Church has been unable to attract young people of the inner city who find more affinity with gangs and the underground economy. Moreover, a great potential power of the Black church is not being realized because its male clergy refuse to empower Black women as preachers and pastors. The key question is whether the Black church remains as close to the Black masses—especially to poor and working classes—as it once was. That closeness determines its strength to facilitate Black protest.

In short, research has shown conclusively that the Black church, Black colleges and other Black community organizations were critical vehicles through which social protest was organized, mobilized and sustained. A truncated class structure was also instrumental to Black protest. It is unclear whether during the twenty-first century these vehicles will continue to be effective tools of Black protest or whether new forces capable of generating protest will step into the vacuum.

In conclusion, I foresee no reason why Black protest should play a lesser role for Black people in the twenty-first century. Social inequality between the races will continue and may even worsen especially for poorer segments of the Black community. Racism will continue to effect the lives of all people of color. If future changes are to materialize, protest will be required. In 1898 as Du Bois glanced toward the dawn of the twentieth century, he declared that in order for Blacks to achieve freedom they would have to protest continuously and energetically. This will become increasingly true for the twenty-first century. The question is whether organizationally, institutionally, and intellectually the Black community will have the wherewithal to engage in the kind of widespread and effective social protest that African Americans have utilized so magnificently. If previous centuries are our guide, then major surprises on the protest front should be expected early in the new millennium.

REFERENCES

Bobo, L. 1997. "The Color Line, the Dilemma, and the Dream: Race Relations in America at the Close of the Twentieth Century." In *Civil Rights and Social Wrongs: Black-White Relations since World War II*, edited by J. Higham, pp. 31–55. University Park. PA: Penn State University Press.

Conley, Dalton. 1999. *Being Black. Living in the Red: Race, Wealth, and Social Policy in America*. Berkeley: University of California Press.

Rustin, Bayard. no date. *Report on Poland*. New York: A. Philip Randolph Institute.

Oliver, Melvin, and Thomas E. Shapiro. 1995. *Black Wealth/White Wealth: A New Perspective on Racial Inequality*. New York: Routledge.

DISCUSSION QUESTIONS

1. What was the Jim Crow regime? How did Black Americans bring an end to the Jim Crow regime?
2. What does Morris mean by *human agency*?
3. What tactics or strategies developed by the grassroots Civil Rights Movement have been employed by other groups?

53

Death on the Roof: Race and Bureaucratic Failure

HARVEY MOLOTCH

Molotch analyzes the federal failure to respond to the devastation caused by Hurricane Katrina, using a sociological lens to do so. His article also reveals the difficulties encountered by trying to work for racial justice within bureaucratic organizations.

Would so many white people struggling for life be ignored for so long? Racism explains some of what went on, but its route was indirect....

One of the race-based explanations is that those left behind are consistently the most deprived. The legacy of slavery, exclusion, and segregation corrals those with least resources into a vulnerable space, natural and economic. Governments go on to reinforce that vulnerability with the way they spend money. Katrina shows how this works in otherwise low-visibility arenas like Army Corps of Engineers flood control projects. Institutional racism, familiar in so many other contexts, does its job on the levee.

But a more subtle explanation involves tracing how race impacts organizations' abilities to deal with unanticipated problems. I will outline one effort. A founding sociological truth is that bureaucratic actors follow rules—something that gets a vast amount of routine accomplished but works against adaptation to change. But sociologists went on to discover that people "bend" those rules, responding for example, to pressures from co-workers or other elements of context. This leads to so-called "informal rules" that affect who does what. The bureaucracy becomes suppler (and less predictable).

This still does not go far enough in recognizing the social ingenuity of bureaucrats. They do what most everyone does when they need to solve problems. They invent as they go. Rules, formal or informal, become a resource people use to explain their own behavior, not simply a force in determining it. Rather than follow-through on dictates, individuals have a sense—gestalt-like—of what to do given the context. This frees up the capacity to innovate, to make the rules come alive "on the fly."

SOURCE: From Website, New York: Social Science Research Council, accessed June 2006. www.ssrc.org

Receptionists in a welfare agency, Don Zimmerman showed, ignore "first-come, first served" when a screaming child makes it impossible for other workers to be heard by their clients.[1] The squeaky wheel gets the grease so that all the other wheels can be kept going. Harried hospital emergency room doctors, whose oath prescribes otherwise, work less hard to save lives of derelicts whose alcoholism they think will lead to eventual fatality anyway.[2]

For good or ill, actors find a way. Sometimes they discover, from the welter of regulations, another rule, a "better" rule to follow. Or they just act under the presumption that others will later see they did the right thing. Indeed, to avoid being what Harold Garfinkel called "judgmental dopes," they *must* use such discretion. Otherwise, their "work to rule" appears incompetent, obstreperous or maybe even insane.[3]

Both a full-on list of rules and the simultaneous capacity to not follow them make society possible. Individuals figure out what to do on an ad hoc basis—mobilizing, elaborating, and finessing the rules as things move along. We are, as the ethnomethodologists reiterate, "artful," and this forms social life, bureaucratic and otherwise.

As with other zones of organized activity, emergencies do involve Weberian protocols. Vehicles, uniforms, and building entrances carry the word "emergency," each bound up with its own routines. Codified triggers do set off responses. The bell rings, the firefighter slides down the pole. At each scale, protocols specify how to fill out the paperwork, who gives the clearance, which budget to use. So when "911" calls for help come in, emergency call-takers require not only that certain bits of information be provided (e.g. nature of problem, whether or not the caller is alone or with others) but be provided in a particular sequence. But that is not the end of the story; call takers, the research shows, also adapt. Indeed, if the call-taker really insists that all the "i"s be dotted, death can happen on the other end of the line.[4]

In the Katrina case we are learning that functionaries, in searching out the rules to apply, did a poor job. They seem to have fallen back on the Weberian default. They became bureaucrats in the formal sense, rather than the bureaucrats who populate much of real life. Federal authorities apparently said the local governments' requests were too vague; they needed specific requisition. The forms had to be filled out in the right way. Competent bureaucrats take it into account when a caller for help is trying to communicate while drowning. To do otherwise is, *in bureaucratic terms*, pathological.

Now back to race. For organizational creativity to happen, there needs to be motivation to look up, think, and find the route. For many people, Katrina would be a no-brainer; something MUST be done, the evident suffering MUST be dealt with. No aspect of protocol can stand in the way. Convention, easy to invoke if you are without a contrary motivation, has to be overcome. What helps?

First, people need a lateral scan that takes in, gestalt-like, the larger context. This includes the suffering of others. The deeper and the more heart-felt, the more energetic will be the search and rescue for the procedures that can work. Ideally, it is as though it is oneself or one's own family in desperate straits.

Second, there needs to be a blink of understanding that others' orientation will be the same, that one will not be alone out on the limb of empathy. You can get away with commandeering fleets of buses, moving funds across budgetary categories, and contacting people out of the sanctioned communication order. Not only can you get away with it, you will—the presumption may be made—later be seen as having done the right thing.

Unprecedented action requires some personal adrenaline within and around the bureaucracy. It happens: a kind of panic of empathy that trumps organizational habit and individual postures. Response to the World Trade Center attack had elements of this. Private companies and public agencies went into action. They put equipment in place and spent a lot of money with contracts worked out on the spot or with no contracts at all. A shared sensibility fueled corporations, bureaucracies, and political units. Given the gargantuan scale and wider consequentiality, it reads in retrospect as organizational heroism—which is what it was.

Anything that inhibits empathy for the victims or weakens the assumption that others share it, undermines the likelihood of effective rescue. The rites, rituals, and sometimes dopey constraints of bureaucratic life will remain the default. Some depict this failure as an inherent feature of bureaucracy, and of government bureaucracy in particular. But this is the case only if one has a naïve sense of what competent people in bureaucracies actually do.

The culprit was not bureaucracy, per se, but the structure of feeling among those responsible and the milieu in which they operated. And now we can see how a little bit of racism can go a long way. It amplifies into the vast realm of business as usual. The default to literalness, always available to the irresponsibly prudent, binds as a force. The rescue workers and the victims alike remain in the iron cage.

White Americans do not completely accept African-Americans. As the sociologist Gary Schulman found when he replicated the Milgram shock experiments, white people are a little more likely to shock black "victims," at least under certain conditions, compared to fellow whites under the same conditions.[5] Certainly, white attitudes toward blacks have become more positive over time. And African-Americans take up prominent U.S. public roles. But for those in charge, the victims of the flood were sufficiently outside to hinder bust-out.

Let's also get back to government incompetence. Incompetent people occupy many positions in all realms. Maybe the wrong person got put in charge of something very important, maybe because it was not anticipated that it would become all that important during their tenure. More common still, people default to routines that no longer work—because that is the world they know and deploying a larger imagination is not their strong suit.

But others are too passionate and imaginative to let it happen; they see too much. They break the mold; they deal with it. The big boss moves in and exercises direct authority. Friends and cronies put aside delicacies of personal loyalty: this must be done. Or those in the next layer just beneath rise up and take control, maybe even sensing that those above will fall into line because the need for innovation and change is so clear.

In this case of Katrina, we had very little of this. Somebody really smart (Karl Rove?) did not press the panic button. Those just below, obviously agitated in many cases, could not assume their actions would be seen as "of course" necessary. They could not surmise, probably quite accurately, that parallel takeovers would make for coordination across different geographic and administrative spheres.

The human capacity for overcoming inertia, including formal emergency inertia, could not come into play. No one with power reached over, pressed down, or pushed up to make the rescue. I think race was at hand.

NOTES

1. Don Zimmerman, "Tasks and troubles: the practical bases of work activities in a public assistance agency" in D. Hansen (ed.) *Explorations in Sociology and Counseling*. New York: Houghton-Mifflin. 1969; Don Zimmerman, "The practicalities of rule use" pp. 285–95 in J. Douglas (ed.) *Understanding Everyday Life*. Chicago: Aldine. 1970.
2. David Sudnow, *Passing On: The Social Organization of Dying*. Englewood Cliffs, NJ: Prentice Hall. 1967.
3. Harold Garfinkel, *Studies in Ethnomethodology*. 1967. Englewood Cliffs, NJ: Prentice Hall.
4. Marilyn Whalen and Don Zimmerman, "Sequential and institutional contexts in calls for Help." *Social Psychology Quarterly*. Vol 30, pp. 172–85. 1987.
5. Gary Schulman, "Race, Sex, and Violence: A Laboratory Test of the Sexual Threat of the Black Male Hypothesis." *American Journal of Sociology*. Vol. 79, No. 5 (Mar., 1974), pp. 1260–1277.

DISCUSSION QUESTIONS

1. What role did bureaucracy play in the disastrous recovery efforts following Hurricane Katrina? How does an understanding of bureaucracy help you interpret this massive federal failure?
2. What does Molotch's argument suggest to you about the difficulties you might encounter in working in organizations where the intent is to address some of our nation's (or your community's) racial problems?

54

Poisoning the Planet: The Struggle for Environmental Justice

DAVID NAGUIB PELLOW AND ROBERT J. BRULLE

Pellow and Brulle's discussion of the environmental justice movement places it within the context of national and international racial politics. In doing so, they also document the activism of people of color in challenging the dumping of toxic waste and pollution in their communities.

One morning in 1987 several African-American activists on Chicago's southeast side gathered to oppose a waste incinerator in their community and, in just a few hours, stopped 57 trucks from entering the area. Eventually arrested, they made a public statement about the problem of pollution in poor communities of color in the United States—a problem known as environmental racism. Hazel Johnson, executive director of the environmental justice group People for Community Recovery (PCR), told this story on several occasions, proud that she and her organization had led the demonstration. Indeed, this was a remarkable mobilization and an impressive act of resistance from a small, economically depressed, and chemically inundated community. This community of 10,000 people, mostly African-American, is surrounded by more than 50 polluting facilities, including landfills, oil refineries, waste lagoons, a sewage treatment plant, cement plants, steel mills, and waste incinerators. Hazel's daughter, Cheryl, who has worked with the organization since its founding, often says, "We call this area the 'Toxic Doughnut' because everywhere you look, 360 degrees around us, we're completely surrounded by toxics on all sides."

THE ENVIRONMENTAL JUSTICE MOVEMENT

People for Community Recovery was at the vanguard of a number of local citizens' groups that formed the movement for environmental justice (EJ). This movement, rooted in community-based politics, has emerged as a significant player at the local, state, national, and, increasingly, global levels. The movement's origins lie in local activism during the late 1970s and early 1980s aimed

SOURCE: From *Contexts*, 6(1): 37–41, 2007.

at combating environmental racism and environmental inequality—the unequal distribution of pollution across the social landscape that unfairly burdens poor neighborhoods and communities of color.

The original aim of the EJ movement was to challenge the disproportionate location of toxic facilities (such as landfills, incinerators, polluting factories, and mines) in or near the borders of economically or politically marginalized communities. Groups like PCR have expanded the movement and, in the process, extended its goals beyond removing existing hazards to include preventing new environmental risks and promoting safe, sustainable, and equitable forms of development. In most cases, these groups contest governmental or industrial practices that threaten human health. The EJ movement has developed a vision for social change centered around the following points:

- All people have the right to protection from environmental harm.
- Environmental threats should be eliminated before there are adverse human health consequences.
- The burden of proof should be shifted from communities, which now need to prove adverse impacts, to corporations, which should prove that a given industrial procedure is safe to humans and the environment.
- Grassroots organizations should challenge environmental inequality through political action.

The movement, which now includes African-American, European-American, Latino, Asian-American/Pacific-Islander, and Native-American communities, is more culturally diverse than both the civil rights and the traditional environmental movements, and combines insights from both causes.

ENVIRONMENTAL INEQUALITIES

Researchers have documented environmental inequalities in the United States since the 1970s, originally emphasizing the connection between income and air pollution. Research in the 1980s extended these early findings, revealing that communities of color were especially likely to be near hazardous waste sites. In 1987, the United Church of Christ Commission on Racial Justice released a groundbreaking national study entitled *Toxic Waste and Race in the United States*, which revealed the intensely unequal distribution of toxic waste sites across the United States. The study boldly concluded that race was the strongest predictor of where such sites were found.

In 1990, sociologist Robert Bullard published *Dumping in Dixie*, the first major study of environmental racism that linked the siting of hazardous facilities to the decades-old practices of spatial segregation in the South. Bullard found that African-American communities were being deliberately selected as sites for the disposal of municipal and hazardous chemical wastes. This was also one of the first studies to examine the social and psychological impacts of environmental pollution in a community of color. For example: across five communities in

Alabama, Louisiana, Texas, and West Virginia, Bullard found that the majority of people felt that their community had been singled out for the location of a toxic facility (55 percent); experienced anger at hosting this facility in their community (74 percent); and yet accepted the idea that the facility would remain in the community (77 percent).

Since 1990, social scientists have documented that exposure to environmental risks is strongly associated with race and socioeconomic status. Like Bullard's *Dumping in Dixie*, many studies have concluded that the link between polluting facilities and communities of color results from the deliberate placement of such facilities in these communities rather than from population-migration patterns. Such communities are systematically targeted for the location of polluting industries and other locally unwanted land uses (LULUs), but residents are fighting back to secure a safe, healthy, and sustainable quality of life. What have they accomplished?

LOCAL STRUGGLES

The EJ movement began in 1982, when hundreds of activists and residents came together to oppose the expansion of a chemical landfill in Warren County, North Carolina. Even though that action failed, it spawned a movement that effectively mobilized people in neighborhoods and small towns facing other LULUs. The EJ movement has had its most profound impact at the local level. Its successes include shutting down large waste incinerators and landfills in Los Angeles and Chicago; preventing polluting operations from being built or expanded, like the chemical plant proposed by the Shintech Corporation near a poor African-American community in Louisiana; securing relocations and home buyouts for residents in polluted communities like Love Canal, New York; Times Beach, Missouri; and Norco, Louisiana; and successfully demanding environmental clean-ups of LULUs such as the North River Sewage Treatment plant in Harlem.

The EJ movement helped stop plans to construct more than 300 garbage incinerators in the United States between 1985 and 1998. The steady expansion of municipal waste incinerators was abruptly reversed after 1990. While the cost of building and maintaining incinerators was certainly on the rise, the political price of incineration was the main factor that reversed this tide. The decline of medical-waste incinerators is even more dramatic.

Sociologist Andrew Szasz has documented the influence of the EJ movement in several hundred communities throughout the United States, showing that organizations such as Hazel Johnson's People for Community Recovery were instrumental in highlighting the dangers associated with chemical waste incinerators in their neighborhoods. EJ organizations, working in local coalitions, have had a number of successes, including shutting down an incinerator that was once the largest municipal waste burner in the Western Hemisphere. The movement has made it extremely difficult for firms to locate incinerators, landfills, and related LULUs anywhere in the nation, and almost any effort to expand existing polluting facilities now faces controversy.

BUILDING INSTITUTIONS

The EJ movement has built up local organizations and regional networks and forged partnerships with existing institutions such as churches, schools, and neighborhood groups. Given the close association between many EJ activists and environmental sociologists, it is not surprising that the movement has notably influenced the university. Research and training centers run by sociologists at several universities and colleges focus on EJ studies, and numerous institutions of higher education offer EJ courses. Bunyan Bryant and Elaine Hockman, searching the World Wide Web in 2002, got 281,000 hits for the phrase "environmental justice course," and they found such courses at more than 60 of the nation's colleges and universities.

EJ activists have built lasting partnerships with university scholars, especially sociologists. For example, Hazel Johnson's organization has worked with scholars at Northwestern University, the University of Wisconsin, and Clark Atlanta University to conduct health surveys of local residents, study local environmental conditions, serve on policy task forces, and testify at public hearings. Working with activists has provided valuable experience and training to future social and physical scientists.

The EJ movement's greatest challenge is to balance its expertise at mobilizing to oppose hazardous technologies and unsustainable development with a coherent vision and policy program that will move communities toward sustainability and better health. Several EJ groups have taken steps in this direction. Some now own and manage housing units, agricultural firms, job-training facilities, farmers' markets, urban gardens, and restaurants. On Chicago's southeast side, PCR partnered with a local university to win a federal grant, with which they taught lead-abatement techniques to community residents who then found employment in environmental industries. These successes should be acknowledged and praised, although they are limited in their socio-ecological impacts and longevity. Even so, EJ activists, scholars, and practitioners would do well to document these projects' trajectories and seek to replicate and adapt their best practices in other locales.

LEGAL GAINS AND LOSSES

The movement has a mixed record in litigation. Early on, EJ activists and attorneys decided to apply civil rights law (Title VI of the 1964 Civil Rights Act) to the environmental arena. Title VI prohibits all government and industry programs and activities that receive federal funds from discriminating against persons based on race, color, or national origin. Unfortunately, the courts have uniformly refused to prohibit government actions on the basis of Title VI without direct evidence of discriminatory intent. The Environmental Protection Agency (EPA) has been of little assistance. Since 1994, when the EPA began accepting Title VI claims, more than 135 have been filed, but none has been formally resolved. Only one federal agency has cited environmental justice concerns to protect a community in a significant legal case: In May 2001, the Nuclear Regulatory Commission denied a permit for a uranium enrichment plant in Louisiana because environmental justice concerns had not been taken into account.

With regard to legal strategies, EJ activist Hazel Johnson learned early on that, while she could trust committed EJ attorneys like Keith Harley of the Chicago Legal Clinic, the courts were often hostile and unforgiving places to make the case for environmental justice. Like other EJ activists disappointed by the legal system, Johnson and PCR have diversified their tactics. For example, they worked with a coalition of activists, scholars, and scientists to present evidence of toxicity in their community to elected officials and policy makers, while also engaging in disruptive protest that targeted government agencies and corporations.

NATIONAL ENVIRONMENTAL POLICY

The EJ movement has been more successful at lobbying high-level elected officials. Most prominently, in February 1994, President Clinton signed Executive Order 12898 requiring all federal agencies to ensure environmental justice in their practices. Appropriately, Hazel Johnson was at Clinton's side as he signed the order. And the Congressional Black Caucus, among its other accomplishments, has maintained one of the strongest environmental voting records of any group in the U.S. Congress.

But under President Bush, the EPA and the White House have not demonstrated a commitment to environmental justice. Even Clinton's much-vaunted Executive Order on Environmental Justice has had a limited effect. In March 2004 and September 2006, the inspector general of the EPA concluded that the agency was not doing an effective job of enforcing environmental justice policy. Specifically, he noted that the agency had no plans, benchmarks, or instruments to evaluate progress toward achieving the goals of Clinton's Order. While President Clinton deserves some of the blame for this, it should be no surprise that things have not improved under the Bush administration. In response, many activists, including those at PCR, have shifted their focus from the national level back to the neighborhood, where their work has a more tangible influence and where polluters are more easily monitored. But in an era of increasing economic and political globalization, this strategy may be limited.

GLOBALIZATION

As economic globalization—defined as the reduction of economic borders to allow the free passage of goods and money anywhere in the world—proceeds largely unchecked by governments, as the United States and other industrialized nations produce larger volumes of hazardous waste, and as the degree of global social inequality also rises, the frequency and intensity of EJ conflicts can only increase. Nations of the global north continue to export toxic waste to both domestic and global "pollution havens" where the price of doing business is much lower, where environmental laws are comparatively lax, and where citizens hold little formal political power.

Movement leaders are well aware of the effects of economic globalization and the international movement of pollution and wastes along the path of least resistance (namely, southward). Collaboration, resource exchange, networking, and joint action have already emerged between EJ groups in the global north and south. In the last decade EJ activists and delegates have traveled to meet and build alliances with colleagues in places like Beijing, Budapest, Cairo, Durban, The Hague, Istanbul, Johannesburg, Mumbai, and Rio de Janeiro. Activist colleagues outside the United States are often doing battle with the same transnational corporations that U.S. activists may be fighting at home. However, it is unclear if these efforts are well financed or if they are leading to enduring action programs across borders. What is certain is that if the EJ movement fails inside the United States, it is likely to fail against transnational firms on foreign territory in the global south.

Although EJ movements exist in other nations, the U.S. movement has been slow to link up with them. If the U.S. EJ movement is to survive, it must go global. The origins and drivers of environmental inequality are global in their reach and effects. Residents and activists in the global north feel a moral obligation to the nations and peoples of the south, as consumers, firms, state agencies, and military actions within northern nations produce social and ecological havoc in Latin America, the Caribbean, Africa, Central and Eastern Europe, and Asia. Going global does not necessarily require activists to leave the United States and travel abroad, because many of the major sources of global economic decision-making power are located in the north (corporate headquarters, the International Monetary Fund, the World Bank, and the White House). The movement must focus on these critical (and nearby) institutions. And while the movement has much more to do in order to build coalitions across various social and geographic boundaries, there are tactics, strategies, and campaigns that have succeeded in doing just that for many years. From transnational activist campaigns to solidarity networks and letter-writing, the profile of environmental justice is becoming more global each year.

After Hazel Johnson's visit to the Earth Summit in Rio de Janeiro in 1992, PCR became part of a global network of activists and scholars researching and combating environmental inequality in North America, South America, Africa, Europe, and Asia. Today, PCR confronts a daunting task. The area of Chicago in which the organization works still suffers from the highest density of landfills per square mile of any place in the nation, and from the industrial chemicals believed to be partly responsible for the elevated rates of asthma and other respiratory ailments in the surrounding neighborhoods. PCR has managed to train local residents in lead-abatement techniques; it has begun negotiations with one of the Big Three auto makers to make its nearby manufacturing plant more ecologically sustainable and amenable to hiring locals, and it is setting up an environmental science laboratory and education facility in the community through a partnership with a major research university.

What can we conclude about the state of the movement for environmental justice? Our diagnosis gives us both hope and concern. While the movement has accomplished a great deal, the political and social realities facing activists (and all of us, for that matter) are brutal. Industrial production of hazardous wastes continues to increase exponentially; the rate of cancers, reproductive illnesses, and respiratory disorders is increasing in communities of color and poor communities; environmental inequalities in urban and rural areas in the United States have remained steady or increased during the 1990s and 2000s; the income gap between the upper classes and the working classes is greater than it has been in decades; the traditional, middle-class, and mainly white environmental movement has grown weaker; and the union-led labor movement is embroiled in internecine battles as it loses membership and influence over politics, making it likely that ordinary citizens will be more concerned about declining wages than environmental protection. How well EJ leaders analyze and respond to these adverse trends will determine the future health of this movement. Indeed, as denizens of this fragile planet, we all need to be concerned with how the EJ movement fares against the institutions that routinely poison the earth and its people.

DISCUSSION QUESTIONS

1. What is the vision of the environmental justice movement and how is an understanding of racism central to this vision?
2. What do you learn from the environmental justice movement about the activism of communities of color?
3. Why do environmental justice activists in the United States have to connect with activists in other nations?

55

Reducing Inequalities

Doing Anti-Racism: Toward an Egalitarian American Society

JACQUELINE JOHNSON, SHARON RUSH, AND JOE FEAGIN

The sociologists in this article discuss the different dimensions of anti-racist work, noting that it must take place at the individual, organizational, and societal levels.

We view both structural and individual change as crucial for creating a new society without racism. Prior efforts to destroy structures and institutions that reinforce a system of racism have generally not cut to the heart of the racist prejudice and discrimination still implemented in the lives of most whites. Serious desegregation efforts by U.S. governmental agencies lasted barely a decade, and weak enforcement of most civil rights laws in states and at the federal level is now a national scandal.... Likewise, efforts that solely target individual racism do not root out the structural embeddedness of racism. Many programs, for example, that stress a liberal ideology of tolerance or color-blindness encourage people to accept individuals, opinions, and cultures that are different from their own, but require little or no work from those in dominant groups to critique and confront systematically their own privileges and power.... In many multicultural programs, whites become just one more "ethnic group" like all the others, rather than the dominant group with great privileges associated with its racial classification (see Van Ausdale and Feagin, 2000).

A nonracist society cannot be achieved if whites continue to deny the reality of the racist society and of racism within themselves. The painful emotional work of actually undoing individual racism must be accomplished in combination with collective efforts for structural change....

EDUCATION AND RE-EDUCATION FOR A NONRACIST SOCIETY

... Eliminating institutional oppressions like racism is essential to achieve an egalitarian society, but institutions are sustained and administered by individuals. Therefore, it is important to emphasize that institutional and individual racism

SOURCE: From *Contemporary Sociology* 29 (January 2000): 95–110. Copyright © 2000 by the American Sociological Association. Reprinted by permission.

are co-dependent. Racist systems construct as "natural" the tendency of individuals to use skin color as a basis for assigning others to in-groups and out-groups (Tajfel 1982).... If these anti-human impulses were not overtly and covertly taught to children, a society would not embed them in its realities....

Educational systems, such as public schools, were seen as primary sites where racial misconceptions could be refuted, resulting in greater racial accord in all social arenas. However, because most attempts at actual desegregation have been controlled by whites, the attempts to deal with institutional racism or re-education were mostly tentative baby steps. Most white desegregationists did not seem to understand, or perhaps did not wish to understand, that racism was deeply embedded in white minds and institutions, including the educational institutions at hand....

Changing the way that the public schools are supported is one way to attack segregation. The current system, where public schools are funded largely by local taxes, guarantees social inequality, segregation, and continued racism across large areas of cities and states. This system reinforces white notions of entitlement and privilege by linking wealth routinely to educational quality. What was once considered overtly racist "white flight" to avoid desegregated schools and neighborhoods has now been reconceptualized, as the socially conscious attempts of white parents to build a "better" life for their children. But, often what makes this life "better" is their limited contact with people of color....

We need new educational approaches and structures if we are to move to the nonracist society. Schools would need to be nonracist, nonsexist, nonclassist, democratic, and egalitarian in structure and process as well as in fundamental values. Children would be taught both through example and ideology that all people—black or white, Anglo or Latino, Asian or non-Asian, boy or girl, man or woman, rich or poor—have a right to grow and develop to their fullest potential. In a nonracist society, all children would be supplied with learning environments of equal quality so that their abilities could be developed....

"DOING" ANTI-RACISM: SOCIAL AND PERSONAL ACTIVISM

... To some degree, most Americans of color are forced routinely to engage in anti-racism work, at least in regard to their own group. These Americans of color may need to expand their activities to include the discrimination faced by other groups of color. But the most challenging task is to move significant numbers of whites into anti-racist actions and activism. This means that whites must move out of their present comfort zones to confront personally the painful and usually emotional work of doing anti-racism every day. We also envision the widespread formation of cross-racial coalitions with others who are devoted to doing anti-racism. Overall, we visualize many white individuals actively, consciously, and consistently working to eliminate racism by rejecting systems of privilege-maintenance in favor of human dignity, mutual respect, and liberty.

Organizational Efforts

African-Americans and other Americans of color have long led the struggles against racism in the United States. They continue to lead that struggle. As we see it, the goal must be to continue that struggle and to recruit more whites to the nonracist cause. Some members of the dominant group, albeit a very small percentage, are moving already toward the ideal nonracist society. Over the last several decades nonracist whites, with other nonracist Americans, have participated in a number of grassroots organizations working against racial oppression....

A next step in a broad nonracist strategy for the United States would be to expand the number of these nonracist organizations and to connect them into a national and international network of all peoples working against systemic racism in this and other societies across the globe....

To effect a genuine move away from racism and toward the nonracist egalitarian utopia, these well-intentioned whites must understand their own racism and that of the society. They must combat institutional racism—the racism built into every facet of American life. Although they may be opposed to discrimination, too many liberal whites are unaware of the demon of white privilege in their own lives. The next step is for them to acknowledge that their own white privilege contributes to the persistence of racism.

Individuals Undoing Racism

... Whiteness and white racism must be carefully learned and maintained over lifetimes. Individual selves and psychologies are shaped by structural realities. Thus, the question arises: What would the nonracist individual in the nonracist society be like? If we can begin to construct such a person, then perhaps we can take real steps toward the utopian society. The emerging nonracist person, like all human beings, is not without faults. Many Americans of color have already moved well down the road to nonracist or nonracist attitudes and actions. It appears that a pressing need today is to create a multitude of whites who can be started along that road. Unlike most whites today, however, a white person committed to the nonracist utopia would be filled with a very deep and lasting respect for all human beings as equals, including those who are physically or culturally different.

Respect, not just tolerance, is the necessary emotional orientation. Because the primary goals of the nonracist utopian society are to eliminate unnecessary suffering and to create the rights to life, liberty, and human happiness so eloquently asserted in the U.S. Declaration of Independence, people therein would be motivated to treat each other equally with dignity and respect. Within this positive energy cycle, everyone will enjoy the equal rights to be free from unnaturally imposed suffering and to be happy. A sceptic might suggest that this sounds good, but ask whether individual Americans, especially individual whites, are up to this difficult task. How do we bring changes in those centrally responsible for racist oppression?

Clearly, being willing to talk candidly about individual and societal racism is one essential step for whites to take in moving toward the nonracist society.

Honestly discussing with Americans of color the realities of racism increases the possibility that whites will move beyond their misunderstandings and fears and begin to put good intentions into the hard work of dismantling racism. This effort requires whites to actively join the struggle by working with African Americans and other people of color side by side, day in and day out....

Moving to the ideal nonracist society will require much work on the cognitive and emotional aspects of contemporary racism. This is perhaps the most difficult task for white Americans. However, whites' identification with oppression on the other side of the color line can develop through at least three different stages: sympathy, empathy, and what might be called transformative insight. The initial stage, sympathy, is important but limited. It usually involves the willingness to set aside some of the racist stereotyping and hostility taught in white communities and the development of a friendly interest in what is happening to the racial other. Empathy is a much more advanced stage, in that it requires the ability to reject distancing stereotypes and a heightened and sustained capacity to see and feel the pain of the racial others. Empathy involves the capacity to sense deeply the character of another's pain and to act on that sensitivity.

Empathizing with victims of racial discrimination is an important and valuable but limited emotional skill. Such empathic feelings are limited because they stem largely from perceiving the reality of anti-black discrimination. The empathic person's energy may be directed mainly at ending those practices. However, this outward focus on blacks' pain may incline empathic whites to avoid the inward reflection necessary to understand the role that their own white privilege plays in maintaining patterns of racism.

Actually crossing the color line provides an opportunity for a more informed understanding of the dynamics of racism that even very liberal whites have not gained from academic studies, civil rights activities, or limited social contacts with black Americans. This third stage of white development we call transformative insight. Transformative insight is more likely to develop in loving and caring interracial relationships. Interpersonal love characterizes most relationships in which people care deeply about each other: parent/child, husband/wife, friend/friend, and so forth.... Transformative insight, or transformative love, is most likely to develop in people who are in loving relationships that challenge institutional norms about power distribution. As a result, whites in such a position come to understand much more about the way in which the racialized hierarchy bequeaths power and privilege. The transformative insight includes a clear understanding of the broad range of privileges that comes from being white in a racist society, privileges a person has whether she or he wants them or not....

A well-intentioned white person has to dig deep to uncover knowledge of this privilege, even though it is obvious to black Americans and other Americans of color. Unfortunately, most whites do not have a caring or loving relationship with even one black person. For that reason, increasing real respectful relations across the color line is essential to the long-term battle against white racism....

Unlearning racism and the essential emotional work to eliminate individual racism are essential steps in becoming an effective nonracist activist. Nonracist activists cannot interact merely within limited social circles. They must actively work to break down existing structures that maintain and reproduce inequality. It is not enough to acknowledge difference or to be tolerant of others. The new anti-racists must acknowledge the real pain and the white privilege embedded in the existing system, must be willing to give up much privilege, and must work actively to create more egalitarian structures. This is tough work—emotionally, spiritually, and physically—but it is ultimately crucial for the construction of a nonracist utopia.

REFERENCES

Carr, Leslie. 1997. *"Color-Blind" Racism*. Thousand Oaks, CA: Sage.

Tajfel, Henri. 1982. *Social Identity and Intergroup Relations*. Cambridge: Cambridge University Press.

Van Ausdale, Debra and Joe Feagin, 2000. *The First R*. Lanham, MD: Rowman & Littlefield.

DISCUSSION QUESTIONS

1. How can schools contribute to building a nonracist society?
2. Why do Johnson, Rush, and Feagin think Whites have to understand their own racism and that of the society?
3. Can you identify a group on your own campus engaged in anti-racist work?

Part IX
Student Exercises

1. Social protests are important ways that people disrupt business as usual and demand that others pay attention. Have you ever participated in a protest? If so, what motivated you to take such action? Was this protest an individual act or part of an organized effort? What was your reaction to the experience? If you have not participated in a protest, why not? You can use your answers to these questions as the basis for a group discussion.
2. Imagine that your college or university announced that they were no longer giving scholarships to students for either economic need or to enhance diversity. Only students with excellent high school records and high SAT or ACT scores would now be eligible for scholarships. Others would have to rely on loans and jobs. Write a brief paper describing the implications of such a policy for your school. What does this assignment suggest about the work of building nonracist institutions in a color-blind era?

Index

Dear Student,

I hope you enjoy reading *Race and Ethnicity in Society: The Changing Landscape,* Second Edition. With every book that I publish, my goal is to enhance your learning experience. If you have any suggestions that you feel would improve this book, I would be delighted to hear from you. All comments will be shared with the authors. My e-mail address is Chris.Caldeira@cengage.com, or you can mail this form (no postage required).
Thank you.

School and address: ____

Department: ____

Instructor's name: ____

1. What I like most about this book is: ______________________________

____

____

2. What I like least about this book is: ______________________________

____

____

3. I would like to say to the authors of this book: ______________________

____

____

4. In the space below, or in an e-mail to Chris.Caldeira@cengage.com, please write specific suggestions for improving this book and anything else you'd care to share about your experience in using this book.

____

____

____

DO NOT STAPLE. PLEASE SEAL WITH TAPE.

FOLD HERE

NO POSTAGE
NECESSARY
IF MAILED
IN THE
UNITED STATES

BUSINESS REPLY MAIL
FIRST-CLASS MAIL PERMIT NO. 34 BELMONT CA

POSTAGE WILL BE PAID BY ADDRESSEE

Attn: Chris Caldeira, Sociology Editor
Wadsworth, Cengage Learning
10 Davis Drive
Belmont, CA 94002-9801

FOLD HERE

OPTIONAL:

Your name: ______________________ Date: ____________

May we quote you, either in promotion for *Race and Ethnicity in Society: The Changing Landscape,* Second Edition, or in future publishing ventures?

Yes: ____________ No: ____________

Sincerely yours,

Elizabeth Higginbotham
Margaret L. Andersen